Mantle Plumes and Their Record in Earth History

In recent years an enormous amount of data have been published related to mantle plumes, both modern and ancient. Some significant questions have arisen. Could the Earth have been more like Mars and Venus during the Archean? Instead of cooling principally by plate tectonics as it does today, did the Archean Earth cool chiefly by rising mantle plumes? How do we identify the effects of mantle plumes in the geologic record? Did plumes have a role in the growth of continents? Are large mantle plume events recorded in the geologic record, and, if so, what were the consequences of these events in terms of our atmosphere, oceans, and biosphere?

Mantle Plumes and Their Record in Earth History provides a timely and comprehensive review of the origin and history of mantle plumes throughout geologic time. This book describes the new and exciting results of the last few years and integrates an immense amount of material from the fields of geology, geophysics, and geochemistry that bear on mantle plumes. Included are chapters on hotspots and mantle upwelling, large igneous provinces (including examples from Mars and Venus), mantle plume generation and melting in plumes, plumes as tracers of mantle processes, plumes and continental growth, Archean mantle plumes, superplumes, and mantle plume events in Earth history and their effect on the atmosphere, oceans, and life in the geologic past.

This book will be valuable as a textbook for advanced undergraduate and graduate courses in geophysics, geochemistry, and geology and will also serve as a reference for researchers in the Earth sciences from a variety of disciplines.

Kent Condie is Professor of Geochemistry at New Mexico Institute of Mining and Technology, where he has taught since 1970. Before then he was at Washington University in St. Louis, Missouri. His textbook *Plate Tectonics and Crustal Evolution*, which is widely used in upper division and graduate courses in the Earth sciences, was first published in 1976 and has gone through four editions – the most recent in 1997. With coauthor Robert Sloan, Condie has also written a beginning textbook in geology entitled *Origin and Evolution of Earth: Principles of Historical Geology*, which was published in 1998. In addition, Condie has written a treatise, *Archean Greenstone Belts* (1981), and has edited two books, *Proterozoic Crustal Evolution* (1992) and *Archean Crustal Evolution* (1994). He is also the author of the popular interactive CD Rom, *Plate Tectonics and How the Earth Works.* Condie's research, primarily dealing with the origin and evolution of continents and the early history of the Earth, has over the years been sponsored chiefly by the U.S. National Science Foundation. He is author or coauthor of more than 250 articles published in scientific journals.

Mantle Plumes and Their Record in Earth History

KENT C. CONDIE

Department of Earth and Environmental Science
New Mexico Institute of Mining and Technology
Socorro, New Mexico

PUBLISHED BY THE PRESS SYNDICATE OF THE UNIVERSITY OF CAMBRIDGE
The Pitt Building, Trumpington Street, Cambridge, United Kingdom

CAMBRIDGE UNIVERSITY PRESS
The Edinburgh Building, Cambridge CB2 2RU, UK
40 West 20th Street, New York, NY 10011-4211, USA
10 Stamford Road, Oakleigh, VIC 3166, Australia
Ruiz de Alarcón 13, 28014 Madrid, Spain
Dock House, The Waterfront, Cape Town 8001, South Africa

http://www.cambridge.org

First published 2001

Printed in the United States of America

Typeface Times New Roman PS 10.5/13 pt. *System* LaTeX 2ε [TB]

A catalog record for this book is available from the British Library.

Library of Congress Cataloging in Publication Data

Condie, Kent C.

Mantle plumes and their record in earth history / Kent C. Condie.

p. cm.

Includes bibliographical references and index.

ISBN 0-521-80604-6 – ISBN 0-521-01472-7 (pbk.)

1. Mantle plumes. I. Title.

QE527.7 .C66 2001

551.1′16 – dc21 2001025504

ISBN 0 521 80604 6 hardback
ISBN 0 521 01472 7 paperback

Contents

Preface

Although plate tectonics and mantle plumes were introduced to geology at the same time in the 1960s and early 1970s by J. Tuzo Wilson and Jason Morgan, unlike plate tectonics, which rapidly collected supporters from the Earth Science community, mantle plumes took a back seat. Yes, Hawaii was an example of a mantle plume and as oceanic plates moved over plumes they leave hotspot tracks. The prevailing attitude was one of "this is fine, but let's now move on to plate tectonics where the real excitement is." For twenty years geoscientists focussed most of their efforts on trying to understand plate tectonics and document examples of it in the geologic record. It was not until the late 1980s that scientists turned some of their attention to mantle plumes, and indeed during the 1990s, when mantle plumes really "became of age", publications dealing with mantle plumes increased exponentially. Why the long period of dormancy for mantle plumes? I believe it was simply because geoscientists were overwhelmed by plate tectonics-a band wagon effect that influenced all of the Earth Sciences.

I think three things brought mantle plumes to the forefront in the nineties. First is high speed computers, which allowed scientists to numerically model mantle processes in reasonable amounts of time with increased accuracy. Models appeared for the production and ascent of mantle plumes, the effects of mantle phase transitions on plumes, and the interaction of plumes with both the continental and oceanic lithosphere. Of course, no matter how sophisticated, models are no better than the assumptions and boundary conditions that go into them. The first models were simple, focussing on Newtonian fluids as analogs for the mantle, with mantle plumes coming from boundary layers with strong thermal gradients. As it became clear that simplistic models were probably far from reality for the mantle, modeling advanced to non-Newtonian fluids, gradients in viscosity and density were included, and finally, we moved from two- to three-dimensional space.

Second, exciting new data from the exploration of Mars and Venus suggested mantle plumes and not plate tectonics were important on these planets. The detailed mapping of the surfaces of both Mars and Venus by the Pathfinder and Magellan Missions returned superb images of the planetary surfaces, which showed gigantic volcanoes, rifts, and domal uplifts, none of which looked like the product of plate tectonics. Geophysical models suggested that many of these features could be produced by mantle plumes, and in some instances, gigantic mantle plumes. If mantle plumes were important on Mars and Venus, why not on Earth?

And third, in the late 1990s, the increased precision of seismic tomography allowed scientists for the first time to begin mapping the Earth's mantle. Truly spectacular color figures began to appear in Nature and Science, almost on a weekly basis, showing what some geoscientists had said all along: the mantle is really quite inhomogeneous and descending slabs probably go all the way to the core-mantle interface. Mantle plumes such as Hawaii and Iceland really do have very deep roots.

Although we have a vast amount of data supporting the idea of mantle plumes, not everyone believes in plumes. One school of thought, led by Don Anderson at Caltech, proposes that modern hotspots do not reflect mantle plumes and they can be explained in other ways more effectively. We must remember that although the database from geophysics and geochemistry is consistent with idea of mantle plumes, other interpretations are possible, although in my opinion, not probable.

Why did I write a book on mantle plumes in the year 2000-2001? Now seemed a good time to bring together under one cover a summary of the truly enormous amount of data that have been published, principally during the 1990s, related to mantle plumes and their role in Earth history. Not only results for modern mantle plumes, but also for mantle plumes in the geologic past. Could the Earth have been more like Mars and Venus during the Archean some 3 Ga? Instead of cooling principally by plate tectonics (subduction) as it does today, did the Archean Earth cool chiefly by rising mantle plumes? How do we identify the effects of mantle plumes in the geologic record? Did plumes have a role in the growth of continents? Are large mantle plume events recorded in the geologic record, and if so, what were the consequences of these events in terms our atmosphere, oceans, and biosphere?

I have approached the subject of mantle plumes in such a way that the book can be used as a university text in an advanced undergraduate or graduate course in geophysics, geochemistry, or geology. The book is also intended as a reference for Earth scientists from a variety of disciplines. It is not intended to be an encyclopedia. Some topics are considered in greater detail than others, and an adequate, but not overwhelming list of references is given for the interested reader to further pursue topics of interest. On occasion, questions are left open-ended, and controversies are highlighted. As an aid to those not familiar with the deluge of geochemical diagrams that have appeared in recent years, I have used only a few geochemical diagrams, and the same diagrams appear in several chapters. For instance, the Th/Ta-La/Yb diagram is introduced in Chapter 5 and then used in subsequent chapters to constrain magma sources and tectonic settings. Basically there are only 5 or 6 geochemical and isotopic diagrams that are applied to both young and old basalts throughout the book. Hence, the reader can refer back to earlier chapters to compare young basalts from given tectonic settings with Archean basalts in later chapters. In some chapters, I have added a section at the end of the chapter, which should be considered as a perspective: what are the outstanding questions and controversies, and where do we go from here?

I am especially grateful to all those who took the time to read and make suggestions for improvement on one or more chapters, including the following: Louise Kellogg,

Geoff Davies, Bonnie Frey, Amy Gibson, Rob Kerrich, Nick Arndt, Andrew Kerr, Richard Ernst, Dallas Abbott, Chris Small, Dave Des Marais, Kirsten Nicolaysen, Bernhard Steinberger, and John Mahoney. I would also like to thank all of those investigators who have so generously allowed me to use figures from their published and unpublished papers. Also, I appreciate the many authors who provided me with preprints of papers in press. I am especially appreciative to those individuals who provided me with electronic copies of figures saving me an immense amount of time in preparing figures. I have learned much about mantle plumes from my timely discussions with Dallas Abbott, who has offered many suggestions and continual encouragement while preparing this book. I would also like to acknowledge Dave Des Marais who has helped me understand how the biosphere-atmosphere-ocean system works and the underlying complexities of the carbon cycle. Some of the ideas proposed in Chapter 9 grew and developed from lengthy email correspondences with Dave.

Kent C. Condie
Socorro, NM

1

Introduction

General Features of Mantle Plumes

A mantle plume is generally considered to be a blob of relatively hot, low-density mantle that rises because of its buoyancy. The existence of mantle plumes in the Earth was first suggested by J. Tuzo Wilson (1963) as an explanation of oceanic island chains, such as the Hawaiian–Emperor chain, that change progressively in age along the chain. Wilson proposed that as a lithospheric plate moves across a fixed hotspot (the mantle plume), volcanism is recorded as a linear array of volcanic seamounts and islands parallel to the direction in which the plate is moving. Morgan (1971) championed the idea of mantle plumes, suggesting that flood basalts formed by melting of plume heads, whereas hotspot volcanic chains were derived from partial melting of plume tails. He also showed that closely spaced hotspots on the same plate had not moved significantly relative to each other and suggested this was evidence that the plumes had come from the core–mantle boundary (Morgan 1972). Morgan noted that some hotspot tracks, like the Mascarene–Chagos–Laccadive track in the Indian Ocean, are traceable to flood basalts and can be used to reconstruct paths of opening ocean basins. Richards, Duncan, and Courtillot (1989) recognized at least 10 flood basalt–hotspot track pairs that formed from mantle plumes in the last 250 Myr.

The first laboratory experiments aimed at understanding mantle plumes better were those of Whitehead and Luther (1975), who showed that plume viscosity has an important effect on the shape of a plume. If a plume has a viscosity greater than its surroundings, it rises as a finger, whereas if it has a lower viscosity, it rises in a mushroom shape with a distinct head and tail. The tail contains a hot fluid that "feeds" the head as it buoyantly rises. Loper and Stacey (1983) developed a theory of flow in plume tails for a case in which the viscosity of a plume is strongly temperature dependent. Because the tail is hot, it has a relatively low viscosity and is quite narrow ($\approx$100 km across). Olson and Singer (1985) developed a theory for the ascent of plume heads that are compositionally distinct from surrounding mantle. They also studied the behavior of plume tails during horizontal shear caused by convective currents. Griffiths and Campbell (1990) were the first to confirm, by experiment and theory, the existence of thermal plume heads and tails and to distinguish between thermal and compositional plumes. In thermal plume heads, the boundary layer around the plume is heated by conduction, becomes buoyant, and rises with, and becomes entrained into, the head. This results in a plume head that reaches a diameter of 1000 km or more, which is two or three times larger than compositional, nonentraining plume heads.

Thermal modeling indicates that, for a layered silicate planet, the layers will cool from the outside inwards, and plumes will be generated at boundaries between layers from heat conducted across the boundaries from greater depths (Davies 1999). Although many details of plumes and their effects are still controversial and debated, the basic theory of mantle plumes is well established, and there is considerable observational evidence to support the plume concept. Only recently, however, has the resolution of seismic tomography improved sufficiently that at least some plumes in the upper mantle can be detected seismically (Li et al. 2000).

Plume Nomenclature

Numerous terms have been applied to mantle plumes, and there is confusion in the use of these terms. Although general agreement has not yet been reached, it is important to standardize the usage for this book. As noted previously, a *mantle plume* is a buoyant mass of material in the mantle that rises because of its buoyancy. On reaching the base of the lithosphere, plumes spread laterally. As suggested by the areal extent of some flood basalts, which are derived by partial melting of plumes, plume heads may reach diameters of 500 to 3000 km (Hill et al. 1992). Plume tails, on the other hand, are typically 100 to 200 km in diameter. Large *hotspots* are the surface manifestation of mantle plumes and are focused zones of melting. They are characterized by high heat flow, variable topographic highs depending on plume depth, and active volcanism. The term *superplume* is used herein to describe plume heads 1500–3000 km in diameter. Expressed in terms of the volume of plume-derived basalt flows, superplumes give rise to erupted volumes exceeding 0.5×10^6 km^3. The term *diapir* has been used to describe some mantle plumes. Herein, *mantle diapir* is used to describe a small mantle plume (<300 km across) that has lost its tail and thus ceases to grow (Herrick 1999). Diapirs may be produced in descending slabs or in the mantle wedge above descending slabs in response to localized thermal gradients. Alternatively, they may form anywhere in the mantle if it is convecting in a hard turbulent regime (Yuen et al. 1993). Another possible production mode is upward "budding" along the tops of superplumes (Sleep 1990).

Subduction of lithospheric slabs into the mantle requires a balancing upward flow. Because slabs are relatively cool, they cool the adjacent mantle as they descend. On the other hand, the mantle between subduction zones, where return flow occurs, is relatively warm and will slowly expand and rise. Larson (1991a) originally used the term *superplume* to describe the large mantle upwelling in the Pacific basin, and some investigators have continued this usage (Maruyama 1994). However, we will retain the more widely employed term *mantle upwelling* to describe these large regions of rising warm mantle between subduction zones. If a large lithospheric plate "protects" a large volume of mantle from the cooling effects of subduction, which happens beneath supercontinents and large ocean basins, a large mantle upwelling may be generated beneath the plate. Two such upwellings occur in the mantle today, one beneath Africa and one in the South Pacific. Such upwellings are typically more than 10,000 km in diameter and may contain many mantle plumes and hotspots.

Plumes may elevate the lithosphere by several hundred meters, producing broad, roughly circular uplifts known as *swells*. In oceanic lithosphere, which is relatively thin, swells may reach diameters of 2000 km with 500–1000 m of relief (Crough 1983; McNutt 1998). Mantle upwellings also raise the lithosphere, producing large swells known as *superswells*, which are more than 10,000 km across (McNutt 1998).

Internal Structure of the Mantle

An Overview

Before beginning a detailed discussion of mantle plumes, we need to review the internal structure of the mantle as revealed by seismology. Our knowledge of the Earth's interior comes chiefly from compressional (P-wave) and shear (S-wave) waves that pass through the Earth in response to earthquakes. Seismic wave velocities vary with pressure (depth), temperature, mineralogy, chemical composition, and the amount of melt present. Three first-order seismic discontinuities that reflect changes in composition or mineralogy divide the Earth into crust, mantle, and core (Fig. 1.1): the Mohorovicic discontinuity, or Moho, defines the base of the crust; the large decrease in seismic wave velocity at the base of the mantle (2900 km) defines the core–mantle interface; and at about 5200 km, a small increase in velocity is the inner-core boundary. Smaller but important velocity changes at 50–200, 410, and 660 km provide a basis for further subdivision of the mantle.

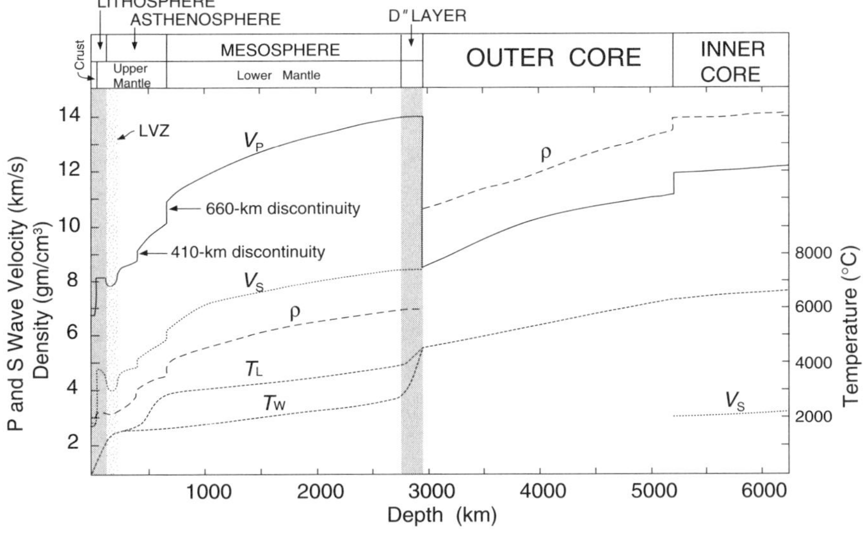

Figure 1.1. Internal structure of the Earth as represented by compressional (V_P) and shear wave velocity (V_S) and calculated density (ρ). Calculated temperature distributions for layered (T_L) and whole-mantle (T_W) convection. Also shown are the two major thermal boundary layers: the lithosphere and D″. After Condie (1997a).

The major regions of the Earth are summarized as follows with reference to Figure 1.1.:

1. The crust consists of the region above the Moho and ranges in thickness from about 3 km at some ocean ridges to about 70 km in collisional orogens.
2. The lithosphere (50–300 km thick) is the strong outer layer of the Earth, including the crust, that reacts to many stresses as a brittle solid. The lithosphere is a mechanical boundary layer that is broken into plates, which are formed at ocean ridges and descend into the mantle at subduction zones. The asthenosphere, extending from the base of the lithosphere to the 660-km discontinuity, is by comparison a weak layer that deforms by creep. A region of low seismic-wave velocity and high attenuation of seismic-wave energy, a low-velocity zone (LVZ), occurs at the top of the asthenosphere and is from 50- to 100-km thick. Significant lateral variations in density and in seismic-wave velocity are common in the mantle at depths of less than 400 km.
3. The upper mantle extends from the Moho to the 660-km seismic discontinuity. It includes the lower part of the lithosphere, the asthenosphere, and two seismic discontinuities that are caused by important solid-state transformations: olivine to wadsleyite at 410 km, and spinel to perovskite + magnesiowustite at 660 km.
4. The lower mantle extends from the 660-km seismic discontinuity to the core–mantle boundary at 2900 km. Between 200 and 250 km above the core–mantle interface, a flattening of velocity and density gradients occurs in a region known as the D″ layer named after the seismic wave used to define the layer. The lower mantle is also referred to as the mesosphere – a region that is strong but relatively passive in terms of deformational processes.
5. The outer core will not transmit S waves and is interpreted as a liquid composed chiefly of molten iron and nickel.
6. The inner core, which extends from the 5200-km discontinuity to the center of the Earth, transmits S waves (although at very low velocities), suggesting that it is near the melting point.

There are two boundary layers in the Earth with steep thermal gradients: the lithosphere at the surface and the D″ layer just above the core (Fig. 1.1). These layers play an extremely important role in cooling and convection in the Earth. In addition, the steep thermal gradient in the D″ layer may be the site at which most mantle plumes are generated.

Considerable uncertainty exists regarding the temperature distribution in the Earth. It is dependent on such features of the Earth's history as (1) the initial temperature distribution, (2) the heat contributed by large impacting bodies, (3) the amount of heat generated as a function of both depth and time, (4) convective and conductive heat loss, and (5) the process of core formation. Most estimates of temperature distribution in the Earth are based on one or a combination of two approaches: models of the Earth's thermal history involving various mechanisms of core formation and models involving redistribution of radioactive heat sources in the Earth by melting and convection processes.

Estimates using various models seem to converge on a temperature at the core–mantle interface of about 4500 ± 500 °C and a temperature at the center of the core of

6700 to 7000 °C. Two examples of calculated temperature distributions in the Earth are shown in Figure 1.1. Both show significant gradients in temperature in the lithosphere. The layered convection model (T_L) also shows a large temperature change near the 660-km discontinuity because this is the boundary between shallow and deep convection systems in this model. The temperature distribution for whole-mantle convection (T_W), which is preferred by most investigators, shows a steep thermal gradient in the D″ layer but no changes at the discontinuities in the upper mantle.

Convection is the dominant mode of heat transfer in the asthenosphere and mesosphere, where an adiabatic gradient is maintained, and thus temperature increases at a very slow rate with increasing depth. On the other hand, conduction is the main way heat is lost from the lithosphere, and temperatures change rapidly with depth and tectonic setting. In this respect, the lithosphere is both a mechanical and a thermal boundary layer in the Earth.

The Lithosphere

The oceanic lithosphere, where thickness is controlled by cooling, can be considered the upper boundary layer with a conductive temperature gradient that overlies a convecting adiabatic interior. Because thickness is dependent on temperature and age, it is sometimes referred to as the thermal lithosphere. Asthenosphere can be converted to oceanic lithosphere simply by cooling. The oceanic lithosphere begins life at ocean ridges as restite left when ocean-ridge basaltic magma is extracted. The progressive thickening of the lithosphere continues until about 70 Myr, and afterwards it remains relatively constant in thickness until subduction. Convective erosion at the base of the oceanic lithosphere may be responsible for maintaining this constant depth.

Unlike oceanic lithosphere, continental lithosphere has a complicated history that probably involves more than one mechanism by which it forms and grows (Condie 1997a). The post-Archean subcontinental lithosphere includes some combination of accreted asthenosphere, remnants of mantle plumes, and remnants of mantle wedges that originally formed above descending plates. Seismic reflection results also suggest that at least some of the continental lithosphere comprises remnants of partially subducted oceanic lithosphere. In northern Scotland, for example, dipping reflectors in the lower lithosphere are thought to represent fragments of now eclogitic oceanic crust (implying deep burial) – a relic of Paleozoic oceanic subduction (Warner et al. 1996). Although basal plume accretion may also have been important in the formation of the thick Archean (>2.5 Ga) lithosphere (Campbell and Griffiths 1992), buoyant subduction must have been important in the Archean as well, and thus oceanic plates may have been plastered beneath the continents, contributing to lithospheric thickening (Condie 1997a).

The Low-Velocity Zone

The seismic low-velocity zone (LVZ) in the upper mantle is characterized by low seismic wave velocities and high electrical conductivity (Condie 1997a). The bottom of the LVZ, sometimes referred to as the Lehmann discontinuity, has been identified

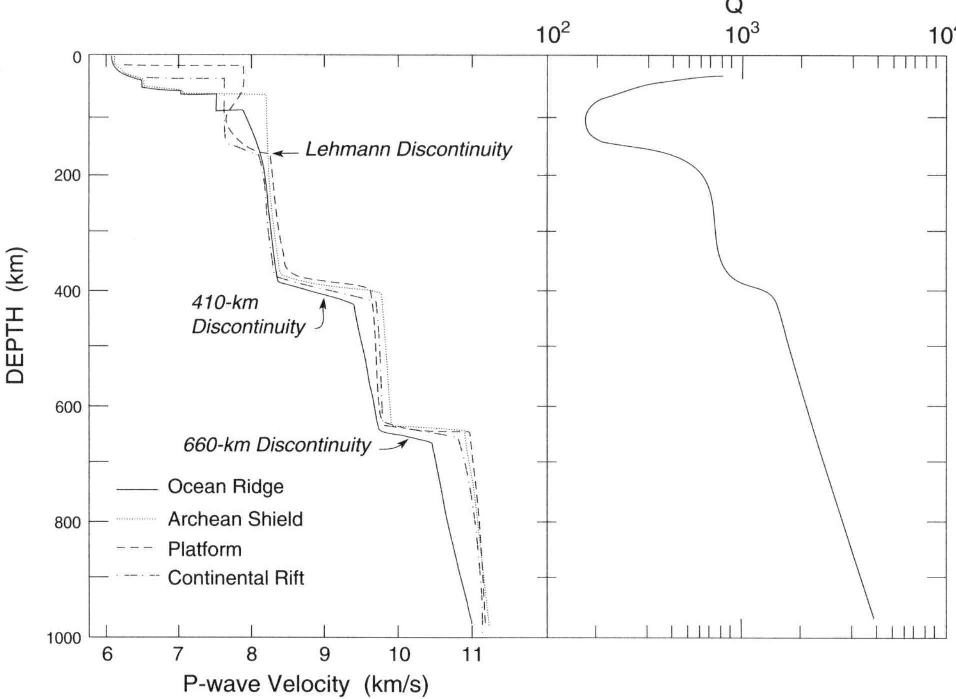

Figure 1.2. P-wave velocity and specific attenuation factor (Q) distribution in the upper mantle. Q varies inversely with seismic wave attenuation such that low values of Q show the greatest attenuation. After Condie (1997a).

from the study of surface wave and S-wave data in continental areas and occurs at depths of 180–220 km (Fig. 1.2) (Gaherty and Jordan 1995). The LVZ plays a major role in plate tectonics, providing a relatively weak, low-viscosity region upon which lithospheric plates move.

Because of the dramatic drop in S-wave velocity and increase in attenuation of seismic energy, interconnected melt along grain boundaries must contribute to producing the LVZ. The probable importance of incipient melting is attested to by the high surface heat flow observed when the LVZ reaches shallow depths, such as beneath ocean ridges and in continental rifts. Experimental results show that incipient melting in the LVZ requires a minor amount of water to depress silicate melting points (Wyllie 1971). With only 0.05–0.1% water in the mantle, partial melting of garnet lherzolite occurs in the appropriate depth range for the LVZ. The source of water in the upper mantle may be from the breakdown of minor phases that contain water such as amphiboles and micas. The theory of elastic wave velocities in two-phase materials indicates that only 1% melt is required to produce the lowest S-wave velocities measured in the LVZ. If, however, melt fractions are interconnected by a network of tubes along grain boundaries, the amount of melting may exceed 5% (Marko 1980). The downward termination of the LVZ probably reflects a combination of the depth at which geotherms pass below the mantle solidus and a rapid decrease in the amount of water available. The width or even the existence of the LVZ depends on the steepness of the geotherms. With steep geotherms, like those characteristic of ocean ridges and continental rifts, the range of penetration of the mantle solidus is large, and hence the LVZ is relatively wide (100–200 km). The gentle geotherms beneath continental platforms, which show a narrow range of intersection with the hydrated mantle solidus, produce a thin or poorly defined LVZ. Beneath Archean shields, geotherms generally do not intersect the mantle solidus; hence, there is no LVZ (Fig. 1.2).

The 410-km Discontinuity

High-pressure experimental studies show that the 410-km seismic discontinuity reflects the breakdown of Mg-rich olivine to the high-pressure Mg–silicate phase wadsleyite at about 14 GPa (Fig. 1.3). Mantle olivine (Fo_{90}) completely transforms to higher density wadsleyite over a less than 300 MPa pressure range at appropriate temperatures for the 410-km discontinuity (Ita and Stixrude 1992). This pressure range is in good agreement with the less than 10-km width of the 410-km discontinuity deduced from seismic data (Vidale et al. 1995). The approximately 6% increase in density observed at this discontinuity suggests that olivine composes 40–60% of the upper mantle as it does in garnet lherzolite xenoliths derived from the upper mantle.

High-pressure experimental data indicate that at depths of 350–450 km, both clino- and orthopyroxene transform into a garnet-structured mineral known as majorite, involving a density increase of about 6% (Christensen 1995). This transition has been petrographically observed as pyroxene exsolution laminae in garnet in mantle xenoliths derived from the Archean lithosphere at depths of 300–400 km (Haggerty and Sautter 1990). It is probable that the increase in velocity gradient beginning at 350 km and leading up to the 410-km discontinuity is caused by the majorite transformation.

Experimental data also indicate that wadsleyite transforms to a more dense spinel-structured phase at depths of 500–550 km. This mineral, referred to as Mg–spinel, has

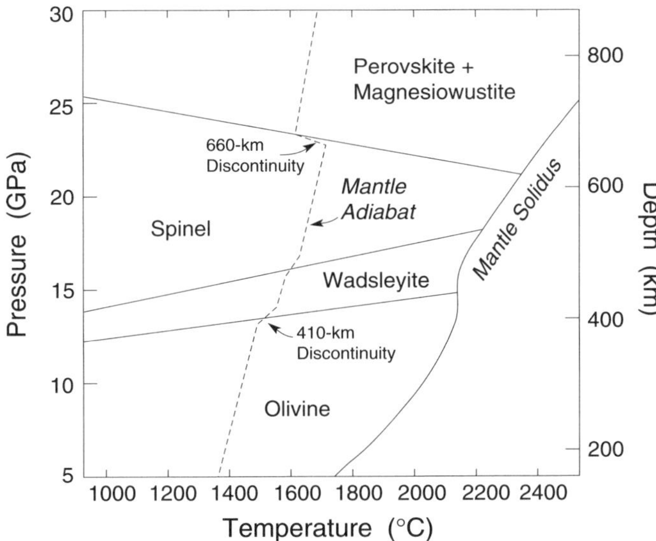

Figure 1.3. High-pressure phase relations for Mg_2SiO_4 in the mantle. After Christensen (1995).

the same composition as Mg-rich olivine but the crystallographic structure of spinel. The small density change (≈2%) associated with the phase change, however, does not generally produce a resolvable seismic discontinuity.

The 660-km Discontinuity

One of the most important considerations related to the style of mantle convection in the Earth is the nature of the 660-km discontinuity (Fig. 1.2). If descending slabs and mantle plumes cannot penetrate this boundary, two-layer mantle convection is favored in which the 660-km discontinuity represents the base of the upper layer. Large increases in both seismic-wave velocity (5–7%) and density (8%) occur at this boundary. High-frequency seismic waves reflected at the boundary suggest that it has a width of only about 5 km but has up to 20 km of relief over distances of hundreds to thousands of kilometers (Wood 1995).

As with the 410-km discontinuity, it is likely that a phase change in Mg_2SiO_4 is responsible for the 660-km discontinuity (Christensen 1995). High-pressure experimental results indicate that spinel transforms to a mixture of perovskite and magnesiowustite at a pressure of about 23 GPa, which can account for the seismic velocity and density increases at this boundary if the rock contains 50–60% Mg–spinel. Mg-perovskite and magnesiowustite are extremely high-density minerals and appear to compose most of the lower mantle.

Unlike the shallower phase transitions, the spinel–perovskite transition has a negative Clapeyron slope in P–T space (Fig. 1.3), and thus the transition may impede slabs from sinking into the deep mantle and also make it difficult for mantle plumes to rise into the upper mantle (Davies 1999). The latent heat associated with phase transitions in descending slabs and rising plumes can deflect phase transitions to shallower depths for positive Clapeyron slopes and to greater depths for negative Clapeyron slopes (Liu 1994) (Fig. 1.4). For a positive slope, like the olivine–wadsleyite transition, the elevated region of the denser phase exerts a strong downward pull on the slab or upward pull on a plume, thus helping drive convection. In contrast, for a negative slope, like the spinel–perovskite transition, the low-density phase is depressed, enhancing a slab's buoyancy and resisting further sinking of the slab. This same reaction may retard a rising

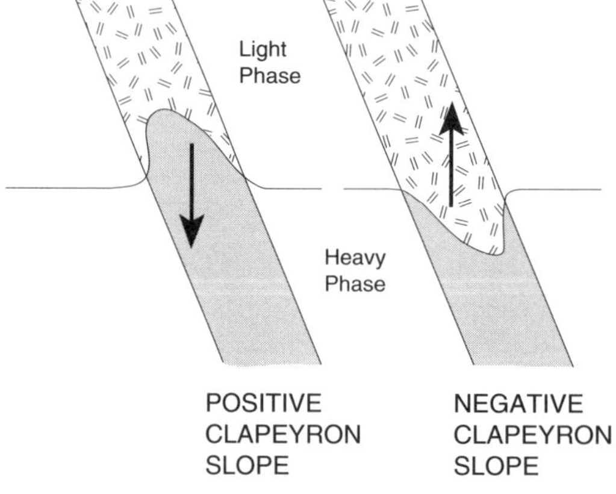

Figure 1.4. Deflection of phase boundaries in subducting slabs (light phase) for reactions with positive and negative Clapeyron slopes.

plume. However, if upward rise of a plume is slow and the temperature is high enough for the phase reaction to proceed, the plume may transform as it rises. Computer models by Davies (1995) suggest that stiff slabs can penetrate the boundary more readily than plume heads, and plume tails are the least able to penetrate it. Some geophysicists have suggested that slabs may locally accumulate at the 660-km discontinuity, culminating in occasional "avalanches" of slabs into the lower mantle (Tackley et al. 1994).

Seismic tomographic images of descending slabs provide an important constraint on the depth of penetration into the mantle. These images indicate that, although some slabs may be delayed at the 660-km discontinuity, all modern slabs eventually sink into the lower mantle. Thus, there is no evidence for layered convection in the Earth in terms of slab distributions in the mantle (van der Hilst et al. 1997).

The Lower Mantle

General Features

High-pressure experimental studies clearly suggest that Mg–perovskite is the dominant phase in the lower mantle. However, it is not clear if the seismic properties of the lower mantle require a change in major element composition at the 660-km phase transition (Wang et al. 1994). Results allow, but do not require, the Fe/Mg ratio of the lower mantle to be greater than that of the upper mantle. If this were the case, the greater density of Fe–perovskite would greatly limit the mass flux across the 660-km discontinuity, and the maintenance of such a chemical difference would favor layered convection.

The isostatic rebound of continents following Pleistocene glaciation, together with gravity data, indicates that the viscosity of the mantle increases with depth by two orders of magnitude and that the largest jump occurs at the 660-km discontinuity (Condie 1997a). This conclusion is in agreement with other geophysical and geochemical observations. For instance, although mantle plumes move upwards relatively fast, it would be impossible for them to survive convective currents in the upper mantle unless they were anchored in a "stiff" lower mantle. Also, only a mantle of relatively high viscosity at depth can account for the small number (two today) of large mantle upwellings. The isolation of geochemical domains in the mantle for billions of years (Chapter 5) can be achieved with a stiff lower mantle that resists mixing.

The D″ Layer

The D″ layer is a region of the mantle within a few hundred kilometers of the core where seismic velocity gradients are anomalously low (Loper and Lay 1995; Montague and Kellogg 2000). Calculations indicate that a relatively small temperature gradient (1–3 °C/km) is necessary to conduct heat from the core into the D″ layer. Because the core diffracts seismic waves, spatial resolution in this layer is poor, and details of its structure are not well known. However, initial seismic results indicate that D″ is a complex region that is vertically and laterally heterogeneous (Kendall and Silver 1996; Lay et al. 1998). Data seem to be equally consistent with a sharp interface between 100 and 300 km above the core–mantle boundary, or small-scale discontinuities that scatter

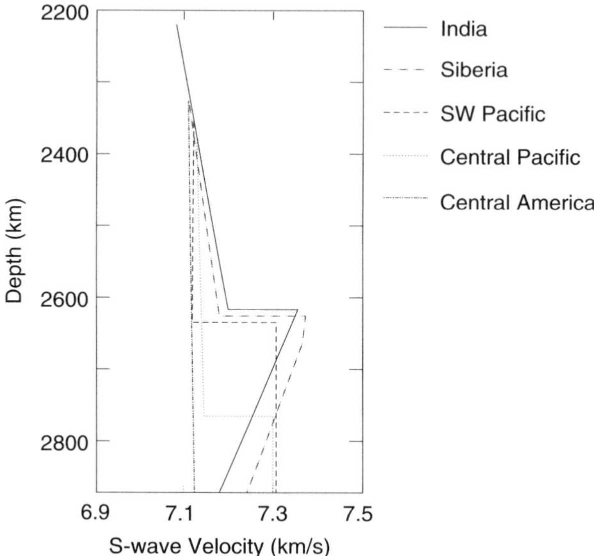

Figure 1.5. S-wave velocity distributions in the D″ layer indicate significant lateral heterogeneity. Data from Knittle and Jeanloz (1991).

seismic waves near the boundary (Fig. 1.5), or both. Estimates of the thickness of the D″ layer suggest that it ranges from approximately 100 to 500 km. Despite the poor resolution, large-scale lateral heterogeneities can be recognized in D″ (Lay et al. 1998; Sidorin et al. 1999). For example, regions beneath circum-Pacific subduction zones have anomalously fast P and S waves, which are interpreted by many to represent lithospheric slabs that have sunk to the base of the mantle. Slow velocities in D″ occur beneath the Central Pacific and correlate both with the surface and core–mantle boundary geoid anomalies and a greater concentration of hotspots (Chapter 2).

There are three possible contributions to the complex seismic structures seen in D″: temperature variations, compositional changes, and mineralogical phase changes. Temperature variations appear to be caused chiefly by the sinking of slabs into D″ (a cooling effect that produces relatively fast velocities) and heat released from the core (causing slow velocities). If a significant gradient in viscosity occurs in D″, convection may take place within the boundary layer (Montague and Kellogg 2000). Mixing of molten iron from the core with high-pressure silicates can lead to compositional changes with corresponding velocity changes. Experiments have shown, for instance, that when liquid iron comes in contact with silicate perovskite at high pressures, these substances react vigorously to produce a mixture of Mg–perovskite, a high-pressure silica polymorph, wustite (FeO), and Fe silicide (FeSi) (Wyession et al. 1998). These experiments also suggest that liquid iron in the outer core will seep into D″ by capillary action and affect a region extending for hundreds of meters above the core–mantle boundary. Recent seismic studies reveal the presence of fuzzy zones at the core–mantle interface, which could be due to intense chemical and physical interactions of the core with mantle silicates (Garnero and Jeanloz 2000). Phase changes, such as the possible breakdown of Mg–perovskite to magnesiowustite and silica, provide the best explanation for the sharp velocity increase seen at about 2600 km (Fig. 1.5). Although chemical segregation may occur in the D″ layer, calculated temperatures are not high enough to melt the perovskite or magnesiowustite phases.

Because the low seismic wave velocities in D″ reflect high temperatures and thus a lowering of mantle viscosity, this layer is commonly thought to be the source of mantle plumes. The lower viscosity will also enhance the flow of material into the base of newly forming plumes, and the lateral flow into plumes is balanced by slow subsidence of the overlying mantle. The model results of Davies and Richards (1992) suggest that a plume could be fed for 100 Myr from a volume of D″ only a few tens of kilometers thick and 500–1000 km in diameter. These results are important for mantle dynamics because they suggest that plumes are fed from the lowermost mantle in contrast to ocean ridges, which are fed from asthenosphere in the uppermost mantle.

The heterogeneous nature of the D″ layer is consistent with the presence of somewhat denser material commonly referred to as "dregs" (Knittle and Jeanloz 1991; Loper and Lay 1995). Slow upward convection of the mantle may pull dense phases such as wustite and FeSi upward from the core–mantle boundary. Eventually, they begin to sink because of their greater density, and these phases may form dregs that accumulate near the base of D″. Modeling suggests that dregs should pile up in regions of mantle upwelling and thin in regions of downwelling with the possibility that parts of D″ could be swept clean of dregs beneath downwellings. This means that dregs must be continually supplied by reactions and upwelling from the core–mantle boundary. Lateral variations in the thickness of D″ caused by lateral dreg movements could account for the large-scale seismic wave velocity variations and the variations in thickness of D″.

Plumes and Convection in the Mantle

Convection in fluids results from buoyancy differences that cause lighter material to rise and denser material to sink. In the Earth, buoyancy differences in the mantle originate at the only two well-documented thermal boundary layers: at the base of the lithosphere and in the D″ layer (Fig. 1.1). Rayleigh–Bernard convection arises because of heating at the base of a fluid. In the Earth, heat is conducted out of the core and warms the D″ layer, lowering its density. Lord Rayleigh was the first to show that convective behavior of a substance is dependent on a dimensionless number known as the Rayleigh number. For a simple homogeneous liquid to convect as it is heated at the base, the Rayleigh number must exceed 2000, and for the convection to be vigorous the Rayleigh number must be the order of 10^5. Irregular turbulent convection (hard turbulence) begins when the Rayleigh number reaches about 10^6, and such convection probably exists in the Earth, where the Rayleigh number is estimated to be about 3×10^6 (Davies 1999). Another factor contributing to mantle convection is lateral motion of subducted slabs as they sink to the bottom of the mantle (Garfunkel et al. 1986). The lithosphere cools by conduction at the Earth's surface, becomes denser than underlying mantle, and begins to sink. As it sinks, new lithosphere is formed at ocean ridges as the asthenosphere rises and undergoes partial melting. In the mantle, convection is driven by a combination of three thermal processes: heating at the bottom by heat loss from the core; heating by internal radioactive sources; and cooling from the top, resulting in sinking of cool lithospheric slabs into the mantle.

Two types of convection are considered for the Earth: layered and whole-mantle convection. In layered convection models, convection occurs separately below and above the 660-km discontinuity, whereas in whole-mantle convection the entire mantle overturns. With the improvement in the resolution of seismic tomographic data, it appears certain that lithospheric slabs descend into the lower mantle, and this requires some style of whole-mantle convection (van der Hilst et al. 1997). Classical pictures of convection in the Earth show convective upcurrents coming from the deep mantle beneath ocean ridges and downcurrents returning at subduction zones. However, it is now clear that ocean ridges are shallow passive features not related to deep convection in the Earth. The low-velocity anomalies under ocean ridges indicate they are shallow features with shallow roots caused by passive upwellings of asthenosphere.

How are plumes generated in the convecting mantle? As discussed more fully in Chapter 4, at thermal boundary layers in the Earth where steep temperature gradients exist, a portion of the layer acquires enough buoyancy to overcome the viscosity of surrounding mantle and begins to rise, forming a plume (Duncan and Richards 1991). Although the only well-established thermal boundary layer in the mantle is the D″ layer above the core, one or more additional thermal boundary layers are possible. Plumes coming from D″ contribute to cooling of the deep mantle and core. The failure of most hotspots to correlate with any particular features on modern plates indicates that the plume and plate modes of mantle cooling are not strongly coupled to each other (Davies 1999). In most instances, plumes appear to rise through the mantle without affecting descending plates. Today, most plumes are concentrated in two large mantle upwellings, suggesting the probability of plume formation is greater where upwellings occur (see Chapter 2). How do mantle plumes survive in a convecting mantle? They survive because they rise much faster (10^6–10^7 yr) than convective overturn (10^9 yr).

Organizational Strategy

In view of all of the different types of information bearing on mantle plumes, organization of this book has been a formidable task. My general approach is to begin with factual data that have been used to support the existence of mantle plumes, including laboratory and numerical experiments, followed by a consideration of the more speculative aspects of plumes and their possible role in Earth evolution. Chapters 2 and 3 focus on hotspots and large igneous provinces, which together constitute the observable database in support of the existence of mantle plumes. Included in these chapters are results from Venus and Mars, which complement the terrestrial database. In Chapter 4 we venture into the mechanisms of how and where mantle plumes are produced and how melting occurs in plumes based on results from laboratory and numerical experiments. Here we also review the only direct evidence for modern mantle plumes: seismic anomalies. Chapter 5 reviews geochemical and isotopic evidence for the existence of mantle plumes and also discusses some of the geochemical complexities encountered in characterizing magma sources in the mantle. Beginning in Chapter 6, I discuss possible roles of mantle plumes in the geologic past, and in this respect, the latter part of the book enters the realm of speculation based on the preserved geologic record. In Chapter 6 we explore the possible role of mantle plumes in the growth of continents, and Chapter 7 shows how it may be feasible to track mantle plumes into the early part of Earth history using lithologic associations and the geochemistry of Archean greenstones.

Continuing into the realm of geoconjecture, in Chapter 8 I discuss the possible role of superplume events in the geologic past as an explanation for episodic continental growth. And lastly, in Chapter 9 we wrap up the plume story with a summary of the possible effects of mantle plumes on the atmosphere, oceans, and life in the geologic past.

2

Hotspots and Mantle Upwellings

Introduction

A significant but volumetrically minor amount of dominantly basaltic volcanism occurs within plates as linear chains of volcanoes that grow older in the directions of plate motion (Wilson 1963; Morgan 1971). Examples of this style of volcanism are the Hawaiian–Emperor chain in the Pacific, the Yellowstone–Snake River plain in the western United States, and the Ninetyeast Ridge in the Indian Ocean. These volcanic tracks appear to form over hotspots, which are believed to be the surface manifestations of mantle plumes (Fig. 2.1). As mentioned in Chapter 1, Wilson (1963) suggested that hotspot tracks form as oceanic crust moves over relatively stationary magma sources in the uppermost mantle. Partial melting of plumes, when they intersect the mantle solidus near the base of the lithosphere, leads to large volumes of magma, which are partially erupted or intruded at or near the Earth's surface. In addition to hotspots, two broad mantle upwellings provide the return flow caused by subduction. The mantle upwellings elevate the Earth's surface up to a few hundred meters, and because they elevate the temperature of the uppermost mantle, they also cause minor but widespread melting, giving rise to volcanism and mafic underplating of the crust. Most of the major hotspots on Earth today occur within the mantle upwellings (Fig. 2.1).

It is commonly thought that mantle plumes and mantle upwellings begin life at the D″ thermal boundary layer just above the core–mantle interface. As we will see in Chapter 8, both upwellings and plumes may play a role in the breakup of supercontinents. In this chapter we will review the major features of mantle hotspots and upwellings and discuss some aspects of their origin.

Hotspot Characteristics

Hotspots are characterized by the following features:

1. In ocean basins, hotspots are overlain by topographic swells with a relief of 500–1000 m and typical widths of 1000–2000 km. These swells are probably indirect manifestations of ascending mantle plumes.

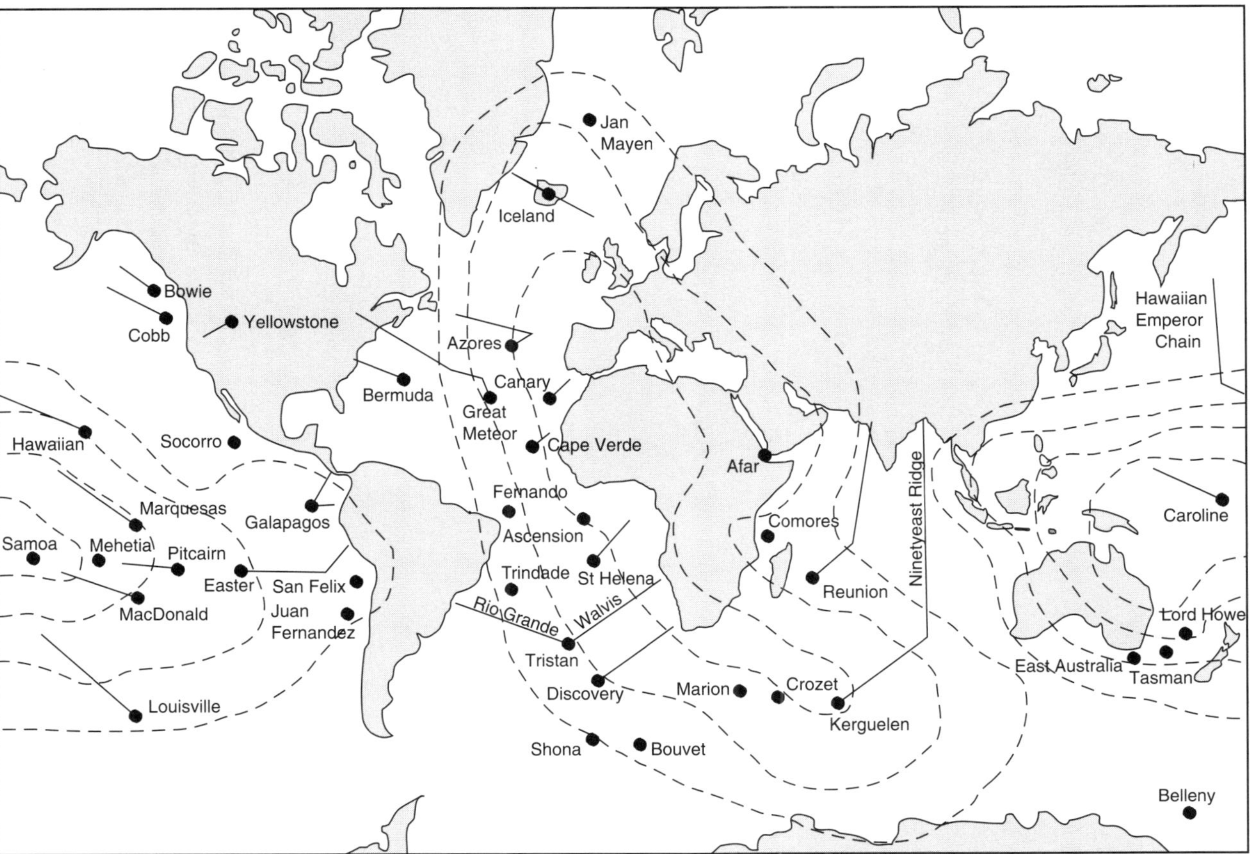

Figure 2.1. Map showing the global distribution of hotspots (solid dots) and hotspot tracks (lines). Dashed lines are geoid anomalies, and the geoid highs over western Africa and the South Pacific mark large mantle upwellings. After Crough 1983 and Duncan (1991).

2. Many hotspots are capped by active or recently active volcanoes. Examples are found in Hawaii and Yellowstone Park in the western United States.
3. Most oceanic hotspots are characterized by intermediate-wavelength gravity highs reflecting a topographic rise centered on the hotspot.
4. One or two ridges of mostly extinct volcanoes lead away from many oceanic hotspots (Fig. 2.1). Similarly, in continental areas, the age of magmatism and deformation may increase with distance from a hotspot. These features are known as hotspot tracks.
5. Most hotspots have high heat flow, probably reflecting a mantle plume source.
6. Most hotspots correlate with positive departures from the average geoid (average sea level).

Somewhere between 40 and 150 active hotspots have been described on the Earth (Morgan 1971; Duncan and Richards 1991). The best documented hotspots have a rather irregular distribution and occur in oceanic and continental areas (Fig. 2.1). Some occur on or near ocean ridges, such as Iceland, St. Helena, and Tristan in the Atlantic basin, whereas others occur near the centers of plates (such as Hawaii). How long do hotspots last? One of the oldest known hotspots is the St. Helena hotspot in the southern Atlantic Ocean, which began to produce basalts about 145 Ma. Many modern hotspots, however, are less than 100 Ma. The number of hotspots seems to correlate with geoid height, and the large number of hotspots in and around Africa and in the Pacific basin fall within the two major geoid highs (Fig. 2.1) (Anderson 1982; Stefanick and Jurdy 1984). The geoid highs appear to reflect processes in the deep mantle, which supports the idea that hotspots are caused by mantle plumes rising from the deep mantle.

Hotspot Tracks

In this section we will examine some of the major hotspot tracks and discuss their structural features, age, and geology. Chains of seamounts and volcanic islands are common in the Pacific basin and include such well-known island chains as the Hawaiian–Emperor, Louisville, Caroline, and Austral–Cook islands (Fig. 2.2). All of these island chains are subparallel to either the Emperor or Hawaiian chains, and the Hawaiian chain is approximately perpendicular to the axis of the East Pacific Rise. Closely spaced volcanoes form aseismic ridges such as the Ninetyeast Ridge in the Indian Ocean and the Walvis and Rio Grande Ridges in the South Atlantic. The life spans of hotspots vary and depend on such parameters as mantle plume size and the tectonic environment into which a plume is emplaced. On the Pacific plate, three volcanic chains were generated by hotspots between 70 and 25 Ma, whereas 12 chains have been generated in the last 25 Myr.

Hawaiian–Emperor Volcanic Chain

The Hawaiian–Emperor volcanic chain is a linear array of volcanic islands, seamounts, and volcanic ridges extending nearly 6000 km across the northern Pacific basin, where it is subducted into the Kurile trench (Fig. 2.2) (Clague and Dalrymple 1989). The chain includes more than 107 volcanoes with a cumulative volume exceeding 10^6 km^3

Figure 2.2. Examples of hotspot tracks on the Pacific plate. Calculated hotspot tracks assume fixed hotspots and show 10-Myr tick marks. Modified after Steinberger and O'Connell (1998), with permission. Copyright © 1998 by the American Geophysical Union.

(Bargar and Jackson 1974). The age of volcanism progresses systematically from modern (at the island of Hawaii) to about 80 Ma at the Detroit seamount, with a kink in the chain at about 43 Ma, which is generally thought to reflect a change in motion of the Pacific plate at this time. Mauna Loa and Kilauea are active volcanoes on Hawaii, and the Hawaiian hotspot is centered some 30–50 km off the southern coast of Hawaii beneath the active Loihi seamount (Moore et al. 1979). It is now widely agreed that the Hawaiian–Emperor chain formed during the last 100 Myr as the Pacific plate moved first north and then west relative to the Hawaiian hotspot. A trail of volcanoes was formed on the ocean floor as each volcanic center was progressively cut off from its source of magma as a new volcano formed behind it (Wilson 1963).

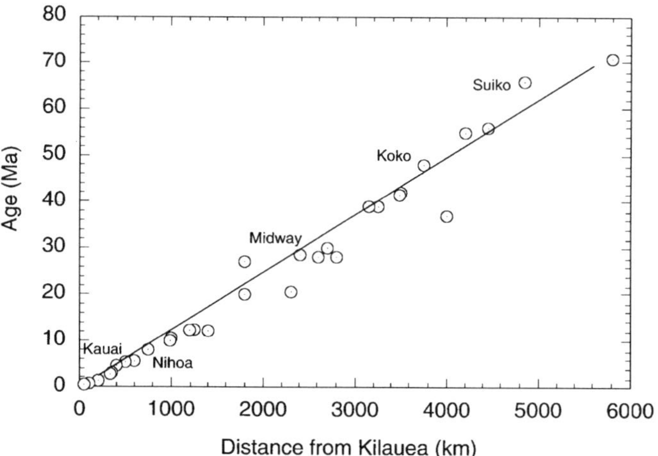

Figure 2.3. Age of volcanoes in the Hawaiian–Emperor chain as a function of distance from Kilauea. Line is a least-squares cubic fit and yields a propagation rate of 8.6 ± 0.2 cm/yr. Modified after Clague and Dalrymple (1989).

As one progresses westward along the Hawaiian chain, there is a gradual transformation of islands to guyots, which are former volcanic islands that were flattened by wave and stream erosion and later subsided (Hamilton 1956). Beginning at the southern end of the chain are active volcanoes such as Kilauea and Mauna Loa on Hawaii followed by eroded volcanic remnants such as Niihau and Nihoa, then by growing atolls like the Midway Islands, and eventually by deeply submerged guyots like Suiko and Detroit (Figs. 2.2 and 2.3). The subsidence of the Hawaiian–Emperor chain with age results from thermal aging of the lithosphere and isostatic response to local loading of growing volcanoes. Depth to the seafloor increases along the chain from about 4.5 km near Hawaii to 5.3 km near the bend in the chain, yielding a subsidence rate of about 0.02 mm/yr.

A distance–age plot for the Hawaiian–Emperor chain using the oldest reliable isotopic ages for basaltic volcanism is shown in Figure 2.3. Applying a York regression to the data from the entire chain yields a volcanic propagation rate of 8.6 ± 0.2 cm/yr (Clague and Dalrymple 1989). There is a change in propagation rate at about 38 Ma from around 9 cm/yr in the Hawaiian segment to about 7 cm/yr in the Emperor segment. It is noteworthy that this rate change does not correspond to the kink in the chain at 43 Ma. There are also short-lived differences in the rate of propagation but with a notable increase in the last 5 Myr. The volume of magma erupted per unit distance (or time) along the track increases significantly from the beginning of the Emperor chain (3000 km) to the end of the Hawaiian chain (Fig. 2.4). The Hawaiian plume is presently producing the greatest volume of lava at the greatest eruption rate in its 80-Myr history. The only section in which the eruption rate does not increase with decreasing age is near 43 Ma, when a change in motion may have occurred in the Pacific plate. The increase in volcanic activity with time in the Hawaiian hotspot is puzzling because, as plumes cool with age, their magmatic productive power should decrease. This is strong evidence that the Hawaiian plume tail is being supplied with new material rising from the deep mantle.

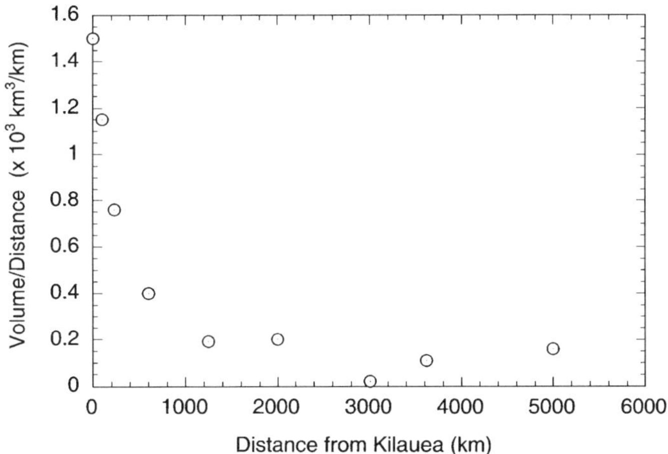

Figure 2.4. Rate of eruption in the Hawaiian–Emperor chain as a function of distance from Kilauea. Note the dramatic increase in rate as the Hawaiian hotspot is approached (1000 to 0 km). Data from Clague and Dalrymple (1989).

Louisville Volcanic Chain

The Louisville volcanic chain is a 4300-km hotspot track in the South Pacific (Fig. 2.2), composed of at least 65 major volcanoes (Lonsdale 1988; Watts et al. 1988). The chain, which is being subducted on its northwest end into the Tonga–Kermadec trench, ranges in age from about 70 Ma to probably present day. The exact location of the Louisville hotspot, however, is still uncertain. Most of the volcanoes in the Louisville chain grew above sea level and later subsided to form guyots, and the age of volcanism progressively decreases to the southeast. The northwestern part of the chain comprises many small guyots 5–10 km in diameter. As with the Hawaiian–Emperor chain, the depth of the guyots increases with progressing age, although the data are somewhat scattered. The regularity of magma eruption along the chain was occasionally interrupted by large volcanoes, which are now 1–3 km high.

The spatial and age pattern of volcanoes in the Louisville chain can be matched to the same plate motion as for the Hawaiian–Emperor chain. However, a change in plate motion at 43 Ma is not as pronounced in the Louisville chain (Fig. 2.2), and the rate of magma production along the two chains is quite different. Before 25 Ma, the Louisville magma production rate was much steadier than the Hawaiian–Emperor rate, but it had a lower total volume of eruption. At about 25 Ma, the magma supply decreased to a small fraction of that in the Hawaiian–Emperor chain such that none of the volcanoes emerged above sea level in the last 11 Myr. Also, the Louisville track is only half as wide as the Hawaiian–Emperor track. Hence, unlike the Hawaiian plume, the Louisville plume must have declined in activity with time.

Easter Volcanic Chain

The Easter volcanic chain in the South Pacific is a near–east-trending chain of volcanic islands and seamounts on the eastern side of the East Pacific Rise (Fig. 2.2). Volcanism ranges from very young at the western end of the chain in the Easter seamounts (about 0.4 Ma) to about 20 Ma at the eastern end. There is, however, some debate as to the location of the hotspot responsible for the Easter chain. Three locations have been proposed for this hotspot (O'Connor et al. 1995): one near the western end of the chain, one at Sala y Gomez Island, and one east of Sala y Gomez Island at about 102° longitude. The most recent isotopic ages, however, clearly indicate that the Easter plume must be located in the vicinity of Sala y Gomez Island, as shown in Figure 2.2. The islands west of Sala, including Easter Island, were formed by magma that was channeled westward from the plume towards the East Pacific Rise (O'Connor et al. 1995).

Austral–Cook and Society Volcanic Chains

The Austral–Cook and Society volcanic chains, north of the Louisville chain, are relatively short chains that roughly parallel the Hawaiian and Louisville chains (Fig. 2.2). Both chains have active volcanoes at the southern ends, which may be manifestations of the hotspots that gave rise to the chains. For the last 5 to 20 Myr, the Pacific plate has moved over these hotspots at an average rate of 11 cm/yr (Juteau and Maury 1999). As the volcanoes age along both hotspot tracks, they are eroded to form guyots, and as these sink, coral reefs around their perimeters have increased in size. Isotopic ages of volcanism generally increase with distance from the hotspots in the Society chain and in the main Austral–Cook chain, consistent with a velocity of the Pacific plate of 11 cm/yr. In the Society chain, there are several islands that have renewed activity at later times (most notably at 380 km), consistent with magma storage over long periods of time (Fig. 2.5a).

The Austral–Cook chain is complicated in that it contains many old, relatively high islands not yet converted to atolls. It also exhibits multiple episodes of volcanism at different sites and uplift of selected islands long after initial immersion by subsidence (Dickinson 1998). There are also young volcanoes far from the McDonald hotspot source, which indicates that a single hotspot will not explain all of the magmatism. McNutt et al. (1997) showed that the Austral–Cook chain is composed of three distinct volcanic chains with ages spanning 34 Myr and with inconsistent age progression. Although the main volcanic chain (Fig. 2.5b) may be produced by plate movement over the McDonald hotspot, the other two chains probably have different sources (McNutt et al. 1997; Dickinson 1998). Seismic tomography studies indicate that the upper mantle in this region is anomalously hot and weak, which suggests unusual amounts of partial melting and broad upwelling. One explanation of the three volcanic chains is that local structures in the lithosphere may have channeled the outflow of widespread melt (perhaps mixed with magmas from the McDonald plume source) into three "hotspots." Shearing between the Pacific plate and the underlying melt may have led to very short age progressions of volcanoes tapping individual melt pockets, as observed in the two short Austral–Cook chains (Aitutaki–Rurutu and Rarotonga, Fig. 2.5b). Alternatively, each of the volcanic chains in the Austral–Cook track may have its own hotspot, perhaps related to relatively small mantle plumes (diapirs) from shallow rather than deep-seated sources (Dickinson 1998).

Continental Hotspot Tracks

Hotspot tracks also occur on the continents, although these tracks are less well defined than in ocean basins owing to the thicker lithosphere (McDougall and Duncan 1988;

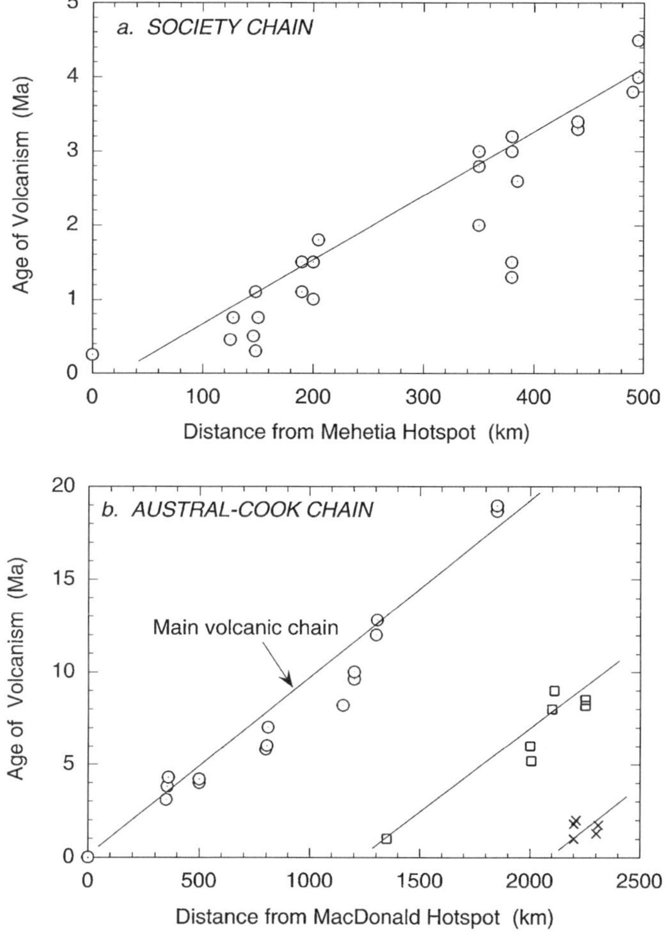

Figure 2.5. Graph showing volcano age versus distance from hotspot source for the Society (a) and Austral–Cook (b) volcanic chains in the South Pacific. Squares and crosses in (b) represent the two short Austral–Cook chains, Aitutaki–Rurutu and Rarotonga, respectively. Modified after Juteau and Maury (1999).

Burke 1996). For example, North America moved northwest over the Great Meteor hotspot in the Atlantic basin between 125 and 80 Ma (van Fossen and Kent 1992). The trajectory of the hotspot is defined by the New England–Corner seamount chain in the North Atlantic (Figs. 2.1 and 2.6) and by Mesozoic kimberlites and alkali igneous complexes in New England and Quebec. Geologic data and paleotemperatures indicate that this region was elevated by at least 4 km as it passed over the hotspot. Dated igneous rocks fall near the calculated position of the hotspot track at the time they formed. In addition, the hotspot appears to have moved south by 11° between 125 and 80 Ma. Based on U/Pb perovskite ages of kimberlites in the Canadian shield, Heaman and Kjarsgaard (2000) have proposed that the Great Meteor hotspot track can be extended to about 200 Ma and that the oldest kimberlite eruption occurred at Rankin Inlet along the northwestern coast of Hudson Bay.

When post-Triassic kimberlites from North and South America and Africa are rotated to their position of origin relative to present Atlantic hotspots, the majority appear to

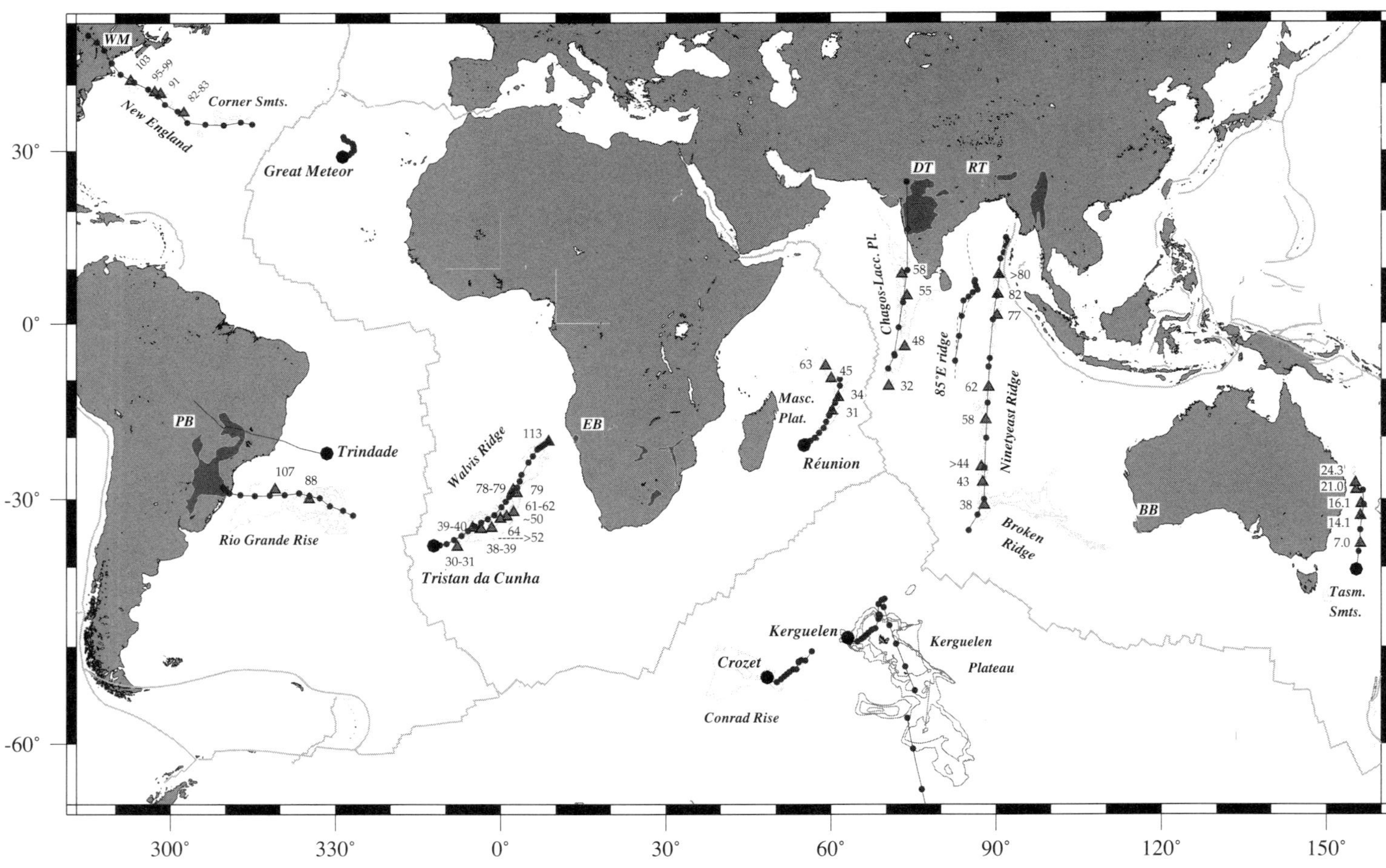

Figure 2.6. Major hotspot tracks in the Atlantic and Indian Oceans showing observed and modeled paths. Black circles = present-day hotspots; triangles = dated volcanoes with ages in Ma; Lines = reconstructed hotspot paths with dots at 5-Myr intervals. WM = White Mountains; DT = Deccan traps; RT = Rajmahal traps; PB = Paraná flood basalts; EB = Etendeka flood basalts; BB = Bunbury basalts. Courtesy of R. D. Muller.

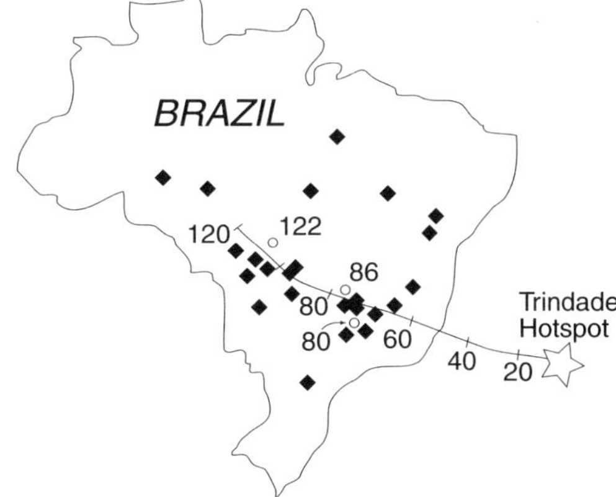

Figure 2.7. Locations of dated kimberlites (open circles) and alluvial diamonds (black diamonds) in Brazil compared with the calculated trajectory of the Trindade hotspot between 120 and 20 Ma. Modified after Crough, Morgan, and Hargraves (1980).

have formed within 5 degrees of a mantle hotspot. As an example, the calculated trajectory of the Trindade hotspot east of Brazil (Fig. 2.7) matches the locations of three dated kimberlites from Brazil and also roughly coincides with the distribution of alluvial diamond deposits, which are derived from nearby kimberlites.

Yellowstone

Of the few continental hotspot tracks that have been identified and characterized with any degree of certainty, the Yellowstone track in the American Cordillera is the most famous (Fig. 2.2). High heat flow, low seismic wave velocities and densities at shallow depth, a southwest-widening topographic swell, and high electrical conductivity at shallow depth beneath Yellowstone National Park in Wyoming are interpreted to reflect a mantle hotspot at this locality (Smith and Braile 1993, 1994). A 600-m high topographic bulge is centered on the Yellowstone caldera and extends across an area 600 km in diameter. Direct evidence of a mantle plume at depth is manifested by anomalously low P-wave velocities that extend to depths of 200 km. The most definitive evidence for the movement of the North American plate over a mantle plume in this area is the progressive increase in age of felsic and mafic volcanism extending southwest from Yellowstone National Park into the Snake River Plain in Idaho (Smith and Braile 1994) (Fig. 2.8). The youngest volcanism in Yellowstone is around 2 Ma followed by volcanism about 6 Ma in the Snake River Plain (Fig. 2.2). Hotspot-related felsic volcanics in northern Nevada and southwest Idaho range to about 16 Ma, and in north–central Nevada both volcanics and gold mineralization in the Carlin District may be related to the Yellowstone hotspot at about 34 Ma (Oppliger et al. 1997). The average rate in the progression of felsic volcanism along the Yellowstone hotspot track in the last 20 Myr is about 4.5 cm/yr, and this can divided into two regimes (Smith and Braile 1994): 3.3 cm/yr between 0 and 8 Ma and 6.1 cm/yr between 16 and 8 Ma (Fig. 2.8). This is faster than the speed of the North American plate of about 2 cm/yr during this time using fixed hotspot models (Gripp and Gordon 1990). Rates of extension due to normal faulting in the Snake River Plain are very similar to those of felsic volcanism, suggesting a similar cause due to movement of the North American plate over the Yellowstone hotspot (Pollitz 1988). Although both the Snake River and Columbia River flood basalts may

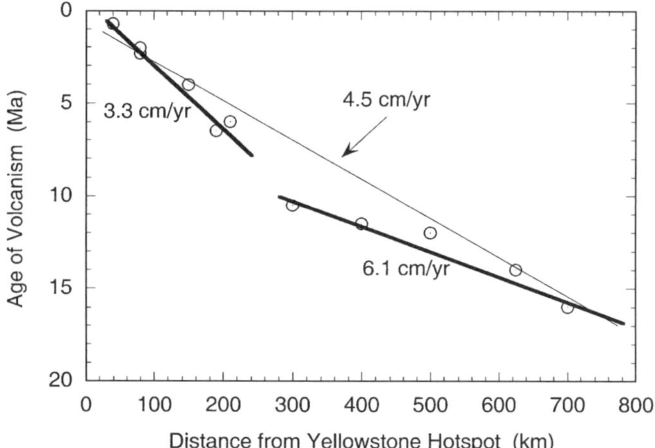

Figure 2.8. Age of volcanism versus distance from the Yellowstone hotspot for the last 20 Myr. The thin line is the linear regression of all of the data, whereas bold lines divide the data into two populations. Modified after Smith and Braile (1993).

have been derived from the Yellowstone plume, only the Snake River basalts lie on the hotspot track. The Columbia River basalts are more fully discussed in Chapter 3.

Recent teleseismic investigations across the Yellowstone hotspot reveal that the swell is held up by mantle of two types: partly molten mantle with low velocities beneath the hotspot track, and depleted mantle with higher velocities beneath the sides of the swell (Humphreys et al. 2000). In addition, the results indicate that the mantle beneath the hotspot track is anisotropic with olivine a axes oriented northwest, consistent with both plate motion and hotspot asthenosphere flow. Imaged velocities are consistent with a mantle plume source only if melt buoyancy within the center of the plume head drives convection, which moves restite to the sides of the plume head.

As indicated by plate reconstructions of the North Pacific, the Yellowstone hotspot may have been long-lived, appearing perhaps as early as 130 Ma in the extreme northwestern Pacific (Johnston and Thorkelson 2000). The earliest recognized magmatism that may be related to the Yellowstone hotspot is the 70-Ma basalts of the Carmacks Group now preserved in an accreted terrane in the Yukon in northern Canada. The voluminous Coast Range basalts erupted about 50 Ma in Washington and British Columbia also appear to have been derived from this hotspot.

Seamount Arrays

Seamounts are extinct volcanoes on the seafloor that are well below sea level. From satellite gravity measurements it is possible to estimate the size and geographic distribution of seamounts in ocean basins (Wessel and Lyons 1997). Most of the large seamounts (>3.5 km tall) occur in hotspot tracks such as the Hawaiian–Emperor and Louisville tracks (Plate 1). However, it is generally agreed that most seamount chains are produced along transform faults or fractures associated with active ocean ridges (Lonsdale 1988). Chapel and Small (1996) recognized two distinct distributions of seamounts in the Pacific. In some cases, seamount chains form at retreating limbs of intersecting spreading centers. Batiza (1982) showed that the volume of magma erupted in non-hotspot submarine volcanoes decreases exponentially away from active ridge crests in a manner similar to depth and average heat flow. Many of these volcanoes are the same age as the crust on which they are constructed and therefore were generated near ridge crests. Some chains of intermediate to small seamounts do not parallel major hotspot tracks, and relatively short chains and clusters of seamounts are quite widespread in the Pacific basin (Plate 1). In some of these, such as the Cobb–Eickelberg chain in the northeastern Pacific (Fig. 2.2), anomalously young pulses of volcanism may reflect rejuvenated magmatism of nonhotspot origin (Keller et al. 1997).

One of the major constraints on the sources of seamount volcanoes is the composition of their volcanics. Are they derived from depleted upper mantle sources similar in composition to ocean-ridge basalts, or do they come from enriched mantle sources typical of the major hotspot volcanoes? The former source is typical of most small seamount chains and clusters in the Pacific basin. On the other hand, seamount volcanics with enriched geochemical characteristics found in larger seamount chains imply mantle plume sources similar to the major hotspot volcanic chains (Batiza 1989).

Although intermediate and small seamounts are more abundant and more widely scattered than large seamounts, many of them also define linear or sublinear trends. Some of these trends are parallel or subparallel to one of the branches of the Hawaiian–Emperor chain (Plate 1). Does this mean that each of these seamount chains has its own small plume and that it reflects the same plate motions as the Hawaiian–Emperor chain? This is a question of widespread interest and importance for which we do not have the answer at present. However, Malamud and Turcotte (1999) have pointed out that a discrepancy between mantle cooling due to subduction (58%) and large mantle plumes (6%) leaves some 36% of the mantle heat flux unaccounted for. These authors showed that the cumulative frequency-size distribution of plume strengths approximates a power law for intermediate and large plumes. If this distribution is extrapolated to smaller sizes, about 5200 small plumes are required to account for the missing heat flux. The authors have suggested that the large population of seamounts on the Pacific seafloor is evidence for the presence of many small plumes. Such plumes would presumably come from depths shallower than the D″ layer at the base of the mantle – perhaps from the 660-km discontinuity.

A nonplume origin has also been suggested for some seamount-island chains by Hieronymus and Bercovici (2000). These authors proposed a model whereby the initial volcanic load on the lithosphere or a local anomaly in the melt source results in magmatic hydrofracturing of the lithosphere at extensional maxima. The propagating fracture is enlarged by magmatic erosion, and it controls the distribution of submarine volcanoes, which are aligned at right angles to the tensile stress field. The ages of volcanism along the chain are not related to plate velocity but depend on the dynamics of fracture propagation and the mechanisms of volcano formation.

Hotspot Swells

One of the important features of hotspots is the elevation of the lithosphere over a hotspot known as a swell. Hotspot swells, which are typically 1000 to 2000 km in

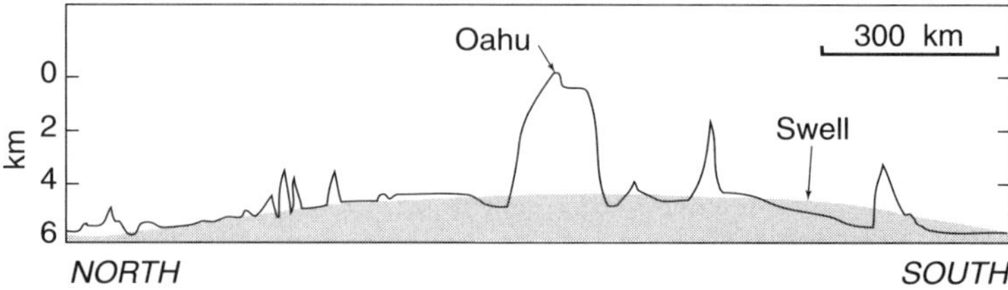

Figure 2.9. Bathymetry of the Hawaiian swell near Oahu with a shaded parabola showing the excess elevation. Modified after Crough (1983).

diameter and several hundred to over 1000 m high, occur in both the oceanic and continental lithosphere (Fig. 2.9) (Crough 1983; Sleep 1992). As the lithosphere moves over a hotspot, the swell moves with the hotspot track, and as the hotspot track ages, the swell subsides at a faster rate than normal ocean crust, which is a feature that must be accounted for in any origin for swells (Fig. 2.10). Detrick and Crough (1978) suggested that swells are produced by thinning of the lithosphere as it rides over a hotspot. Because a mantle plume is hotter and less dense than the lithosphere, replacement of the lower part of the lithosphere with plume material causes isostatic uplift of the surface of the plate. After the lithosphere moves off the hotspot and cools, it thickens to its normal thickness and the swell subsides. This simple thermal model, however, has two problems. First, it requires unrealistically high rates of heating of the lithosphere, and second, a W-shaped heat flow anomaly on the Hawaiian swell is not explained (Von Herzen et al. 1989). In a more sophisticated model using three-dimensional variable-viscosity convection and parameterization of melting rate, Ribe and Christensen (1999) showed that the Hawaiian swell is supported chiefly by the thermal buoyancy flux of the plume.

Jordan (1979) proposed that swells form because of underplating of compositionally lighter mantle residue left after the extraction of basaltic melts from the plume. Supporting this idea are seismic studies in hotspot chains suggesting that significant

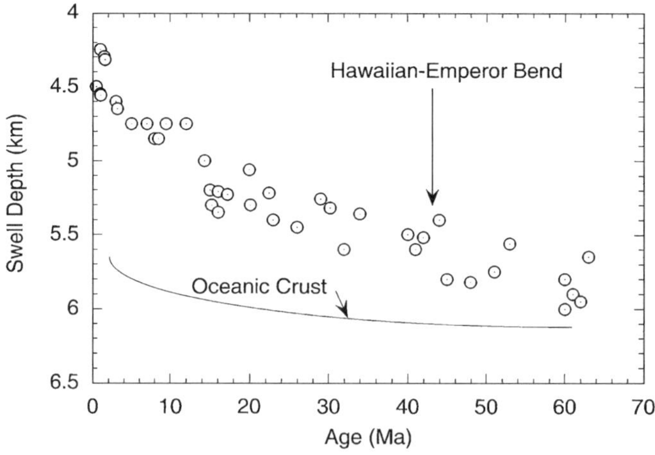

Figure 2.10. Depth of the seafloor as a function of volcano age for the Hawaiian–Emperor chain. Line is the predicted depth of normal seafloor as it ages in moving away from a spreading center. Modified from Clague and Dalrymple (1989).

crustal thickening due to mafic intrusions accompanies hotspot volcanism (Morgan et al. 1995). Because these intrusions come from the melted part of the plume, when they are extracted they leave a large volume of buoyant restite in the plume source (restite is the solid residue that remains after partial melting). In the model of Morgan et al. (1995), as the lithosphere moves off a plume, the swell subsides at a rate faster than the subsidence of the seafloor because the rates of melt production and residual lithosphere underplating fall off faster than seafloor subsidence. For rapidly moving plates, the buoyant swell root is dragged away from the hotspot by the plate, causing further spreading and thinning of the swell root with age. In contrast to the model of Morgan et al. (1995), the numerical model of Ribe and Christensen (1999) indicates that the buoyancy caused by the restite cannot exceed 33% of the total buoyancy flux for the Hawaiian plume, and thus depletion plays a secondary role to thermal buoyancy, at least in this case.

Hotspot Volcanoes

Hotspot volcanoes evolve through stages from their birth on the seafloor to their extinction far out in a hotspot chain (Juteau and Maury 1999). Of all the oceanic volcanoes, Hawaiian volcanoes are best known and are often used as the "type model" for oceanic volcano development (MacDonald and Katsura 1964). During the earliest submarine volcanic stage, known as the initial stage, one or more volcanic centers develop above a hotspot, and volcanoes grow rapidly to a kilometer or more in height (Fig. 2.11, Stage 1). As volcanoes grow, their slopes can increase to 15° or more. The base of the initial volcanoes chiefly comprises flows with ropy surfaces similar to ocean-ridge lavas but differing from these lavas in often being alkalic in composition. Above about 93-km depth, bulbous and tabular pillow flows dominate together with hyaloclastic breccias formed from fragmentation of the flows. Flows are often erupted from radial rift zones, the trends of which are controlled by stresses in the lithosphere or by old fractures. At depths less than 500 m, pillow basalts are replaced by hyaloclastic breccias, and when a volcano emerges above sea level, strombolian cones and associated subaerial lavas dominate. Hydrothermal springs are associated with the submarine volcanism.

During the second or shield stage (Fig. 2.11, Stage 2), tholeiitic basalt flows build a shield volcano in a relatively short time span of less than 1 Myr (Jackson et al. 1972).

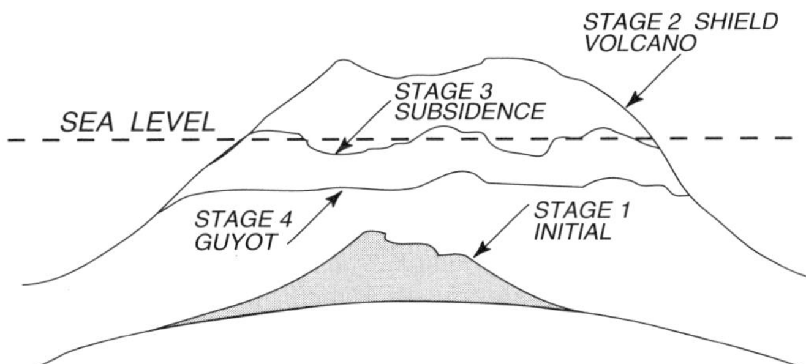

Figure 2.11. Stages in the evolution of a typical hotspot volcano: 1) initial stage, 2) shield stage, 3) subsidence stage, and 4) the guyot stage.

Most of the mass of hotspot volcanoes is formed during the shield stage, when typically more than 95% of a volcanic island is produced (Macdonald and Katsura 1964; Clague and Dalrymple 1989). During this stage, large calderas develop similar to those on Hawaii. Submarine sedimentary aprons and landslides form on the slopes of the shield volcanoes. During the latter part of the shield stage, a thin veneer of alkali lavas and pyroclastic rocks may cover the shield volcanoes and fill the calderas, as found for instance on Mauna Kea and related volcanoes in Hawaii. After a few million years of volcanic quiescence and erosion, very small volumes of alkali volcanics are erupted at isolated vents. If this evolution is typical of hotspot shield volcanoes, it would appear that they are composed almost entirely of tholeiitic basalts and that only very small volumes of alkali volcanics erupted during the initial stage and then again at the termination of volcanic activity in the shield or early subsidence stage.

During the subsidence stage (Fig. 2.11, Stage 3), volcanic islands begin to subside, and reef limestones are deposited around their margins. It is during the early part of this stage that volcanism wanes and eventually terminates as a volcano moves off its parental mantle plume. Most of the subsidence is due to an increase in depth of the ocean floor as oceanic lithosphere cools while it moves away from a hotspot and from an ocean ridge (Juteau and Maury 1999). During the subsidence stage, barrier reefs commonly develop around the margins of volcanoes, producing large lagoons or atolls.

During the last stage in the development of a hotspot volcano, known as the guyot stage (Fig. 2.11, Stage 4), a now largely or completely extinct volcanic center continues to subside to great depths, retaining the relatively flat top produced by earlier erosion. A variety of pelagic sediments may be deposited on guyots as they sink.

Hotspot Magma Composition

Both the tholeiitic and alkaline series are represented in hotspot igneous rocks. During the shield stage of hotspot volcano evolution, tholeiites greatly dominate in some chains as illustrated, again, by the Hawaiian shield volcanoes. Intermediate and felsic lavas resulting from prolonged fractional crystallization during the shield stage are uncommon (Juteau and Maury 1999). This is probably due to the very rapid growth of volcanoes, which results in a very short residence time of basaltic magmas in shallow fractionation chambers. In contrast, intermediate to felsic lavas together with associated alkali basalts are common, although in small volumes, during the initial and postshield stages of volcano development. The reason for this time progression is more fully discussed in Chapter 4. Hotspot (and perhaps nonhotspot) volcanoes in the South Pacific differ from the classic Hawaiian examples in that tholeiites are absent or rare, and members of the alkali series dominate throughout the development of the volcanoes, including the shield stage.

Fractional crystallization and progressive melting are recorded in hotspot volcanics, although the former often leaves the strongest imprint. Fractional crystallization leads to a strong decrease in the concentration of compatible elements (elements partitioned chiefly into the solid phases) and variable increases in incompatible elements (elements partitioned into the liquid phase) during magmatic processes

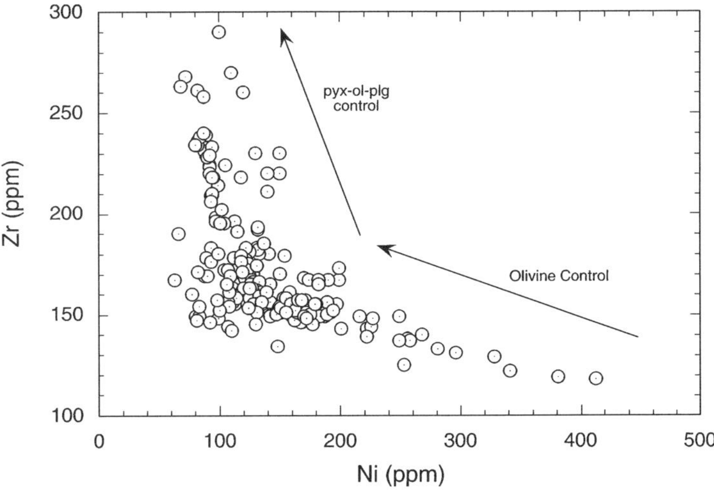

Figure 2.12. Zr–Ni distributions in tholeiites from Kilauea. Also shown are control lines for olivine and olivine + pyroxene + plagioclase. Data from www.georoc.mpch-mainz.gwdg.de.

(Rollinson 1993). In contrast, decreasing degrees of partial melting generally result in notable increases in the concentration of incompatible elements in a magma, with very little change in compatible elements, which remain in the restite mineral assemblage. In addition, gaps between magma groups are common during progressive melting.

Partial melting and fractional crystallization often can be distinguished by plotting the concentration of a compatible versus an incompatible element (Cocherie 1986). For instance, on a Zr versus Ni plot (Zr is incompatible and Ni is compatible) of tholeiites from Kilauea in Hawaii, we see a moderate increase in Zr with a large decrease in Ni, which is interpreted to reflect progressive fractional crystallization (Fig. 2.12). A notable kink in the trend at about 150 ppm Ni reflects a change in mineral control, from olivine to a mixture of olivine + pyroxenes + plagioclase. The onset of crystallization of pyroxenes and plagioclase decreases the rate at which Ni is removed from the melt, and thus the mineral control line steepens and Zr increases more rapidly, as shown in Figure 2.12.

In contrast to Kilauea, the alkaline volcanics from the island of Tubuai in the Austral–Cook hotspot track show several populations on a Zr–Ni plot (Caroff et al. 1997) (Fig. 2.13). Alkali basalts and basanites lie on an olivine control line, suggesting they are related by fractional crystallization involving olivine removal similar to the tholeiites from Kilauea. The nephenlinites and analcitites, however, show little variation in Ni, but considerable variation in Zr, which is consistent with an origin by progressive batch melting. The Zr "gap" between these two igneous rock groups at about 400 ppm of Zr is also consistent with a partial melting model. Caroff et al. (1997) have shown that the Tubuai nephelinites can be produced by 1.5–3% melting, and the analcitites by 3–6% melting, of a homogeneous mantle source (see batch melting vector in Fig. 2.13). The Tubuai lavas appear to reflect two stages of partial melting. During the first stage, 5–8% melting of a plume head produced a basanite parent magma (with more than 300 ppm of Ni), which later underwent fractional crystallization of olivine to produce a linear array of basanites and alkali basalts. Later, the melting zone may have migrated into the lithosphere, which is lower in Ni and possibly metasomatically enriched in incompatible

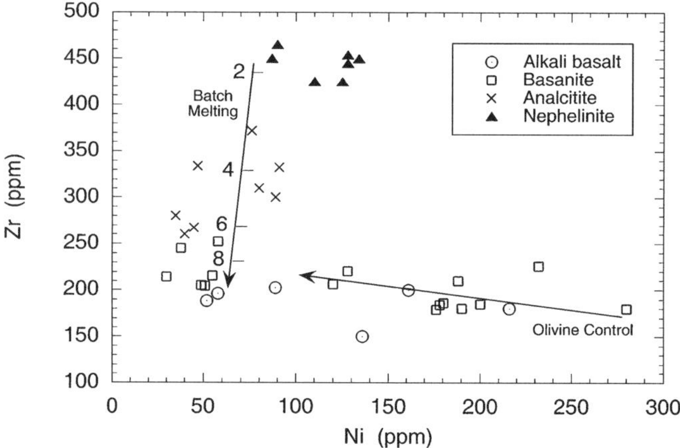

Figure 2.13. Zr–Ni distributions in volcanic rocks from Tubuai in the Austral–Cook volcanic chain. Data from Caroff et al. (1997). Numbers indicate percentage of batch melting of a garnet lherzolite source. Olivine control line is for fractional crystallization.

elements. Small degrees of melting of the mantle lithosphere gave rise sequentially to the nephelinite and analcitite lavas.

Seismicity and Tectonics of Hotspots

Hotspots with active volcanoes are seismically active, and we will use two modern hotspots, Hawaii and Yellowstone, to illustrate hotspot seismicity.

Hawaii

On Hawaii, seismicity is concentrated around the active volcanoes Mauna Loa and Kilauea as well as the Loihi seamount (Fig. 2.14). At Kilauea, earthquakes occur in

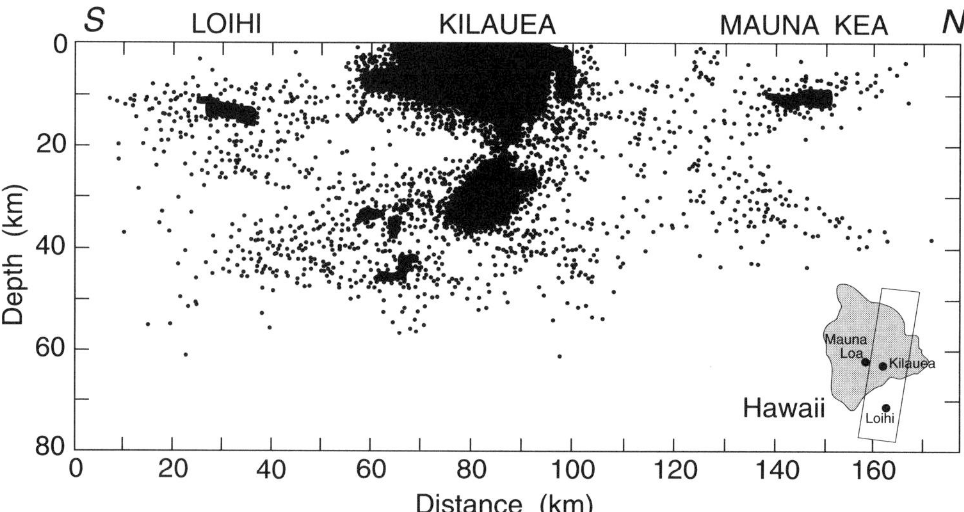

Figure 2.14. Cross section showing earthquake hypocenters in eastern Hawaii from 1960 through 1984. Inset map shows locations of earthquakes included. After Klein et al. (1987).

one of two locations: at shallow depths (1–4 km) under the rift system or along the vertical magma conduit beneath the volcano. They form tight clusters and usually occur in swarms (Klein et al. 1987; Klein and Koyanagi 1989). The base of the shallow earthquakes at about 15 km approximately coincides with the Moho beneath Hawaii. These earthquakes, which are chiefly the result of eruptions or intrusions, rarely have magnitudes over 4 and appear to be caused by stresses resulting from volume changes in magma reservoirs and conduits. Flanking earthquakes are more diffuse and extend to depths of about 60 km (base of the lithosphere). These earthquakes occur at more uniform rates and have larger magnitudes (up to 7) than the rift–conduit-related earthquakes. This implies that stresses causing the deeper earthquakes are more uniform than stresses producing the shallower earthquakes.

Earthquake hypocenters less than 20 km deep along Kilauea's conduit define a vertical seismic zone no larger than 4 km in diameter, whereas those from greater depths define a zone at least 13 km in diameter that plunges to the south (Fig. 2.14). The actual magma conduits are probably much thinner than the seismic zones surrounding them. Most earthquakes less than 20 km deep occur in swarms accompanying recharge of shallow magma chambers and are extensional, whereas earthquakes more than 20 km deep are nearly continuous in time and reflect compressional stress regimes (Klein 1982). The gap in seismicity at 15–20 km is interpreted as a zone of neutral stress and may represent the brittle–ductile transition. Mantle earthquakes (more than 20 km deep) result chiefly from stresses from the overlying volcanic pile, and especially the weight of the volcanics, which depress the lithosphere.

A generalized representation of magma production beneath Kilauea based on seismic and petrologic data is shown in Figure 2.15. Magma is produced by partial melting in a broad zone (20–30 km across) in the top of the Hawaiian plume. Small batches of magma escape from the plume and collect in intermediate chambers at a depth of 20 to 30 km, where fractional crystallization removes olivine and other phases

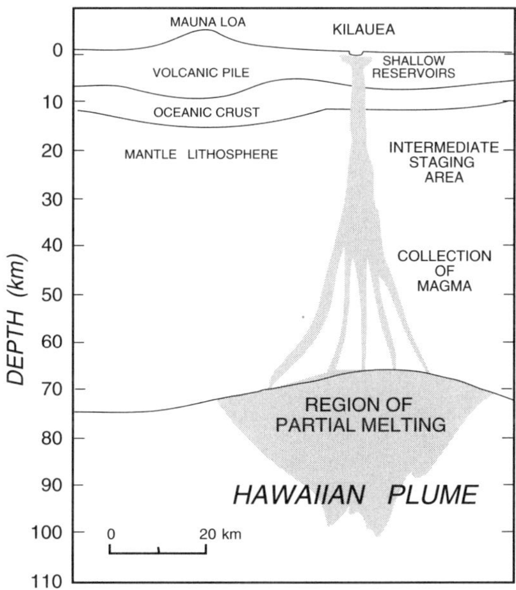

Figure 2.15. Schematic cross section of the lithosphere beneath Kilauea in Hawaii showing melting in the Hawaiian mantle plume, magma transport, and levels of magma collection before eruption. Modified after Wright (1971).

from the magma (Wright 1971). The large concentration of earthquakes at this depth reflects extensional deformation associated with magma collection. Magmas then move to a shallow reservoir (5–10 km deep), where more fractional crystallization occurs involving removal of large amounts of pyroxenes and plagioclase.

Yellowstone

The Yellowstone hotspot is characterized by intense swarms of shallow earthquakes with occasional intermediate earthquakes (Smith and Braile 1993, 1994) (Fig. 2.16). All of these reflect extensional stresses related to the movements of magma at shallow depths (<10 km). Lateral variations in earthquake depths appear to reflect variation in depth of the brittle–ductile transition. Outside the Yellowstone caldera, focal depths extend to more than 15 km, whereas inside the caldera they are less than 10 km (Fig. 2.17).

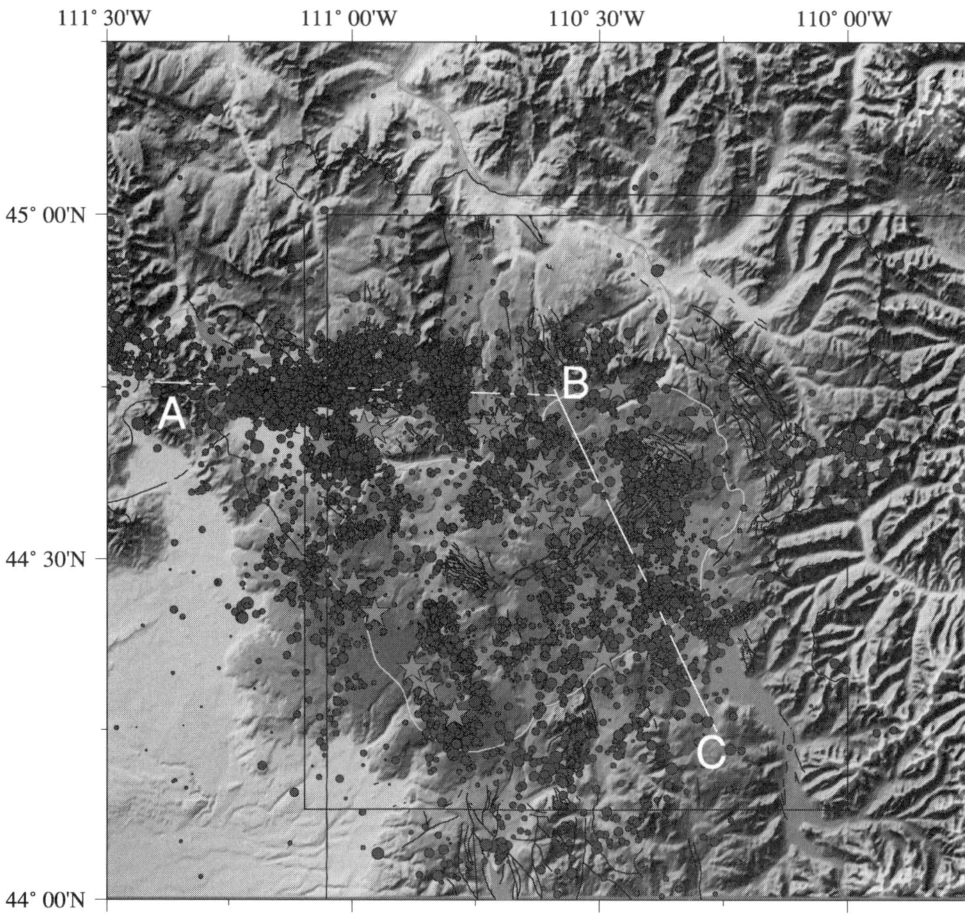

Figure 2.16. Map showing seismicity during the period 1973–1996 for Yellowstone National Park and vicinity. Epicenters from University of Utah Seismograph Station files. Stars indicate post-caldera volcanic events. From *Windows into the Earth: The Geologic Story of Yellowstone and Grand Teton National Parks*, by Robert B. Smith and Lee J. Siegel, Copyright © 2000 by Robert B. Smith and Lee J. Siegel. Used by permission of Oxford University Press, Inc.

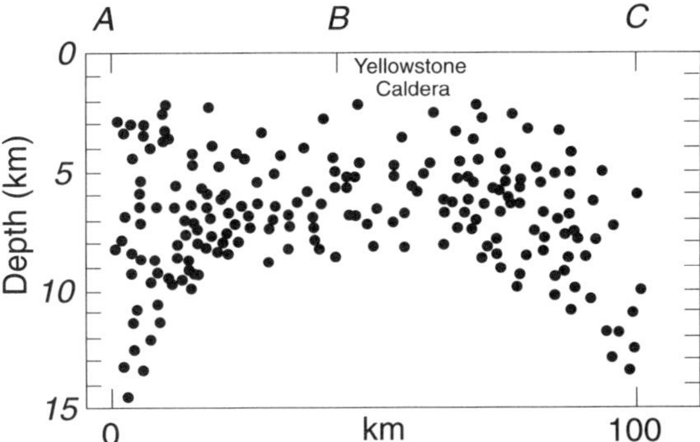

Figure 2.17. Cross section showing earthquake hypocenters in Yellowstone National Park 1973–1996. A–B and B–C section locations are shown in Figure 2.16. Modified after Miller and Smith (1999).

Inside the caldera, the crust is heated to a ductile state at 350–450 °C, a temperature at which the crust is incapable of supporting shear stresses. The upper crustal layer, with typical thicknesses of 10–15 km ($V_p = 5.9$–6.1 km/s) is absent or very thin in the Yellowstone area. This layer appears to have been replaced with igneous rocks related to the Yellowstone hotspot. The Yellowstone Plateau (comprising most of Yellowstone National Park, Fig. 2.16) has tectonically risen by up to 100 cm between 1923 and 1984, reflecting transport of magma into the middle or upper crust, pressurization of a deep hydrothermal system, or both (Dzurisin et al. 1990). In addition, the seismic wave velocities in the upper mantle beneath Yellowstone and the Snake River volcanic plain to the southwest are anomalously low by up to 3% extending to depths of 200 km. Such low velocities are consistent with partial melting or storage (or both) of plume-derived magmas in the lithosphere along the Yellowstone hotspot track.

During the precaldera phase in Yellowstone Park, plume-derived basaltic magmas were transported through the mantle lithosphere into the lower crust, where they collected in holding chambers at depths of 30–40 km (Smith and Braile 1994). The combined effects of heating and possible partial melting of the lithosphere at this location reduced P-wave velocities to as low as 7.6 km/s. Continued transport of heat into the crust by basaltic magmas and volatiles raised crustal temperatures sufficiently to cause partial melting, leading to the production of large volumes of felsic magma. The felsic magma collected in a vast, shallow magma chamber, and a combination of thermal expansion and magma transport resulted in significant crustal uplift. This culminated in felsic volcanism, the eruption of voluminous rhyolite ash flows, and the formation of the Yellowstone caldera by later collapse.

Plume–Hotspot Relationships

Wilson (1963) proposed a relatively fixed mantle plume beneath a moving plate to explain the linear age progression in oceanic volcanic chains such as the Hawaiian–Emperor chain. Morgan (1971) later showed that hotspot volcanic chains could be formed in the Pacific basin by the movement of rigid lithospheric plates over fixed hotspots. The remarkable coincidence of the major Pacific hotspot volcanic chains and the calculated paths strongly supports the conclusion that the plumes responsible for

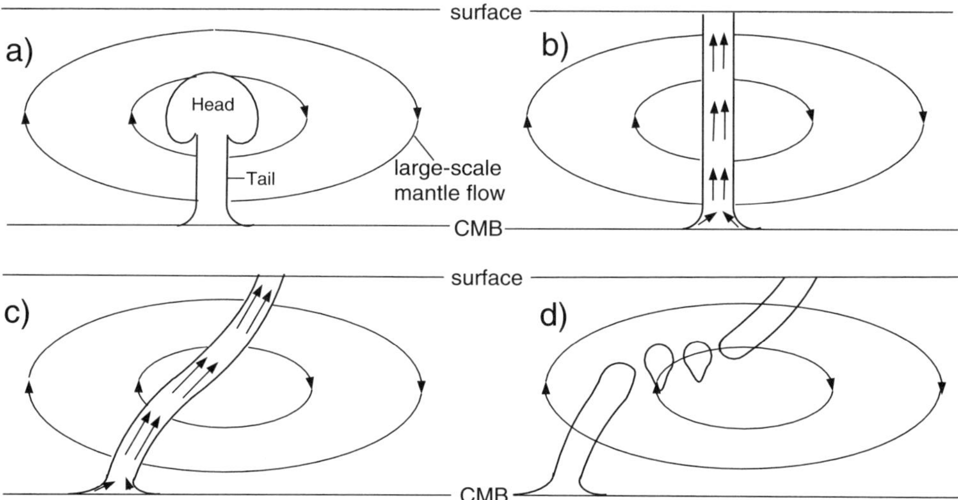

Figure 2.18. Four potential stages in the evolution of a mantle plume after it impinges on the lithosphere. CMB = core-mantle boundary. Modified after Steinberger and O'Connell (1998).

these hotspots have remained approximately fixed relative to each other for the last 50 Myr or more (Steinberger and O'Connell 1998).

Because plume heads should rise rapidly in the mantle, they will rise along vertical trajectories with a straight tail from the source. Although the material in the tail will rise much faster than surrounding mantle, compared with a plume head, the tail will not have as much buoyancy contrast owing to its smaller size. Consequently, plume tails may be advected in the upper mantle, and thus hotspots may move as their tails are deformed (Richards and Griffiths 1988). In general, four stages are recognized in the lifetime of a plume (Whitehead 1982) (Fig. 2.18). The first is the rising stage (a), which consists of the plume head and tail. In the second stage (b), the plume head is flattened and undergoes partial melting. The result of this is that portions of the plume head are added to the lithosphere either as magmatic additions or accretion of restite left from the melting. Some of the plume restite may also be mixed into the upper mantle. The net result of stage 2 is that only the tail remains attached to the hotspot. During the third stage (c), the plume tail is distorted by mantle flow. If the distortion becomes significant (>60° from the vertical), the tail may become unstable in the upper mantle and split into separate segments, as shown in (d).

The hotspots that produced the major volcanic chains in the Pacific basin appear to have remained approximately fixed relative to each other. To remain fixed, it would appear that the tails of the Pacific hotspots cannot be significantly distorted by mantle advection. Steinberger and O'Connell (1998) and Steinberger (2000) have shown that significant distortion and segmentation of plume tails can be avoided if the asthenosphere has a viscosity significantly lower than the lower mantle from which plumes are derived. Although plumes drift in the upper mantle in response to advection of the asthenosphere, a low viscosity of the asthenosphere ensures a high rising velocity of plume tails, which prevents their becoming strongly tilted. The Steinberger and O'Connell model also shows that, in the vicinity of ocean ridges, hotspots tend to be advected towards the ridge, thus explaining why some hotspots occur at (Iceland)

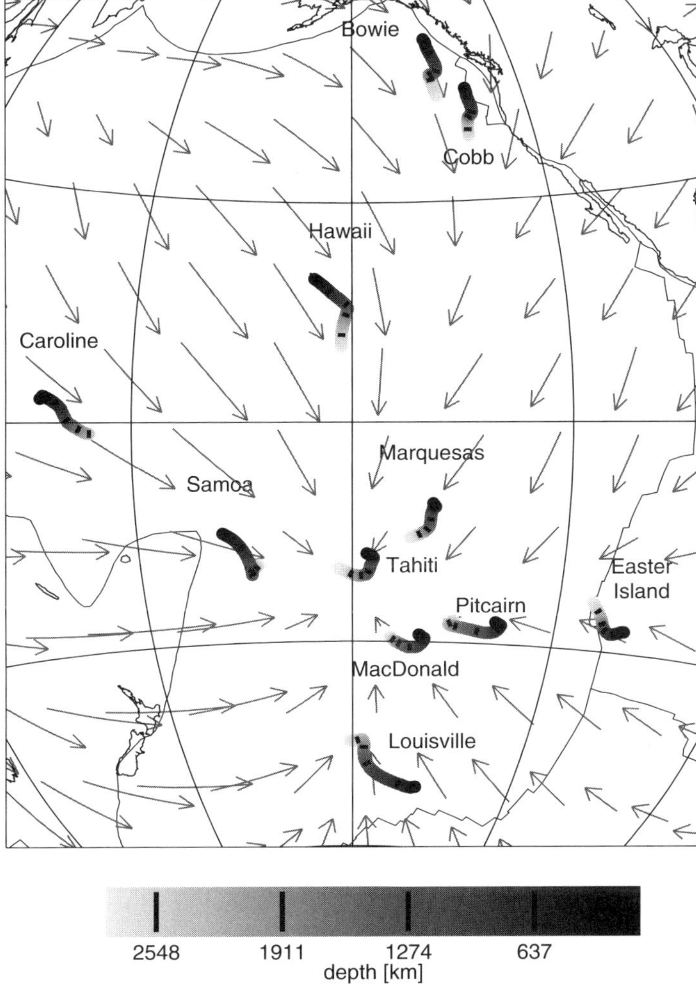

Figure 2.19. Calculated shapes of plume tails as a function of depth in the Pacific basin for the last 68 Myr. Arrows indicate mantle flow at 2700-km depth, and arrow lengths indicate the amount of displacement of plume tails occurring in the last 68 Myr from constant flow. Courtesy of Bernhard Steinberger.

or near (Tristan) ridges. The source regions of plumes in the deep mantle tend to be advected towards large upwellings such as those in the South Pacific and Africa, which is a feature of the model consistent with a large number of plumes concentrated in such upwellings (Steinberger 2000) (Figs. 2.1 and 2.19). Older plumes tend to be tilted more than younger ones. This occurs because the bottoms of the plumes are advected towards the mantle upwellings, whereas plume heads are influenced chiefly by mantle flow at shallow depths, which may be in a different direction. Because of the coherence of flow beneath the Pacific region, the relative motion of hotspots is quite small, which is an outcome of the model that helps reconcile the observation that apparently stationary plumes originate from a convecting lower mantle.

Hotspots may interact with lithospheric plates in a variety of ways, some of which are illustrated in Figure 2.20. If oceanic plate motions relative to hotspots are small, large

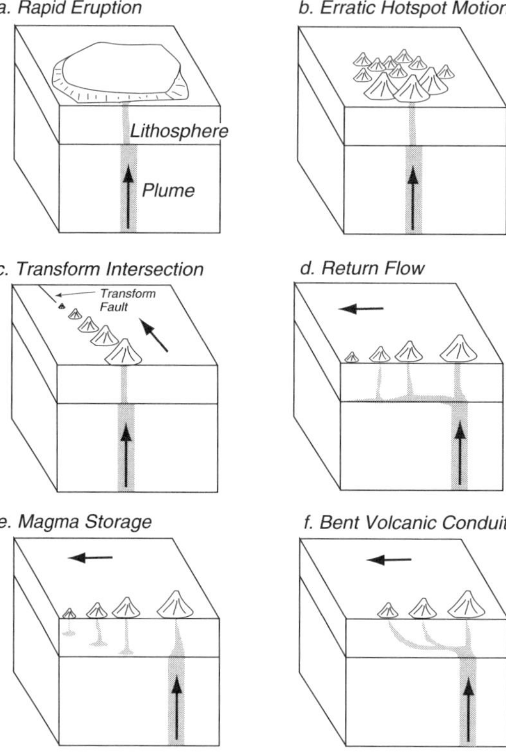

Figure 2.20. Examples of possible interactions of lithospheric plates and hotspots. Modified after Condie (1997a).

amounts of magma are erupted, forming a large island with thick crust, as exemplified by Iceland, and oceanic plateaus such as the Shatsky rise and the Ontong Java plateau in the South Pacific (a). If plate motion is erratic in direction and rate, irregular clusters of volcanoes may form at hotspots (b), as characterized perhaps by the Mid-Pacific Mountains (Fig. 3.1). Volcanic chains that cross transform faults often show cross trends of volcanoes parallel to the faults. Such is the case along the Hawaiian ridge where it crosses the Molokai and Murray fracture zones in the northeastern Pacific. Because transform faults are zones of weakness, magma may be injected into them as they pass over hotspots (c). Another interaction is possible if a plume breaks out at the base of the lithosphere and flows from the hotspot back along the weakened hotspot track (d). This could result in irregularities in the age of volcanism along the hotspot trace, as observed in some island chains such as the Cook–Austral chain. Because the lower lithosphere is near its melting temperature, magmas may be stored here for considerable lengths of time. This can lead to eruption of magma (generally small amounts) along the volcanic chain after it passes over the hotspot (e). When plume conduits are bent to more than 60°, the conduit may segment, and the tail may rise and form a new conduit (f). If a magma conduit is not disconnected from the plume, eruption may continue in volcanoes that have passed over the hotspot.

Very active hotspots occur only on fast-moving plates, whereas weak hotspots occur on both fast- and slow-moving plates (Sleep 1990). This distribution partly results from the effect of plumes on plate size and velocity. For instance, a strong hotspot beneath a slow-moving plate may weaken the plate and cause it to break up and increase the velocity of the new plates (Morgan 1981). This may have occurred during breakup of Gondwana near the Réunion hotspot in the southern Indian Ocean, which reflects a large plume that arose beneath the slow-moving Gondwana supercontinent.

Plume–Ridge Interactions

Although plume–ridge interactions are well established from geochemical and isotopic data (Schilling 1991), these interactions are also verified from seafloor morphology, gravity, seismic tomography, laboratory experiments, and fluid dynamic models (Kincaid et al. 1996). When hotspots form near an ocean ridge, they may interact with the ridge, and in some instances, such as the Shona and Discovery hotspots near the Mid-Atlantic Ridge, they may "capture" a ridge segment (Small 1995). Many hotspot–ridge interactions show evidence of distinct ridge jumps in the direction of the hotspot as the ridge attempts to relocate on the hotspot. If ocean ridges are not fixed, they will sometimes collide with hotspots, as happened with the Iceland hotspot about 15 Ma (Vink 1984). If a hotspot reflects a large plume, as in the case of Iceland, the ridge will be "captured" by the hotspot, and continuous ridge jumps are required to keep it on the hotspot. When ocean ridges are carried across small plumes, as is the case for most of the hotspots in the Atlantic basin, they are not captured. In fact, most near-ridge hotspots reflect small plumes with relatively small buoyancy flux (e.g., Azores, Galápagos, Tristan da Cunha, Juan de Fuca, Bowie), probably not capable of capturing ridges (Sleep 1990).

One of the first well-documented examples of ridge–plume interaction was that of the Mid-Atlantic Ridge and the Iceland plume. Schilling (1973, 1991) has demonstrated a progressive change in chemical composition of lavas along the Reykjanes segment of the ridge in approaching the southern coastline of Iceland. This is shown in Figure 2.21 for the (La/Sm)n ratio. Schilling interpreted the progressive change in the (La/Sm)n ratio as representing mixing of depleted mantle tapped at the ocean ridge and a relatively

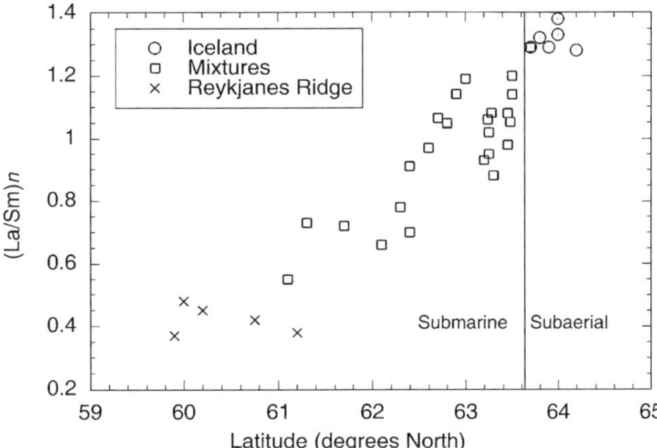

Figure 2.21. Plot of (La/Sm)n ratio versus latitude in basalts along the Mid-Atlantic ridge progressing northward into Iceland showing the effects of mixing of ocean–ridge and plume mantle sources. (La/Yb)n denotes the primitive-mantle normalized ratio. After Schilling (1973).

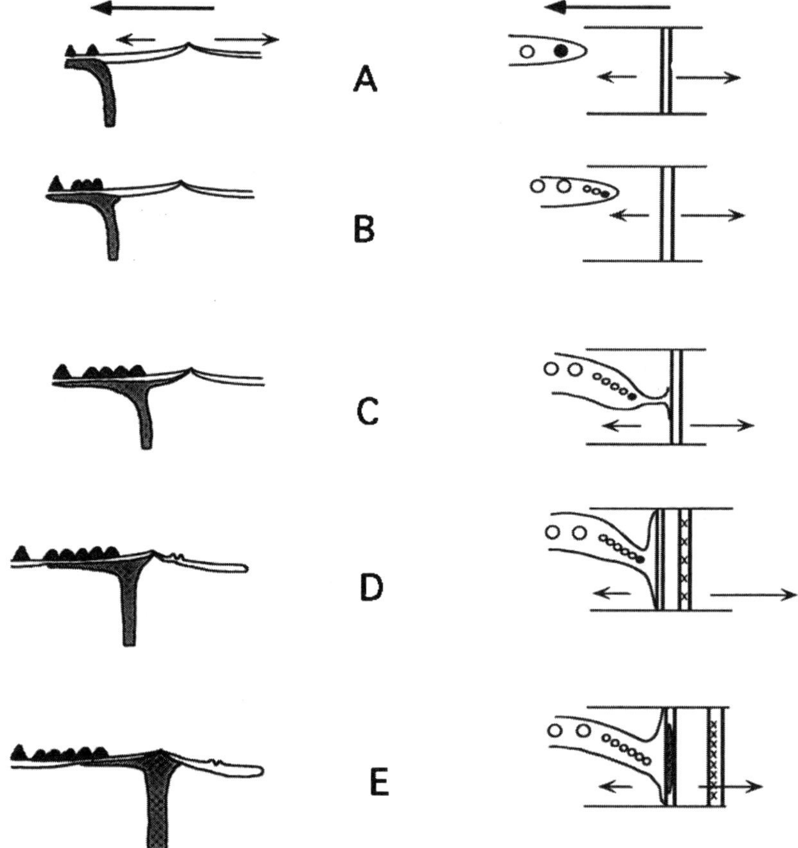

Figure 2.22. Diagrams showing possible interactions of an ocean ridge and a mantle plume as a spreading center migrates toward a hotspot. Cross section (left) and map (right) views are shown. Solid circle, present hotspot location; open circles, hotspot track; bold arrow, plate motion; small arrows, spreading directions. Modified from Small (1995), with permission. Copyright © 1995 by the American Geophysical Union.

enriched component in the plume source beneath Iceland. He suggested that some of the plume is diverted along the axis of the ridge, where it mixes with depleted mantle in varying degrees. Recent studies of helium isotopes in basalts derived from the Galápagos plume also support this idea for the Galapágos hotspot inasmuch as plume helium rich in ^{3}He becomes diluted with upper mantle helium (rich in ^{4}He) as it mixes with mantle beneath the Galápagos spreading center (Kurz and Geist 1999).

As an ocean ridge approaches a plume, various ridge–plume interactions are possible, as illustrated in Figure 2.22 (Small 1995). In case A, there is no connection between the plume and the ridge system. In B, the plume is close enough to the ridge that some plume material is forced "uphill" towards the ridge along the base of the lithosphere. In C, a significant amount of the plume becomes incorporated in the spreading center. As the flux of material in the plume to the spreading center increases, it is distributed along the axis of the ridge. Finally in stage D, the lithosphere between the plume and the ridge is sufficiently heated and weakened to allow a ridge jump as the spreading center moves abruptly towards the plume. During this stage, a much larger fraction of the plume flux enters the spreading center than goes into the hotspot islands, and eventually production of hotspot islands ceases. In stage E, the plume is centered beneath the ridge, and the extinct volcanic chain and the abandoned ridge segment are rifted away in opposite directions from the spreading center.

Numerical and laboratory models are also important for a better understanding of plume–ridge interactions. For ridge-centered plumes such as Iceland, the relationships between plume size, plume flux, and spreading rate were first studied in laboratory tank experiments by Feighner and Richards (1995) and further developed in numerical models (Feighner et al. 1995; Ribe et al. 1995). The numerical models of Ribe (1996) for off-ridge plumes and stationary ridges established the scaling laws for the dependence of plume size on a range of variables, including plume flux, spreading rate, plume–ridge distance, and lithospheric thickening with age. These results and those of Ito et al. (1997) show that plume diameter and plume–ridge distance correlate with the ratio of plume flux to spreading rate for stationary ocean ridges. In the case of migrating ridges, the distance of plume–ridge interaction is reduced when the ridge migrates toward the plume because of excess drag on the plate. In contrast, when a ridge migrates away from a plume, the plume–ridge interaction is enhanced because of reduced drag on the plate. The models further suggest that mantle plumes may spread along the base of the lithosphere parallel to ocean ridges and thus may comprise a significant fraction of the oceanic lithosphere.

The Hotspot Reference Frame

As originally recognized by Morgan (1972), if hotspots have remained approximately fixed relative to plates, the age distribution of igneous rocks in hotspot tracks should provide a record of absolute plate motions during the breakup of Pangea in the last 200 Myr. This is known as the hotspot reference frame. Another reference frame for plate motions is the paleomagnetic reference frame provided by rocks that have acquired magnetization in the Earth's magnetic field at different times in the geologic past. By measuring the remanent magnetization in rocks and determining the age of magnetization from isotopic dating, an apparent polar wandering path can be calculated that records latitudinal motions of plates. In addition, relative plate motions can be determined from the correlation of the magnetic stripes on the seafloor, which record the production rate of new lithosphere at ocean ridges. This method provides only relative motions because ocean ridges are also moving over the mantle as they generate new lithosphere. It is unlikely that plumes are precisely fixed with time, and some interplume motion is expected because of viscous coupling between the base of lithospheric plates and the convecting asthenosphere, as described above (Fig. 2.22). If these motions are small or constant over large regions, however, interplume velocities may be considerably less than plate velocities, and thus hotspot tracks can be used to estimate plate motions.

The amount of interplume motion can be estimated by comparing the geometry and ages of volcanism along hotspot tracks with reconstructions of past plate motions based on relative plate motions (Jarrard and Clague 1977; Duncan 1991). If hotspots under all plates are fixed relative to each other, the calculated motions of plates should follow hotspot tracks observed on these plates from volcanic chains. Any deviation of the predicted plate movements from actual hotspot tracks indicates the magnitude and directions of interplume drift. An example of this approach is given in Figure 2.6 for some of the major hotspots in the Atlantic and Indian Oceans (Muller et al. 1993). The present-day positions of the hotspots are inferred from modern or recently active volcanism and high heat flow. The modeled hotspot tracks calculated in 5-Myr intervals show a remarkable agreement with the observed tracks and thus indicate little motion of the hotspots in the Atlantic and Indian Oceans relative to each other.

The calculated hotspot track verifies that the New England–Corner seamount chain in the North Atlantic was produced above the Great Meteor hotspot, which is now separated from the chain by the modern Mid-Atlantic Ridge (Fig. 2.6). In the South Atlantic, the Walvis Ridge and Rio Grande Rise have formed on opposite sides of the Tristan da Cunha hotspot as the Atlantic basin opened beginning about 130 Ma. In the Indian Ocean, the Mascarene–Chagos–Laccadive hotspot track emanates from the Réunion hotspot east of Madagascar and is broken by the young Carlsberg Ridge in the western Indian Ocean. This track leads to the Deccan traps in India, strongly suggesting that these flood basalts were produced by the Réunion hotspot. The reconstruction suggests that the Kerguelen hotspot in the southern Indian Ocean formed the Ninetyeast Ridge after 100 Ma, but previously the hotspot was on the Antarctic plate and formed the south Kerguelen Plateau and Broken Ridge. Although the Kerguelen hotspot is never coincident with the Rajmahal traps in northeast India, if the initial mantle plume affected a 2000-km-diameter area, which seems likely (White and McKenzie 1989), the Rajmahal basalts may have been derived from the Kerguelen hotspot.

The Eightyfive-East Ridge (Southeast of India) matches well with the present-day location of the Crozet hotspot beneath the Conrad Rise southwest of the Kerguelen plateau (Fig. 2.6). The large gap between the southern end of the Eightyfive-East Ridge and the Conrad Rise segments of the hotspot track was caused by a major jump of the Mid-Indian Ridge about 65 Ma. This jump transferred the Crozet hotspot (on the east end of the Conrad Rise) from the Indian to the Antarctic plate, and thus no trace of the hotspot is left on the Indian plate south of southern termination of the Eightyfive-East Ridge.

A similar exercise for the Pacific basin shows that major hotspots within the basin have remained approximately fixed relative to each other for the last 100 Myr (Morgan 1971; Steinberger and O'Connell 1998). The classic examples are the Hawaiian–Emperor and Louisville hotspot tracks, the calculated trajectories of which fall on top of, or very close to, the observed tracks (Fig. 2.2). Also showing good agreement are the Line–Tuamotu and Austral–Cook chains in the central South Pacific, the Caroline chain in the extreme western Pacific, and the Cobb–Eickelberg and Kodiak–Bowie chains in the northeastern Pacific. The results strongly suggest that the Pacific hotspots have remained stationary relative to each other for the last 100 Myr. Also, the almost perfect fit of volcanic chains in the Pacific plate with a pole of rotation at 70°N, 101°W and a rate of rotation about this pole of about 1 deg/Myr for the last 10 Myr suggests that Pacific hotspots have remained fixed relative to each other over this time.

What about possible motion between Pacific and Atlantic–Indian hotspots? Studies comparing these two groups of hotspots have reported significant discrepancies between the predicted and actual hotspot tracks (Molnar and Stock 1987; Acton and

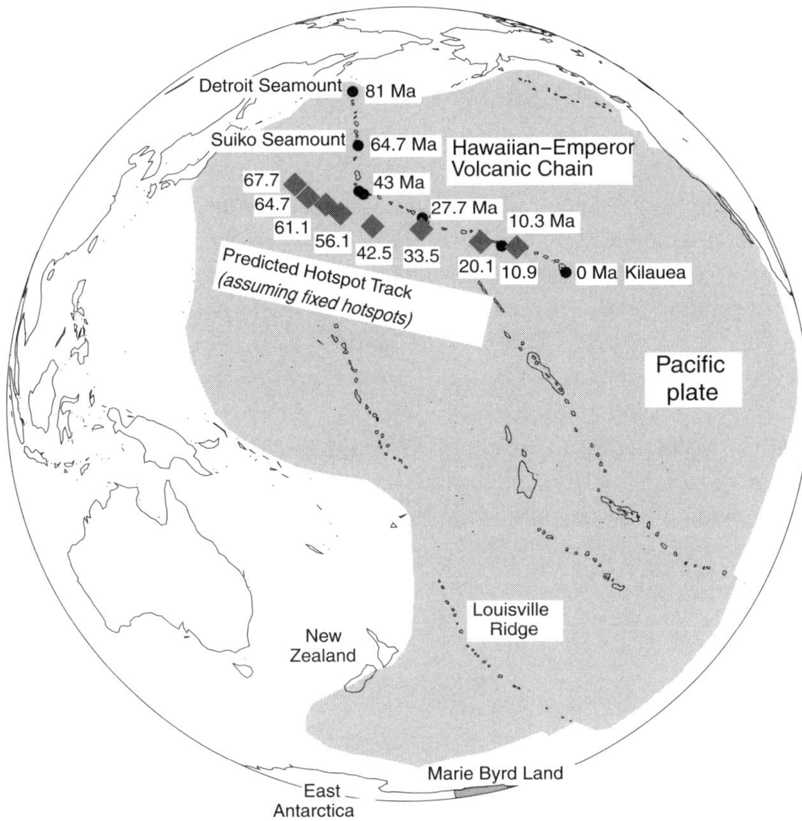

Figure 2.23. Map of the Pacific region showing the Hawaiian–Emperor chain with ages of volcanism. Also shown is the calculated hotspot track on the assumption that the Hawaiian hotspot was fixed relative to Atlantic–Indian hotspots for the last 100 Myr. Modified from DiVenere and Kent (1999), with permission. Copyright ©1999 by Elsevier Science.

Gordon 1994; DiVenere and Kent 1999). This discrepancy is illustrated in Figure 2.23 for the Hawaiian–Emperor chain assuming this hotspot is fixed relative to the Atlantic–Indian hotspots. The discrepancy is particularly striking prior to the 43-Ma bend in the chain with an offset between the predicted and observed positions of the track at 65 Ma of about 1600 km. Paleomagnetic studies of hotspot volcanics have also documented relative motion between the Atlantic–Indian and Pacific Ocean hotspots. Van Fossen and Kent (1992) showed that during the Cretaceous, all Atlantic hotspots moved southward as a coherent group, but only the Louisville hotspot in the Pacific moved southward. Tarduno and Gee (1995) compared paleomagnetic paleolatitudes of Pacific and Atlantic hotspots and concluded that there must have been large-scale motions between the two hotspot suites. DiVenere and Kent (1999) showed that Cenozoic relative motions between East and West Antarctica can account for only about 20% of the apparent motion between the Hawaiian–Emperor chain and Atlantic–Indian hotspots (25 mm/yr). Thus, it would appear that inter-hemispheric relative motion between Atlantic–Indian and Pacific hotspots is about 20 mm/yr for the last 65 Myr. This rate of interplate hotspot motion is comparable to average plate velocities.

True Polar Wander

True polar wander is relative movement between the Earth's spin axis and the mantle. This can be caused by movement of the entire lithosphere as a single unit at the low-velocity zone at the base of the lithosphere relative to the Earth's rotational pole. The amount of true polar wander in the Earth, if any, can be estimated by comparison of plate motions calculated from paleomagnetic and hotspot reference frames if it is assumed hotspots have remained fixed in that portion of the globe being analyzed (Courtillot and Besse 1987). To do this, paleomagnetic pole positions of continents are compiled and corrected for rotation relative to a single continent (usually Africa for the last 200 Myr) using the dated magnetic anomalies on the seafloor to reconstruct past plate motions. From the results, a composite polar wander path for all plates is obtained. This path is then compared with a hotspot track as viewed from the fixed continent. The true polar wander path is calculated from the angular rotation that shifts the mean paleomagnetic pole of a certain age to the North Pole, and then the hotspot pole of the same age is calculated by applying the same rotational offset. Estimates of true polar wander using this approach agree well with astronomical estimates, suggesting the Earth's rotational axis has moved at a rate of less than 1 deg/Myr during most of the Tertiary (Andrews 1985).

Hotspot Origin

Four basic models have been presented to explain hotspot volcanic chains with progressive changes in age of volcanism (Clague and Dalrymple 1989): propagating fractures, local convection, shearing between the lithosphere and asthenosphere, and mantle plumes. Of these, only the mantle plume model has survived the test of time. In the propagating fracture model, the lithosphere breaks and as a fracture propagates laterally, and magma is tapped from the asthenosphere, producing a chain of volcanoes (Green 1971; Turcotte and Oxburgh 1978). The most significant problem with this model is that most hotspot volcanoes do not tap the depleted asthenosphere but have enriched magma compositions suggestive of deep mantle sources. In the convection model proposed by Wilson (1963), a hotspot chain is formed over the top of two upwellings with diverging flow directions, which is similar to what goes on at an ocean ridge. As the lithosphere moves across the source conduit leading from the deep central core of one upwelling cell to the surface, the hotspot track is produced on the overriding plate. Several other perturbations of the upwelling model have been proposed (Clague and Dalrymple 1989), all of which involve upwelling convection cells beneath hotspot ridges. Not only is there no evidence for long linear convection cells beneath hotspot ridges, but most of these models can neither account for the composition of magma sources as dictated by trace elements and isotopes nor can explain hotspot swells. According to the shear melting idea of Shaw (1973), hotspots are formed by plate motions and are regulated by various feedback mechanisms. Shearing between the lithosphere and asthenosphere results in a rise in temperature, causing a hotspot. Little if any evidence, however, has been reported to support such a model.

Hotspot volcanoes are known to develop on swells. Heat flow increases over these swells, and the lithosphere is thinner, suggesting that it is being pushed upwards by buoyant mantle. The only model that adequately explains this relationship is the mantle plume model. In this case, a mantle plume rising from the deep mantle reaches the base of the lithosphere and pushes it upwards. The plume also has a higher potential temperature than surrounding mantle; hence, the plume undergoes decompression melting, producing magmas that rise to form hotspot volcanic chains as the lithosphere moves over the hotspot. Chemical and isotopic data, as discussed in Chapter 5, also support a mantle plume source for hotspot magmas.

Venusian Hotspots

Large volcanic structures on Venus occur primarily in the equatorial region of the planet and are typically associated with topographic rises (Head et al. 1992; Stofan et al. 1995). Some of the better known rises thought to result from mantle plumes are Beta Regio, Dione Regio, and Themis Regio (Fig. 2.24). The rises on Venus have diameters of the order 1000 km or more, which are similar to terrestrial hotspot diameters. Venusian rises show a wide range in topographic and morphologic characteristics as well as in their gravity signatures (Smrekar and Phillips 1991). Many have large shield volcanoes and lava plains indicative of voluminous volcanism. Others, such as western Eistla (Fig. 2.24), are characterized by small (<50-km diameter) volcanic edifices. Over half of the Venusian rises have coronae (small circular uplifts) on or near their flanks, which imply the presence of numerous small mantle plumes associated with, or derived from, a larger superplume at depth. Evidence for extensional deformation is present at all Venusian rises, which range from small grabens to large zones defined by fractures and major rift systems (Stofan et al. 1995).

The amount of material erupted from terrestrial and Venusian hotspot volcanoes ranges over three orders of magnitude, suggesting similar ranges of melting in plumes on both planets. The distribution of swell volume as a function of swell height shows that Venusian swells overlap with their terrestrial counterparts that form on crust with ages of 90 Ma or greater (Fig. 2.25). Although volumes of hotspot swells on Venus are not significantly different from those on Earth, the ratio of gravity to topographic relief of the Venusian examples is a factor of 2–4 higher than on Earth (Smrekar and Phillips 1991). It is generally agreed that Venus has a thick, motionless lithosphere (Turcotte 1993), yet hotspot swells and volcanoes have failed to grow as large as the Tharsis rise

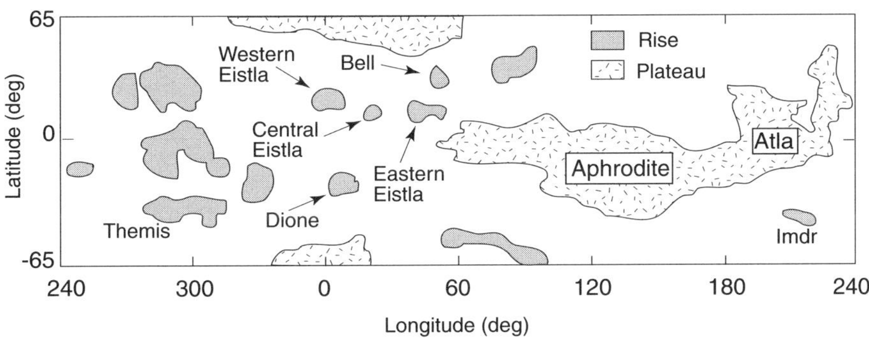

Figure 2.24. Map of Venus showing some of the major topographic rises. After Stefan et al. (1995).

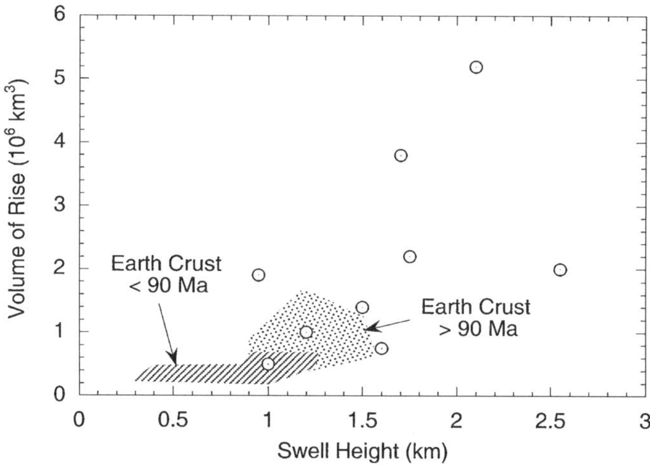

Figure 2.25. Plot showing rise volume versus swell height for Venusian rises compared with terrestrial rises and plateaus. After Stefan et al. (1995).

and Olympus Mons on Mars (Chapter 3). In fact, even with moving plates on Earth, some terrestrial and Venusian hotspots have grown to comparable sizes (Fig. 2.25). Although the timing of hotspot formation on Venus is not well known, the overlap in size range of terrestrial and Venusian hotspot swells suggests that the time-integrated plume strengths are comparable on both planets (Stofan et al. 1995).

Mantle Upwellings

Introduction

To balance the subducted slabs that sink into the mantle there must be an upward return flow. Because subducted plates are relatively cool, they decrease the temperature of nearby mantle, thus leaving relatively warm mantle in the regions between subduction zones. These broad, warm regions of mantle, known as mantle upwellings, are relatively buoyant and rise, providing the return flow. Mantle upwellings elevate Earth's surface up to a few hundred meters, producing "superswells," and because they elevate the temperature of the uppermost mantle, widespread small degrees of melting occur near the tops of upwellings giving rise to volcanism and to mafic underplating of the crust. Another characteristic of upwellings is that they contain most of the modern hotspots and hence most of the modern mantle plumes. A possible explanation for this is that deep flow in a mantle upwelling, just above the core–mantle boundary, sweeps material upwards, thickening the D″ layer. This promotes instability in D″, which leads to the formation of new plumes that rise within the upwelling. That plumes with the highest buoyancy flux are concentrated in the Pacific upwelling (Ribe and Valpine 1994) is consistent with this idea. Davaille (1999) has shown experimentally that upwellings with constituent plumes (hotspots) can form with only a small amount of density layering ($\approx$1%) in the mantle.

Today, two large upwellings occur in the mantle, one beneath the African plate and one beneath the Pacific plate (Maruyama 1994). In this section, we will review some of the characteristics of these upwellings.

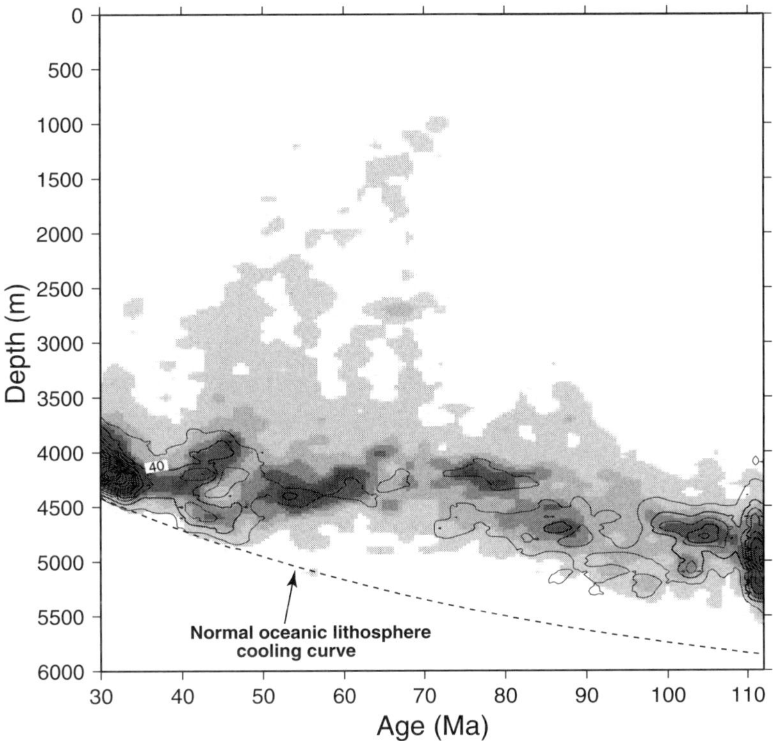

Figure 2.26. Depth–age plot for lithosphere over the Pacific superswell. Shading for all data and contours excludes regions around major hotspot chains to avoid the influence of hotspots. Courtesy of Marcia McNutt.

Superswells

Superswells are broad, oval upwarps of the lithosphere with up to a kilometer or more of relief, and they extend for many thousands of kilometers (Fig. 2.1). The roots of superswells can be imaged by seismic tomography, and their topography and geoid anomalies are large enough to provide important constraints on dynamic mantle flow beneath plates (McNutt 1998).

Perhaps the most unusual feature of superswells is that their depth below sea level does not follow the normal cooling path of oceanic lithosphere over the last 70 Myr (Fig. 2.26). This means that superswells subside more slowly than average oceanic lithosphere as it ages and moves away from spreading centers. Possible reasons for the slow subsidence of superswells include thickening of the crust, adding sediments to the top of the crust, or the changing of uppermost mantle mineral assemblages to less dense rocks. Although crust is locally thickened beneath oceanic plateaus and oceanic volcanoes, widespread thickening of the oceanic crust in superswells is not observed (Caress et al. 1995; McNutt 1998). Sediment cover also does not thicken over either the African or Pacific upwellings. Partial melting of the upper mantle to produce basaltic magmas should leave a buoyant restite because garnet is preferentially removed during melting of mantle rocks (Jordan 1979). Calculations by McNutt (1998) indicate that, although this buoyant restite contributes to the uplift in the Pacific superswell (about 20%), it cannot explain all of it. The only mechanism that quantitatively seems

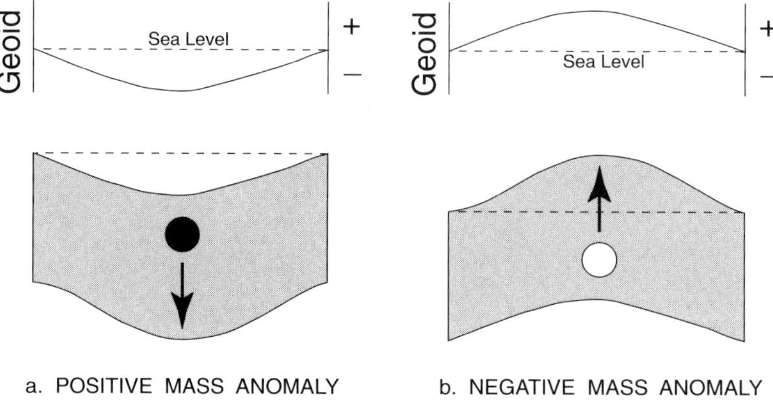

Figure 2.27. Effects of positive (a) and negative (b) mass anomalies on the geoid.

capable of explaining superswells is dynamic uplift caused by upwelling of a convecting mantle (Davies and Pribac 1993). Uplift models of Lithgow-Bertelloni and Silver (1998) reproduce the dynamic topography over the African superswell remarkably well. The African upwelling is most prominent in the D″ layer above the core, which strongly suggests that this layer is the source for the upwelling.

Geoid Anomalies

The geoid is the gravitational equipotential surface of the Earth that coincides with sea level in oceanic areas. Because gravitational potential decreases inversely with distance to source mass, whereas gravitational acceleration decreases inversely with the square of the distance, the geoid provides a long-range probe into the Earth. Deviations of the geoid from an idealized hydrostatic ellipsoid are known as geoid anomalies.

If the mantle were rigid, a positive mass anomaly in the upper mantle would produce a positive geoid anomaly. However, as illustrated in Fig. 2.27(a), the mantle is not rigid but deforms as a ductile solid, and thus a positive mass anomaly causes a downward flow in the mantle, which in turn forms a depression on the surface of the planet (Hager et al. 1985). Because it results in the replacement of rock by air or seawater, the depression can be regarded as a mass deficiency relative to a laterally uniform Earth, and this results in a negative contribution to the geoid. If the mass anomaly is near the surface, then the mass deficiency caused by the surface depression is opposite to but larger than the mass anomaly; thus, the net geoid anomaly will be negative and much smaller than either of its contributors. The mass anomaly will also cause a small deflection at the bottom of the mantle, which makes a small negative contribution to the geoid. In a general case for deflection of both top and bottom surfaces, the combined mass deficiencies will balance the mass excess, and the net geoid anomaly will be negative (Fig. 2.27(a)). On the other hand, if the mass anomaly is near the bottom surface, it will be deflected most and will contribute the most to the anomaly. In the Earth, positive mass anomalies are caused by the sinking of subducted plates into the mantle.

Now if we consider a mantle plume, we are dealing with a negative mass anomaly in the mantle. In this case, the upward flow in the plume raises both the surface of the Earth and the core–mantle interface and produces a positive geoid anomaly because air or water is being replaced by rock (Fig. 2.27(b)). Geoid anomalies caused by mantle plumes are associated with hotspot swells, and they are generally less than 1000 km across and typically a few meters high. The size of most geoid anomalies associated with hotspots compared with their swell heights indicates rather shallow compensation depths of 50–70 km (McNutt 1998). In most instances, these depths probably correspond to the depths of the corresponding hotspots.

In contrast to small geoid anomalies associated with mantle plumes, the Pacific and African upwellings are characterized by large positive geoid anomalies. The Earth's shape deviates by up to about 100 m from that of an ideal rotating Earth, and thus geoid anomalies can be contoured much like a topographic map. In the case of mantle upwellings, the upward flow in the mantle should raise the surface of the Earth and the core-mantle interface and produce a positive geoid anomaly, as described above. This is what is observed if the effect of sinking slabs is subtracted from the geoid (Fig. 2.28(a)). In this model, whole-mantle convection is assumed, and the lower mantle has a viscosity 10 times that of the upper mantle (Hager et al. 1985). The resulting pattern shows two major geoid anomalies at low spherical harmonic degrees: one in the western Pacific and the other beneath Africa. These coincide with, and are presumably caused by, the two large mantle upwellings originating from the core–mantle interface. Supporting this interpretation is the correlation of hotspots and seismic velocity anomalies (discussed below) with the geoid highs (Richards et al. 1988).

The core–mantle boundary is also raised beneath the upwellings, as predicted by the model (Fig. 2.28(b)); that is the upper and lower surfaces of the mantle (+crust) are deformed in the same direction and in the same geographic locations. The order of 3 km of relief occurs on the core–mantle anomalies. These dynamically induced deformations at both the top and bottom of the convecting mantle are caused by substantial mass anomalies.

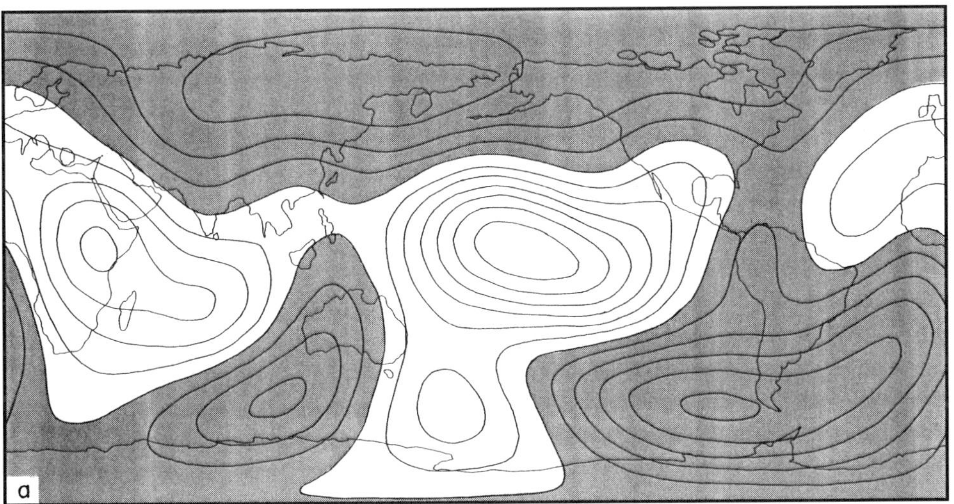

Figure 2.28. (a) The Earth geoid for spherical harmonics 2–6 obtained by subtracting the effects of subducted slabs. Geoid lows are shaded gray. The contour interval is 200 m, and density contrast is 2.3 g/cm^3. Modified after Hager et al. (1985).

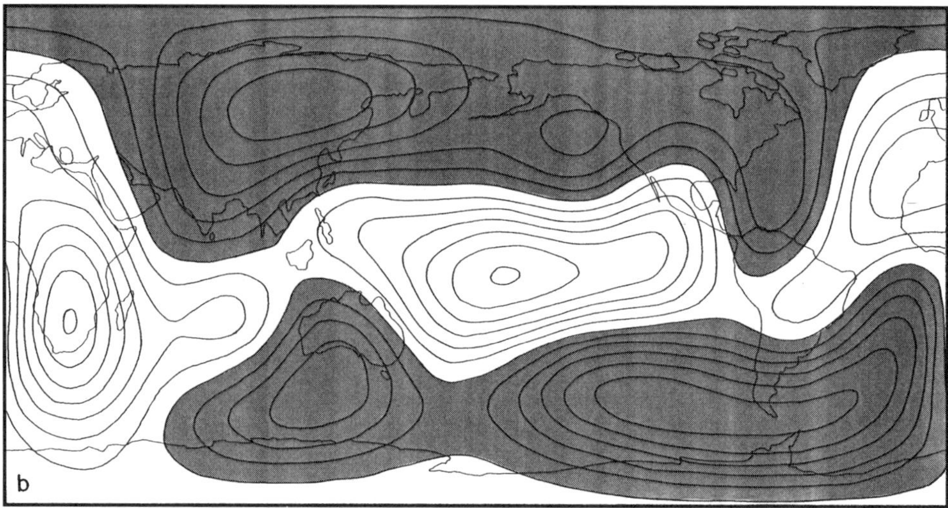

Figure 2.28. (b) Calculated topography on the core–mantle interface for spherical harmonics 2–3 and a density contrast of 4.5 g/cm^3 across the interface. The contour interval is 500 m, and lows are shaded gray. After Hager et al. (1985). Note the similarity of the surface and core–mantle interface geoid anomaly maps.

Seismic-Wave and Density Anomalies

Despite the increased resolution of seismic tomography in the past few years, there is still disagreement on the interpretation of published P- and S-wave velocity distributions in the mantle. However, most or all investigators agree on two observations: (1) regions of anomalously high velocity correlate with downgoing slabs, and (2) two broad slow-velocity anomalies, which may permeate most or all of the mantle, occur beneath Africa and the Pacific basin (Masters et al. 1996; Grand et al. 1997; Ishii and Tromp 1999; Ritsema et al. 1999; Tackley 2000). The details of these anomalies and their lateral and vertical distribution are not as yet clearly established. A recent model of Ishii and Tromp (1999) based on a large collection of free-oscillation data and free-air gravity anomaly constraints is shown in Plate 2. Also shown are calculated densities at equivalent depths. The seismic low-velocity anomalies, which are generally interpreted to reflect relatively hot mantle upwellings, can be tracked from the D″ layer at the base of the mantle to at least midmantle depths (1800 km) beneath both Africa and the South Pacific. At shallower depths, however, only the South Pacific anomaly continues to the base of the lithosphere. If real, the slow-velocity anomaly beneath the Pacific seems to bifurcate into two zones in the upper mantle, one centered near Antarctica and the other beneath northwestern Canada (Plate 2). In the deep mantle beneath Africa, the seismic-wave anomaly underlies southern Africa (2800 km). However, at depths of less than 1000 km, the anomaly either fades or becomes thinner and jogs northeast beneath the East African rift system (Ritsema et al. 1999).

Fading or bifurcation of mantle upwellings is expected when they encounter a region of lower viscosity such as the upper mantle (Lithgow-Bertelloni and Silver 1998). These features may result from one of the following: (1) plate migration over the last 100 Myr, (2) rapid change in mantle viscosity at about 1000 km depth, or (3) movement of an

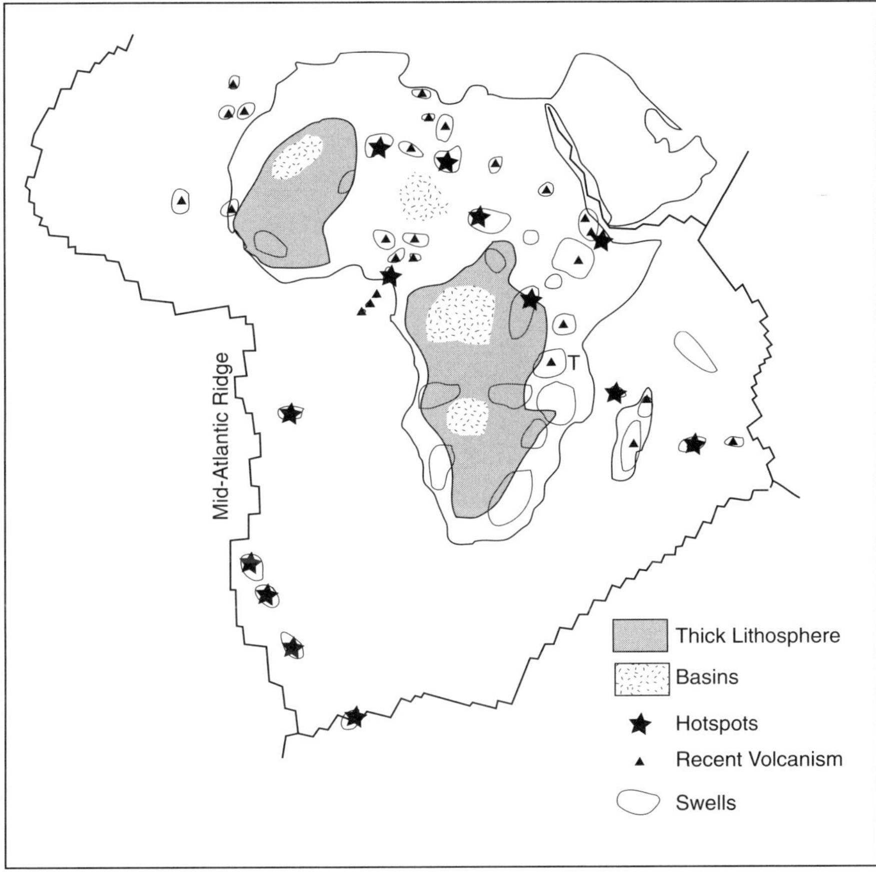

Figure 2.29. Map showing major swells, basins, hotspots, and recently active volcanic sites on the African plate for the last 30 Myr. Also shown are lithospheric roots more than 200 km thick beneath Archean cratons. This thick lithosphere correlates with the distribution of Archean cratons. T denotes Tanzania swell. Data from Burke (1996) and Ritsema and van Heijst (2000).

upwelling caused by convective currents in the mantle. The first explanation is least likely because there is no evidence that plates are "connected" to mantle upwellings. Supporting convective current motions in the deep mantle (explanation 3) as being important are the models of Steinberger (2000), which show that plume conduits are offset in the same direction as the upwelling beneath Africa. From the 660-km discontinuity upward, anomalously fast velocities occur beneath Africa, suggesting deep roots to the lithosphere (Plate 2). High-velocity Raleigh-wave anomalies also have been reported beneath Archean cratons in Africa, confirming the existence of thick lithosphere extending to depths of at least 300 km (Fig. 2.29) (Ritsema and van Heijst 2000).

Lateral variations in density at a given depth are relatively small throughout most of the mantle. Major exceptions are in the deep mantle beneath the Pacific and Africa and again at shallow levels, where densities are anomalously high (Plate 2). The observation that density is not correlated with seismic-wave velocity in much of the mantle, and especially in the transition zone, suggests that lateral variations in seismic-wave velocities may not be entirely due to temperature variations (Ishii and Tromp 2000). The two most prominent density anomalies beneath Africa and the Pacific occur in the D″ layer just above the core–mantle interface. This finding may mean that dense material piles up beneath upwellings because it is too heavy to be entrained and carried upward. The dense material can be descended slabs, or metal from the outer core, or both. In either case, compositional as well as thermal heterogeneity is required to explain lateral variations in the lowermost mantle.

The Pacific Upwelling

One of the characteristic features of the Pacific upwelling is widespread volcanism. Approximately 14% of the presently active hotspots occur in an area in the Pacific representing less than 5% of the total global area (Sleep 1990). Seamounts are also more abundant in the part of the Pacific overlying the upwelling (Plate 1). As an example, the cumulative number of seamounts more than 200 m high is an order of magnitude greater than those that occur off the upwelling (McNutt 1998). Because seamount volume is roughly proportional to seamount height, the total volume of seamounts of all sizes indicates that volcanism on the Pacific upwelling is an order of magnitude greater than it is off the upwelling. This increase in volcanism applies also to the production rate of oceanic plateaus and islands on the central part of the Pacific plate.

The African Upwelling

The African plate has not moved much during the last 30 Myr, but has remained centered on the African mantle upwelling (Burke 1996). A line of seamounts leading away from the St. Helena hotspot in the South Atlantic records a migration rate of only 20 mm/yr for the last 19 Myr, which is interpreted as the absolute motion of the African plate (O'Connor et al. 1999). This suggests a significant deceleration (33%) of the African plate in the last 20 or 30 Myr. Because the African plate is surrounded chiefly by spreading centers (with only minor subduction on the north), subduction is not a major driving force for plate motion. During the last 30 Myr, basins and swells on the order on 200 to 2000 km across have developed in Africa (Burke and Wilson 1972; Burke 1996; Fig. 2.29). This basin and swell topography is unique to Africa on the present-day Earth. Many of the swells appear to have developed over young mantle plumes, as indicated by recent volcanic activity and high heat flow (Fig. 2.29; Burke 1996). There is a striking difference in swell height between continental and oceanic areas. Continental swells have a median diameter of about 300 km and a median height of about 500 m in contrast to oceanic swells with corresponding values of 150 km and 1800 m (Herrick 1999). This probably reflects a combination of erosion and a thicker lithosphere on the continent. Modeling shows that thicker lithosphere results in wider and shorter swells, which is consistent with small mantle plumes (100–200 km across) at the base of the lithosphere (Janes and Squyres 1993).

Whereas only large, strong plumes coming from the deep mantle can produce surface uplift and volcanism in a moving plate, Burke (1996) has suggested that mantle diapirs (small plumes), perhaps coming from shallower levels, may produce these features in a near-stationary plate. The source, and even the reality, of such diapirs is problematic. Perhaps they could form by "budding" from the tops of superplumes. Alternatively, they may come from the 660-km seismic discontinuity or other thermal or compositional anomalies in the middle mantle. Herrick (1999) has proposed that diapirs, both on Earth and on Venus, become detached from their tails during hard turbulence in the mantle.

Seismic-wave velocity results show that lithosphere up to 300 km thick underlies the Archean cratons in Africa (Fig. 2.29; Ritsema and van Heijst 2000). As with most Archean cratons, this finding indicates that the African cratonic mantle lithosphere has not been significantly thinned by rifting or plume erosion in the past 200 Myr. Interestingly, three of the four major young basins in Africa occur over this thick lithosphere. The thick lithospheric roots provide constraints on the distribution of mantle plumes beneath Africa because plumes colliding with the base of thick lithosphere should not produce swells at the surface (Ritsema and van Heijst 2000). Consistent with this prediction, of the six swells presumed to be of plume origin beneath the African continent, five occur in areas of thin lithosphere, and one occurs near the margin of thick lithosphere. An S-wave anomaly beneath the swell in Tanzania in East Africa (Fig. 2.29) is the first direct evidence of a mantle plume head beneath African swells today (Nyblade et al. 2000).

Descending Slabs and Mantle Upwellings

Although most descending lithospheric slabs have little intense seismicity in the transition zone of the upper mantle, some young slabs, like the Tonga–Kermadec slab in the South Pacific, show intense seismicity in this zone. The cause of such high seismicity is not well understood, but the descent of the Tonga–Kermadec slab along the western edge of the Pacific mantle upwelling may have some bearing on the explanation. Increases in slab seismicity with depth are generally related to an impenetrable chemical boundary, a jump in mantle viscosity, or to a force associated with the solid-state reaction at the 660-km discontinuity. Gurnis et al. (2000), however, showed that density of the upper mantle can also be important in both the intensity of seismicity and depth of slab penetration. In the case of Tonga, there is no evidence for penetration of the slab into the lower mantle as there should be if subduction started some 50 Ma. For long-lived subducting slabs, the lower mantle tends to pull slabs to great depths. Using tomographic inversions of mantle seismic structure, Gurnis et al. (2000) have shown that, for the young Tonga slab, the lower mantle actually pushes upward, and this can explain the intense deformation in the slab at shallow depths. Why the upward push? The Tonga–Kermadec slab descends on top of the Pacific mantle upwelling. This is the first study showing that modern mantle upwellings directly affect plate tectonics.

Geotectonic Bipolarity

Within the past 180 Myr, the entire oceanic lithosphere beneath the present oceans has been extracted from the mantle at ocean ridges, and a corresponding amount has been subducted into the mantle. This growth and recycling is a manifestation of large-scale circulation in the mantle. During this time, the Pacific plate, which remained

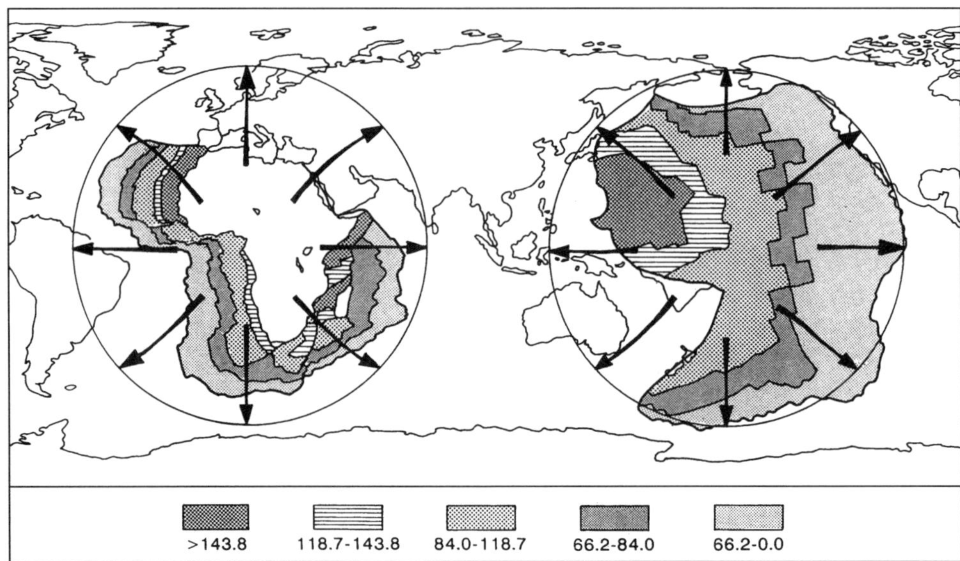

Figure 2.30. Geotectonic bipolarity as revealed by simultaneous concentric growth of the African and Pacific plates in the last 180 Myr. Age of the ocean floor in Ma is shown by patterns, and arrows show horizontal flow in the mantle beneath the plates. Courtesy of Nazario Pavoni.

approximately symmetrical with respect to the equator and today is centered over the Pacific mantle upwelling, has grown from a small plate to a very large one (Fig. 2.30). The roughly concentric growth pattern of the plate shows diverging movements of neighboring plates away from the Pacific plate as it grew. This growth appears to reflect a regular, diverging flow pattern in the mantle upwelling beneath the Pacific plate. The evolution in size and shape of the plate as it grew constrained the size of the upwelling for the last 180 Myr (Pavoni 1997). Concurrent with growth of the Pacific plate, a comparable growth pattern occurred in the African plate centered over the African mantle upwelling (Fig. 2.30). This is documented by the breakup of Gondwana and Laurasia beginning some 180 Ma. Relative to the African plate, the North American plate moved northwestward, the South American plate to the west, the Antarctic plate to the south, and the Australian plate to the east.

The bipolar growth of the African and Pacific plates in the last 180 Myr is consistent with mantle upwelling beneath each plate with descending flow at 70–90° distance from the central pole of each plate.

Plumes in Perspective

Although the mantle plume model is widely accepted for hotspot chains, there are many different kinds of volcanic chains as well as non-chains on the seafloor. Some chains have obvious hotspots, whereas others do not. Some show an age progression along the chain, and others do not. Some oceanic volcanoes and seamounts occur in clusters, and still others occur as single isolated units. Some chains chiefly contain large volcanoes, some small volcanoes, and others contain both small and large volcanoes. In many volcanic islands and seamounts, the shield volcanic stage involves eruption of chiefly tholeiitic basalts, whereas in others, alkali basalts and their derivatives dominate the shield stage. Although the roots of some mantle plumes can be tracked with seismic tomography into the deep mantle, it does not follow that all mantle plumes come from the D″ layer above the core. Increasing refinement of seismic-wave tomographic data is necessary to see if plumes are also produced at other levels in the mantle. As we will see in Chapter 5, the geochemically depleted character of some oceanic island basalts indicates they entrained a deep, depleted mantle domain unlike the shallow, depleted source of ocean-ridge basalts. What is the origin and age of this deep, depleted domain, and how has it remained relatively isolated in a convecting Earth? There is also much work to do in understanding the origin of the large diversity of volcanic structures on the seafloor, many of which may not be related to plumes.

We are only beginning to understand mantle upwellings. Greater resolution of seismic data is needed to enhance our understanding further of the shapes and bifurcations of upwellings, their relationships to mantle plumes, and the complexities of interactions of both plumes and upwellings with the base of the lithosphere.

3

Large Igneous Provinces

Introduction

Large igneous provinces, commonly referred to as LIPs, are voluminous occurrences of dominantly mafic igneous rock not directly related to plate tectonic processes (Carlson 1991; Coffin and Eldholm 1994). Included as LIPs are oceanic plateaus, continental flood basalts, passive margin volcanics, ocean-basin flood basalts, submarine ridges, giant dyke (and sill) swarms, and some large layered intrusions. Major LIPs less than about 250 Ma are shown in Figure 3.1. Next to mafic magmas emplaced at ocean ridges, LIPs are the most significant accumulations of mafic igneous rock on and near the Earth's surface. As with hotspot volcanics, LIPs are generally thought to have a mantle plume origin and to account for 5–10% of the heat and magma extracted from the mantle (Davies 1988; Sleep 1990). However, unlike ocean-ridge basalts, which appear to have been continuously erupted throughout geologic time, LIPs may be episodic, which is an intriguing feature of these rocks discussed in Chapter 8.

Volcanic LIPs are dominated by thick, laterally extensive basalt flows, some which have areal distributions of more than 10^5 km^2 and volumes greater than or equal to 10^6 km^3 (Table 3.1). At depth, layered mafic intrusions and sills and dykes may be important. In flood basalt provinces, sills are common and may contribute a significant fraction of the total igneous volume. In some LIPs, felsic and intermediate igneous rocks produced by fractional crystallization or partial melting of crustal rocks are associated with initial and late stages of LIP development (Campbell and Griffiths 1990). Unlike most ocean-ridge volcanics, LIPs contain both tholeiitic and alkalic components, although the former generally dominate. In some rifted continental margin LIPs, such as the Tertiary volcanic province along the coast of Norway and the Paraná–Etendeka flood basalts on the margins of the South Atlantic, felsic to intermediate lavas are relatively common (Eldholm et al. 1989).

Large igneous provinces occur in both continental and oceanic settings in intraplate locations, on present and former plate boundaries, and within and along the edges of continents (Fig. 3.1). Continental flood basalts are chiefly tholeiitic flows and intrusive equivalents erupted on continental crust over short periods of time on the order of 10^5–10^6 years. Erupted chiefly from fissures, they comprise laterally extensive flows.

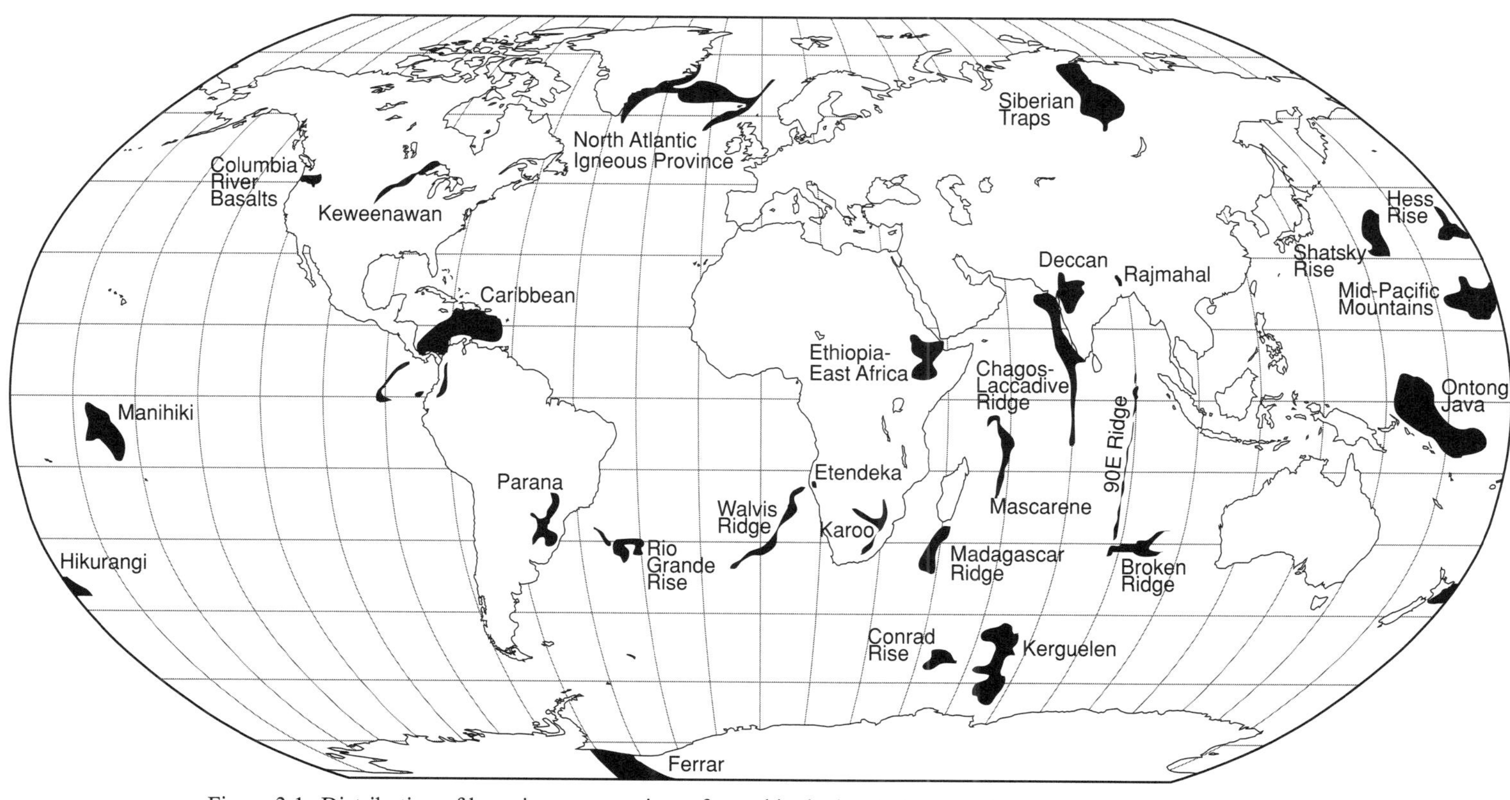

Figure 3.1. Distribution of large igneous provinces formed in the last 250 Myr. After Coffin and Eldholm (1994).

Table 3.1. ***Ages, Volumes, and Eruption Rates of Large Igneous Provinces Compared with Oceanic Crust***

	Age (Ma)	Volume ($\times 10^6$ km^3)	Duration (Myr)	Average Eruption Rate (km^3/yr)
Siberian traps	250	4.0	9 (1)	0.5 (4.0)[a]
Karoo	178–195	2.5	17	0.15
Paraná–Etendeka	127–138 (131–135)	2.0 (0.8)	11 (5)	0.18 (0.16)
Deccan	64.5–65.5	8.2	1	8.2
Ontong Java	88–93; 120–124 (121–123)	44–57 (45)	30 (2)	1.5 (23)
Columbia River	14.5–17.5 (15.5–16.5)	0.175	3 (1)	(0.06) 0.2
North Atlantic	40–61 (58–61)	6.6	20 (3)	0.3 (2.2)
Iceland	0–15	9	15	0.6
Ethiopia/East Africa	15–23; 28–32 (30)	1 (0.7)	15 (1)	0.07 (0.7)
Kerguelen	110–114	10–15	4	3.5
Bushveld Complex	2060	0.384	0.06	6
Oceanic Crust	0–180	1950	180	10.8

Chief references: Carlson (1991); Coffin and Eldholm (1994); Cawthorne and Walraven (1998).
[a] Values in parentheses refer to age, duration, or eruption rate of most of the volcanism.

Passive-margin volcanics, found along the trailing edges of rifted continental margins, contain large volumes of volcanic–intrusive mafic rocks associated with uplift along the rifted margins (Roberts et al. 1984; Eldholm 1991). In ocean basins, the largest LIPs are oceanic plateaus, which are enormous, relatively flat-topped plateaus that rise 2 km or more above surrounding seafloor. Composed chiefly of mafic volcanics and intrusives, they are thicker and generally older than surrounding oceanic crust. Extrusive volcanics may be solely submarine volcanics or, in a few cases like Kerguelen in the southern Indian Ocean, a combination of submarine and subaerial volcanics. Large submarine volcanic ridges are another variety of LIPs. Some of these ridges are hotspot tracks related to continental flood basalts (Walvis Ridge, Ninetyeast Ridge; Fig. 3.1), some are remnants or pieces of oceanic plateaus, and still others may be fossilized ocean ridges. The least well known of the LIPs are ocean-basin flood basalts, which are thick submarine flows and associated sills erupted on oceanic crust. They differ from oceanic plateaus in that they are considerably thinner and may in some instances represent the incipient formation of oceanic plateaus.

Short-lived magmatism characteristic of LIPs is documented by high-precision isotopic dating of flood basalts and rifted continental-margin basalts. For instance, a combination of $^{40}Ar/^{39}Ar$ dating and magnetostratigraphy has shown that basalts of the Deccan and the Siberian traps were erupted in ≤ 1 Myr (Courtillot et al. 1986; Duncan and Richards 1991; Campbell et al. 1992). Such transient volcanic events are often attributed to melting in mantle plume heads. As with oceanic hotspots, when continental lithosphere moves over hotspots, it may leave a track beginning with the eruption of flood basalts. Examples are the North Atlantic Igneous Province (including Iceland), the Ninetyeast Ridge and Rajmahal traps in India, and the Chagos–Laccadive Ridge, and the Deccan traps in western India (Fig. 3.1). Not all flood basalts, however, can be related to preserved hotspot tracks, as evidenced by the Siberian traps.

Large igneous provinces occur in many different tectonic settings. These include on or near ocean ridges (Iceland), at plate triple junctions (Shatsky Rise), on old oceanic crust (Manihiki), along passive plate margins (North and South Atlantic margins), and on continental cratons (Siberian and Deccan traps). Large igneous provinces that form near spreading centers, such as Iceland and perhaps Ontong Java, have basalts that are chemically distinct from ocean-ridge basalts, although mixing can occur between plume and ocean-ridge mantle sources.

Characteristics of Flood Basalts

Continental flood basalt provinces are constructed chiefly of thick basaltic lavas ranging between 20 and 100 m thick and erupted on gentle slopes (Tolan et al. 1989). Single flows are up to hundreds of kilometers long and some have volumes of thousands of cubic kilometers (Self et al. 1997). Most are compound pahoehoe flows, and pillow lavas are found near the base of some flows (Swanson and Wright 1980). The pillows are pahoehoe lobes that formed as lava flowed into lakes or shallow marine environments. In contrast, aa flows are relatively rare in most provinces. Most of the flow packages consist of multiple, distinct lobes often many kilometers in length (Fig. 3.2). The margins of flows are usually not visible because they are larger in scale than most outcrops. Most sheet lobes, as illustrated by those in the Columbia River basalts, do not have flat tops but instead have gentle hills and swales of 1–5 m amplitude and 10–50 m wavelength (Fig. 3.2). Tumuli can be very large, up to 1000 m long and 10–20 m high, and are often associated with lava breakouts. The compound nature of flows in flood basalts, with small lobes and toes at the bases of single flows, indicates they were emplaced as a series of lobes separated in space and time (Self et al. 1997). Lava lobes often inflate owing to blockages or changes in gradient in deep lava channels that force lava upwards.

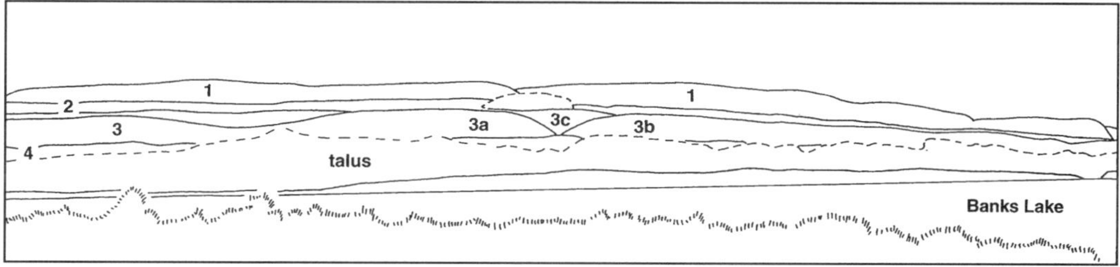

Figure 3.2. Sheet flows in the Columbia River basalts. Section is 220-m high and 3 km across and shows four sheet flow lobes near Banks Lake, Washington. From Self, Thordarson, and Keszthely (1997), with permission. Copyright ©1997 by the American Geophysical Union.

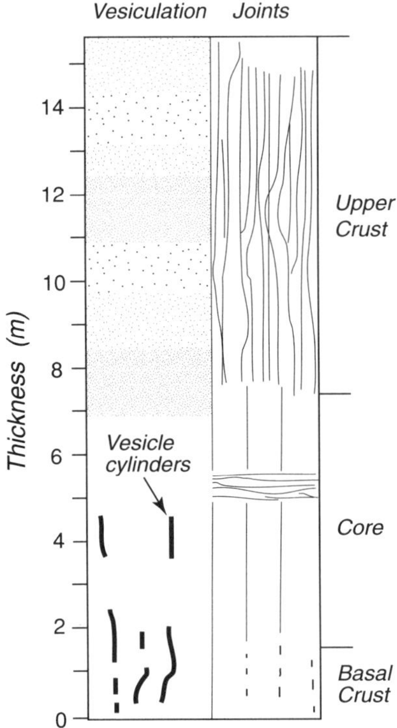

Figure 3.3. Idealized cross section of a lava flow from the Columbia River basalts. Modified after Self et al. (1997).

Lava flows from the Columbia River Province are commonly divisible into three zones (Fig. 3.3): upper crust, core, and basal crust. This tripartite division can be traced in single flows for tens of kilometers. The upper crust contains numerous vesicles, has closely spaced irregular cooling joints, and comprises 40–50% of the total flow thickness. Glass contents can vary greatly from 10 to nearly 90%. Vesicularity usually decreases, whereas vesicle size increases downward in the flow. The lava core, which makes up 40–60% of the flow thickness, has very few vesicles, regular jointing, and is composed chiefly of crystalline rock ($\geq$90% crystals). Typical of the core are volatile-rich magma veins that rise as vesicle cylinders through the lava. The basal crust, which comprises less than 10% of the flow, is less than 1 m thick and contains 50–90% glass. This crust is only slightly vesicular and may have poorly developed, platy jointing.

Most flood basalts come from fissure eruptions rather than from volcanoes (Martin 1989). In the Roza linear vent system of the Columbia River basalts, small dykes and spatter cones define the vent system at the surface, which is about 5 km wide and extends for at least 150 km along strike (Swanson et al. 1975). From studies of Hawaiian fissures and lava fountains, it is likely that only one or two segments of the Roza fissure (each several kilometers in length) were active at a given time.

Detailed studies of basaltic flows in flood basalt provinces, are generally interpreted to reflect eruption of inflating sheet flows (Self et al. 1997). Although estimated eruption rates are relatively high for basalt flows ($\approx$4000 m^3/s), they are not enormous catastrophic eruptions, such as those that may occur above subduction zones. More likely, flood basalts are erupted gradually as pahoehoe sheet flows that accumulate to form massive flow fields.

LIP Eruption Rates

Estimates of the volume, age, and eruption rates of some major LIPs are given in Table 3.1. The large volume of igneous rock found in oceanic plateaus (such as Ontong Java) indicates these plateaus represent major additions to the oceanic crust. Because of the possibility of extensive underplating by mafic magmas, it is not clear if oceanic plateaus occupy larger volumes than flood basalt counterparts on the continents (Schubert and Sandwell 1989). In addition to lavas, large volumes of intrusive igneous rock occur within both continental and oceanic plateaus, as evidenced by uplifted sections of older plateaus that have been accreted to the continents (Kerr et al. 1998). In major flood basalts, such as the North Atlantic and Paraná provinces, extensive sill and dyke complexes are associated with the eruptions (Emeleus 1991).

Because the amount of intrusive rock equals or probably exceeds the volcanic component, the eruption rates in Table 3.1 are minimal, for they are based only on exposed and preserved portions of LIPs. Variable amounts of lava have been removed by erosion from flood basalts, which again contributes to errors in volume estimates. Another factor important in estimating eruption rates is the duration of volcanism. Because most of the LIPs are not precisely dated, the duration intervals given are upper limits. Although the total duration of some plateau volcanism is quite long, most of the volcanism may have occurred in a short time interval. For instance, in Ontong Java the total duration is on the order of 30 Myr, but most of the volume of the plateau was formed in about 4 Myr. Taking these uncertainties collectively, we see that the eruption rates of LIPS compiled in Table 3.1 are, at best, lower limits. It is probable that the actual magma extraction rate during the main eruptive events is two to three times the rate given, making LIPs comparable or larger than eruption rates at ocean ridges (about 11 km^3/yr).

A major difference between oceanic and continental LIPs may be the total duration of magmatism. With few exceptions, continental flood basalts appear to have been erupted in short periods of time of 1 Myr or less (Courtillot et al. 1988; Renne and Basu 1991; Hofmann et al. 1997). In contrast, oceanic plateaus that have been precisely dated commonly show two or three pulses of magmatism extended over 30–40 Myr (Revillon et al. 2000). For instance, isotopic ages from Ontong Java show major magma episodes at 120 and 90 Ma (Birkhold et al. 1999), and those from the Caribbean plateau show major episodes at 90–88 and 76 Ma with a possible third episode at 55 Ma (Revillon et al. 2000).

Crustal Structure of Oceanic Plateaus

Seismic Structure

High-quality seismic refraction data are available from several oceanic plateaus, Ontong Java in the South Pacific, Kerguelen in the southern Indian Ocean, and the Caribbean plateau (Operto and Charvis 1996; Gladezenko et al. 1997; Richardson et al. 2000) (Fig. 3.1). Seismic cross sections of Kerguelen and Ontong Java are shown in Figure 3.4 compared with average oceanic crust in the Pacific and with average continental crust. Ontong Java averages about 33 km to the Moho, whereas in Kerguelen the Moho depth is only about 24 km. Both Ontong Java and Kerguelen have a thin sediment

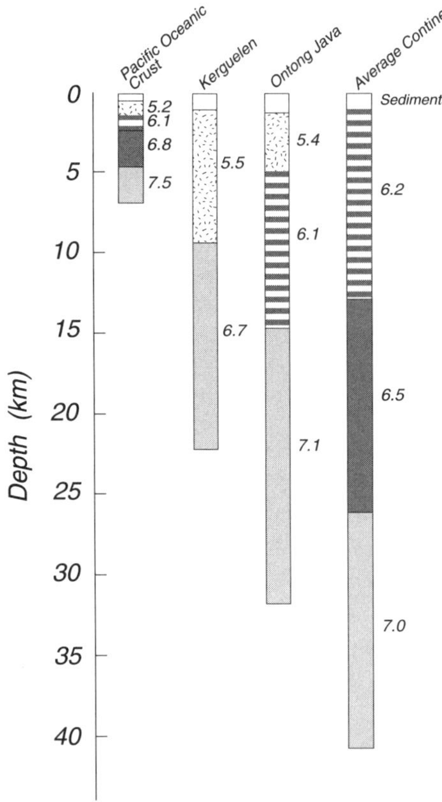

Figure 3.4. Seismic structure of the Kerguelen and Ontong Java oceanic plateaus compared with average crust in the Pacific basin and average continental crust. Velocities given in km/s. Data from Operto and Charvis (1996), Gladezenko et al. (1997), and Condie (1997a).

layer on top underlain by an upper crustal layer with P-wave velocities averaging about 5.5 km/s. In the case of Ontong Java, a midcrustal layer is also recognized with velocities of about 6.1 km/s. In both cases the lower crust ranges from 12 to 17 km thick with velocities averaging 6.7 km/s in Kerguelen and 7.1 km/s in Ontong Java. In addition, the Kerguelen section shows a highly reflective zone in the lowermost 5 km just above the Moho. These oceanic plateaus are 3.5–4 times thicker than oceanic crust but are not as thick as average continental crust.

Seismic reflection profiles and refraction data from the Caribbean oceanic plateau provide constraints of crustal structure, as illustrated by the east–west crustal section in Figure 3.5 (Mauffret and Leroy 1997). Beginning in the west, the Upper Nicaragua Rise is part of the Central American continental crust. The Lower Nicaragua Rise and Hess escarpment are underlain by the Caribbean plateau, which comprises two seismic layers. The upper layer (2V), which is on the order of 4 km thick, is probably composed chiefly of submarine basalt flows underlain by a zone of mixed flows, sills, and dykes. A prominent reflector (B″) separates these units. Although variable, the 3V layer reaches thicknesses of 16 km and is characterized by high velocities in the lower part (7.2–7.4 km/s). The Haiti subbasin is underlain by thin crust and probably represents a rift. The Beata Ridge, erupted 80–75 Ma, is the elevated eastern flank of this basin, and gabbros of the underlying layer 3V are exposed along the submarine escarpment (Case et al. 1990; Revillon et al. 2000). East of Beata Ridge, the topmost layer contains

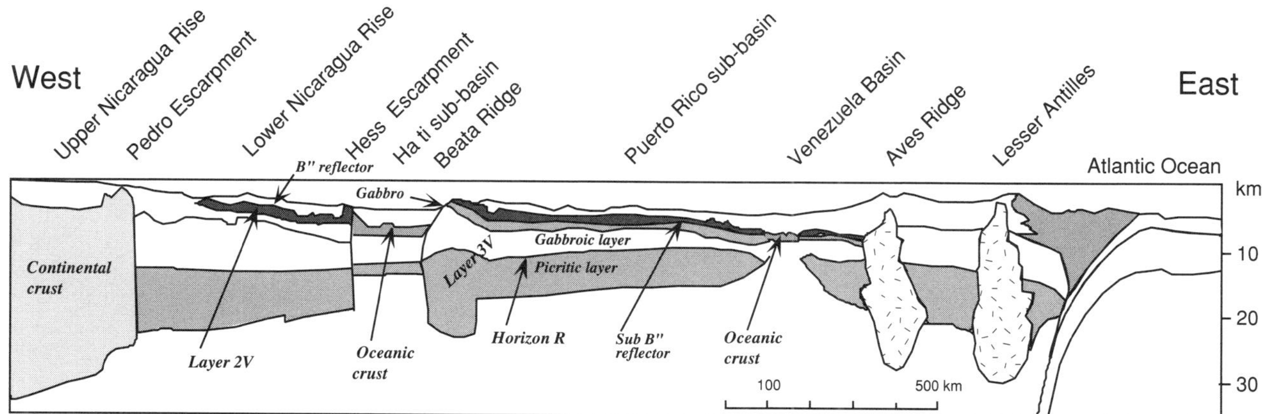

Figure 3.5. Crustal section of the Caribbean plate. Modified after Mauffret and Leroy (1997).

sediments intruded with basaltic sills that make strong seismic reflectors. South of Puerto Rico, the Puerto Rico subbasin is characterized by relatively thin crust ($\approx$10 km) and young volcanic cones that form prominent topographic features on the seafloor. Like the Haiti subbasin, the Venezuela basin appears to be a rifted basin in the Caribbean plateau underlain by thin crust. The eastern margin of the Caribbean plateau has thick crust ($\geq$20 km) owing, in part, to the construction of the Aves Ridge and Lesser Antilles arcs along the edge of the plateau.

What do the various seismic velocity layers in oceanic plateaus represent? Hussong et al. (1979) suggested that oceanic plateaus are simply "expanded" oceanic crust. However, expanding the Pacific oceanic crustal layers (Fig. 3.4) does not reproduce the seismic layering observed in oceanic plateaus (Gladczenko et al. 1997). From drilled and obducted lavas on Ontong Java and Caribbean plateaus, it is likely the upper crustal layers ($V_p = 5.5$ km/s) are largely mafic volcanics and associated sills and dykes. The 4.5-km thick upper layer in Ontong Java may be represented by an uplifted 4-km-thick section exposed on Malaita in the Solomon Islands (Petterson 1995). The upper part of the Caribbean plateau is preserved as obducted thrust sheets in Colombia and Ecuador and on islands around the Caribbean basin. The 6.1-km/s layer beneath Ontong Java, although not drilled, is generally interpreted as mafic intrusives composed dominantly of dykes and sills and probably some layered complexes.

When compared with exposed portions of the Caribbean oceanic plateau in Colombia (Kerr et al. 1998), the seismic profiles are consistent with an upper zone composed chiefly of layered gabbros with a lower layer of picrites and mafic cumulates. Uplifted portions of the Caribbean plateau contain several layered mafic–ultramafic intrusions that have been interpreted as fossil crustal magma chambers that fed overlying basalts (Kerr et al. 1998; Revillon et al. 2000).

Composition of the Deep Crust

The composition of the deep crustal layer beneath oceanic plateaus can be constrained from laboratory measurements of seismic wave velocities, from lower crustal xenoliths, and from obducted fragments of plateaus that expose the lower crust. Mafic magmatic underplates should yield velocities on the order of 7.0–7.5 km/s (White et al. 1992), which is inconsistent with the observed velocities of 6.7 to 7.1 km/s found at the Ontong Java, Kerguelen, and Caribbean plateaus. This is confirmed by the studies of White and McKenzie (1995), who showed that high-MgO melts typical of the axial regions of plumes should generate thick crust with high seismic wave velocities. Mafic granulites have acceptable velocities to explain the observed lower crustal velocities beneath oceanic plateaus, and these granulites occur as xenoliths in Kerguelen basalts (Gregoire et al. 1998). Thermobarometric studies of mafic granulite xenoliths indicate they come from depths of 15–25 km for pyroxene granulites and 25–45 km for garnet granulites. Hence, pyroxene granulites may be the dominant rock type in the lower crust of oceanic plateaus. Garnet granulites, on the other hand, seem to come from the upper mantle, although not from depths great enough to make eclogite.

During collision of oceanic plateaus with continents, slices of the crust in a plateau can be obducted on to the continent and provide access to partial sections of the lower

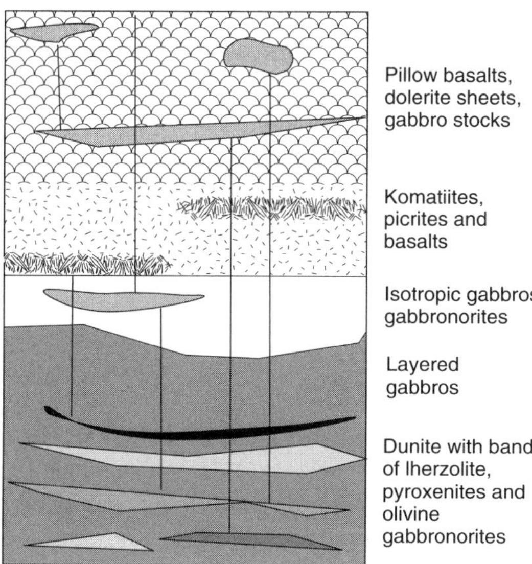

Figure 3.6. Schematic section of the Caribbean oceanic plateau based on exposed sections in Colombia and elsewhere. Modified from Kerr et al. (1998), with permission. Copyright © 1998 by the American Geophysical Union.

crust of oceanic plateaus (Abbott and Mooney 1995; Saunders et al. 1996). The only accreted oceanic plateau for which deep crustal levels have been recognized is the Caribbean plateau. Based on field work in Colombia and adjacent areas in the Caribbean and on geochemical studies of volcanic rocks, Kerr et al. (1998) have constructed a diagrammatic section of the Caribbean oceanic plateau (Fig. 3.6). The upper part of the plateau comprises pillow basalts and related sills and contains no sheeted dykes typical of ophiolites. This suggests that the magma supply greatly exceeded the extension necessary to accommodate it, supporting a mantle plume source. The basalt layer is underlain by a layer composed chiefly of picrite and basaltic komatiite, which, in turn, is underlain by a layer of gabbro and norite. Layered gabbros are cumulates from fractional crystallization, whereas isotropic gabbros appear to have crystallized in local magma chambers. The basal portion of the Caribbean plateau, which is 5–8 km thick, is composed chiefly of dunite and lherzolite with intrusions of mafic igneous rock. The ultramafic rocks, which typically show cumulus textures, are probably cumulates from fractional crystallization of picrite and komatiite magmas injected into magma chambers at mid-to-deep crustal levels. From the available seismic data combined with the mineralogy of obducted slices, the Caribbean plateau appears to be no thicker than about 30 km and in most instances no thicker than 20 km (Fig. 3.7).

Lithospheric Roots

Three-dimensional seismic tomographic inversion studies indicate the presence of a low-velocity root beneath Ontong Java extending as deep as 300 km with S-wave velocities reduced by at least 5% (Richardson et al. 2000). It is likely that this low-velocity root remained attached to the plateau as it drifted over the asthenosphere since its formation in the Cretaceous. Similar deep, low-velocity roots have been reported beneath the Paraná and Deccan flood basalts, and these have been related to plume heads (VanDecar et al. 1995; Kennett and Widiyantoro, 1999). The low-velocity zone beneath

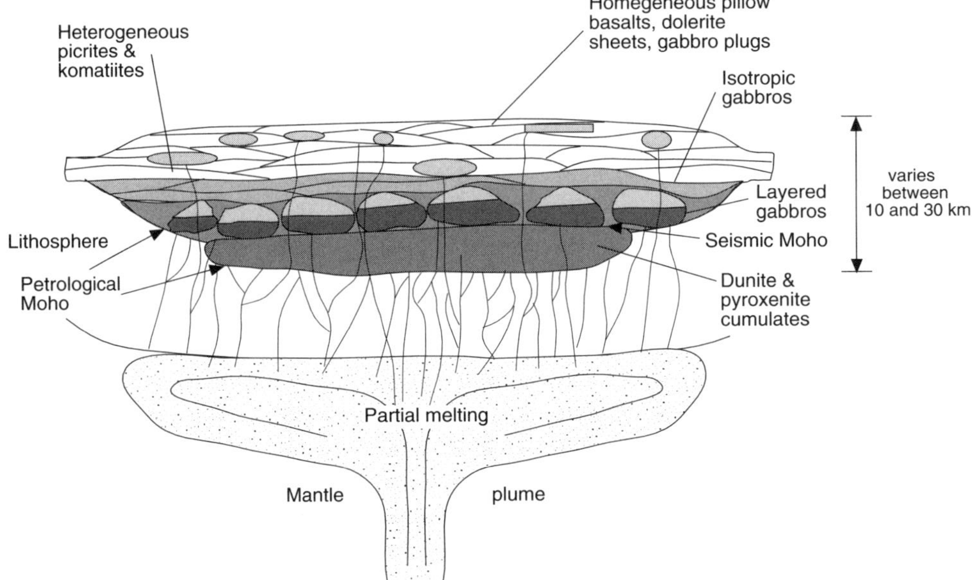

Figure 3.7. Hypothetical cross section of the Caribbean oceanic plateau showing possible relationship to a mantle plume source. Modified from Kerr et al. (1998), with permission. Copyright ©1998 by Elsevier Science.

Ontong, however, differs from the Paraná and Deccan low-velocity zones in that it is much larger (1200 km across compared with 300 km for the others), the S-wave velocity drop is greater (5% compared with 1.5–2.4%), and the anomaly is attached to an oceanic rather than a continental plate. Richardson et al. (2000) indicated that the Ontong Java root cannot be thermally maintained; hence, it must represent a compositional anomaly carried along with the moving Pacific plate. An intriguing possibility is that the low-velocity root represents the deformed head of a mantle plume. If so, however, there is no agreed upon hotspot track leading to the abandoned plume tail.

Examples of Large Igneous Provinces

Columbia River Basalts

Although most investigators agree that the Columbia River basalts, which erupted 16 Ma, were related to the Yellowstone hotspot, the nature of this relationship is uncertain (Hooper 1997). The major problem is that the Columbia River basalt vents do not lie on the Yellowstone hotspot track 16 Ma when they were erupted (Fig. 3.8). A second problem is that these basalts began erupting in eastern Oregon and Washington at the same time (17–15 Ma) that basalts erupted in the Snake River plain, which is on the Yellowstone hotspot track some 300 km to the south. This has led to controversy and debate over whether the Columbia River basalts are related to the Yellowstone hotspot. Most investigators agree that the Columbia River magmas were derived from several different magma reservoirs and that each reservoir underwent variable degrees of fractional crystallization (Carlson and Hart 1987; Hooper and Hawkesworth 1993). Multiple

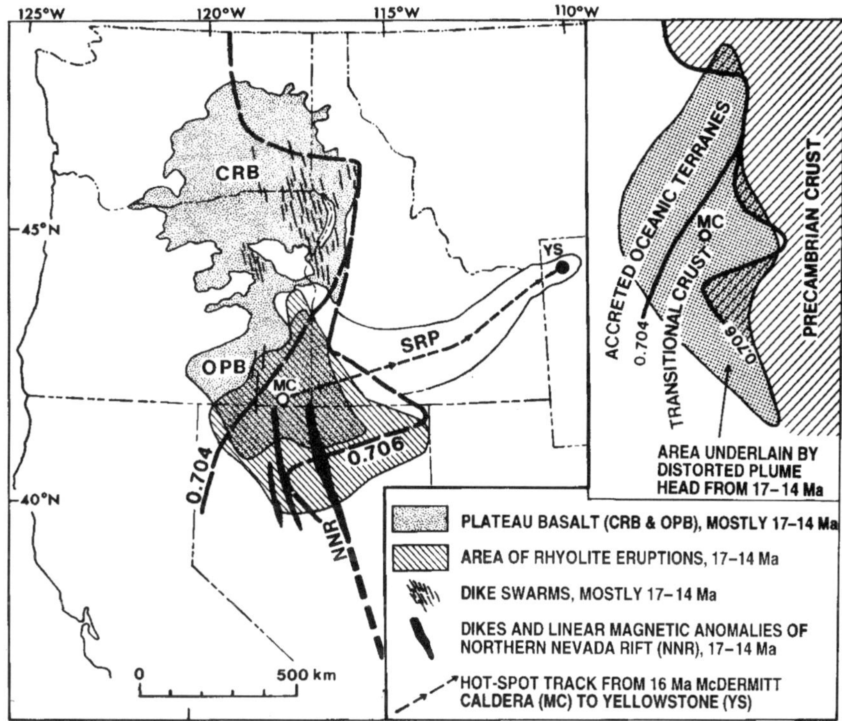

Figure 3.8. Distribution of volcanic rocks associated with the Yellowstone hotspot (YS) 17–14 Ma. CRB = Columbia River basalts; SRP = Snake River Plain; OPB = Oregon Plateau basalts. Initial $^{87}Sr/^{86}Sr$ isotope ratios noted by 0.704 and 0.706 contours. Modified after Camp (1995), with permission. Copyright ©1995 by the Geological Society of America.

sources were involved in magma production involving contributions from depleted and enriched mantle, subcontinental lithosphere, and probably some contamination by the continental crust (Carlson 1984; Carlson and Hart 1987; Hooper 1997). It appears that more than 85% of the estimated 175,000 km^3 of lava constituting the Columbia River plateau was erupted in the remarkably short time interval of 16.5–15.5 Ma (Table 3.1) (Tolan et al. 1989).

Several plume and nonplume models have been proposed to explain the Columbia River basalts, the most important of which are briefly discussed below.

1. ***Back-arc rifting model.*** This model involves magma production in the back-arc region behind the Cascades arc (Carlson and Hart 1987). The model does not explain the intensity of volcanism over such a short time interval nor does the amount of allowed extension in this model (<1%) appear adequate for such a large volume of magma production (Hooper 1997).
2. ***Induced plume model.*** Dickinson (1997) has suggested that Columbia River basaltic magmas were generated from torsional forces along the Pacific–North American plate boundary. Shearing along the edge of the North American plate perturbs the shallow mantle sufficiently to induce plume activity at deeper levels. The induced mantle plume(s) moves upward along a zone of weakness in the

lithosphere on the western edge of the Archean craton, and it is in this zone that decompression melting occurs. The stresses that caused the plume to form are supplied by the impact of a new shear zone along the eastern edge of the Pacific plate. This model also accounts for the synchroneity of Columbia River basalt volcanism with volcanism in the Transverse Ranges and in the borderlands of California.

3. ***Deflection of the Yellowstone plume.*** Geist and Richards (1993) suggested that interactions between the Yellowstone plume and the descending Farallon plate deflected the plume northwards about 17 Ma. The subduction zone was reconfigured when a "window" opened in the descending plate and the plume broke through, producing the Columbia River basalts (Fig. 3.9). Afterwards, the plume readjusted to the vertical plume tail, producing the hotspot track leading along the Snake River Plain (Fig. 3.8). This also may be the reason that the total volume of flood basalts in the Columbia River plateau is so small (an order of less than most other flood basalts; Table 3.1).
4. ***Deformation of the Yellowstone plume head.*** Camp (1995) has suggested that the Yellowstone plume rose near the boundary of the thick root of the Archean

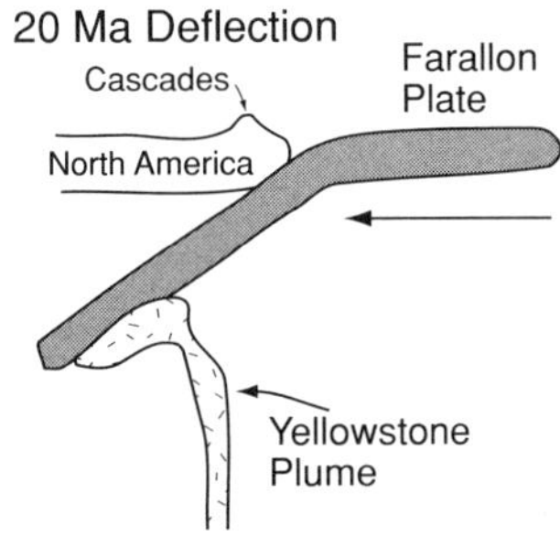

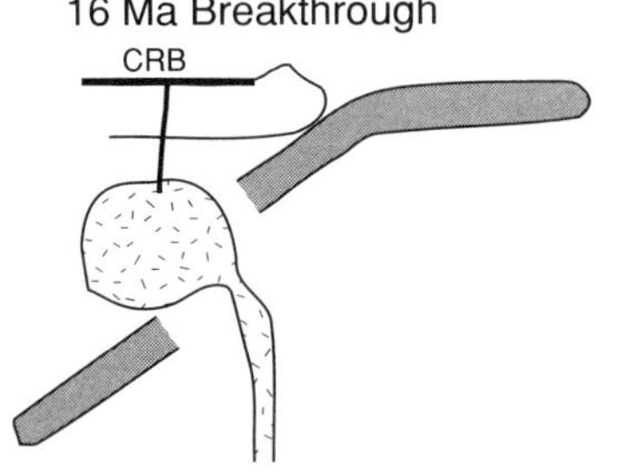

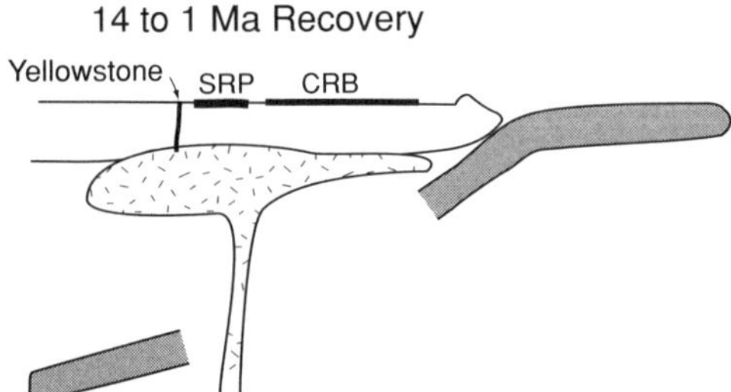

Figure 3.9. Schematic cross sections showing breakthrough of the Farallon plate by the Yellowstone mantle plume. At 20 Ma, plume is deflected northeast by the Farallon plate; at 16 Ma, plate breakthrough and formation of the Columbia River basalts; at 14–1 Ma, plume-related volcanism along the Snake River Plain and in Yellowstone Park. CRB = Columbia River basalts; SRP = Snake River Plain. After Geist and Richards (1993).

craton and the Pacific plate and was deformed against this boundary (Fig. 3.8). As the thickened plume head rose adiabatically beneath the adjacent oceanic lithosphere of the Pacific plate, the pressure release resulted in voluminous melting. The sequence of melting is reflected in the time sequence of basalts erupted in eastern Oregon and Washington. The Imaha basalt is derived from the plume head, and the Grande Ronde basalt comes from a mixture of plume head and oceanic lithosphere. Finally, the late Saddle Mountains basalts come from partial melting of the continental lithosphere in the overriding North American plate. From 17 to 14.5 Ma, the high eruption rate may have been due to the rapid ascent of magma along weak boundaries of the recently accreted terranes. Magma was transported to the north along dyke systems, thus accounting for the offset of major eruption from the plume head. At 14 Ma, the Precambrian lithosphere of the North American plate began to override the northern end of the plume head, whereas the southern end (located over the plume tail) breached the craton, and the Snake River Plain began to form as the North American plate continued its southwestern movement. The overriding of the northern part of the plume shut off the volcanism in the Columbia River plateau.

North Atlantic Igneous Province

The North Atlantic Igneous Province is one of the largest LIPs known and extends on the west from Ellesmere Island in Canada southeastward across Greenland and Iceland to the rifted margin of northwestern Europe (Fig. 3.10). When both subaerial and submarine volcanic and intrusive rocks are considered, the province has an area of at least 1.3×10^6 km^2 and a volume of 6.6×10^6 km^3 (Eldholm and Grue 1994). Most or all of the activity in this large province is related to the still-active Iceland hotspot (Saunders et al. 1997; Johnston and Thorkelson 2000; Fitton et al. 2000). In

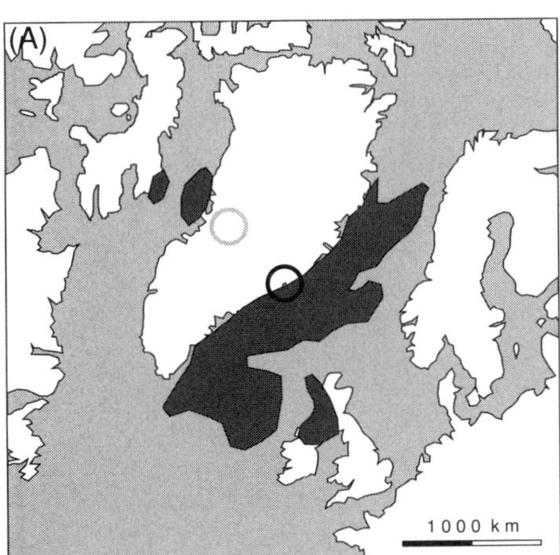

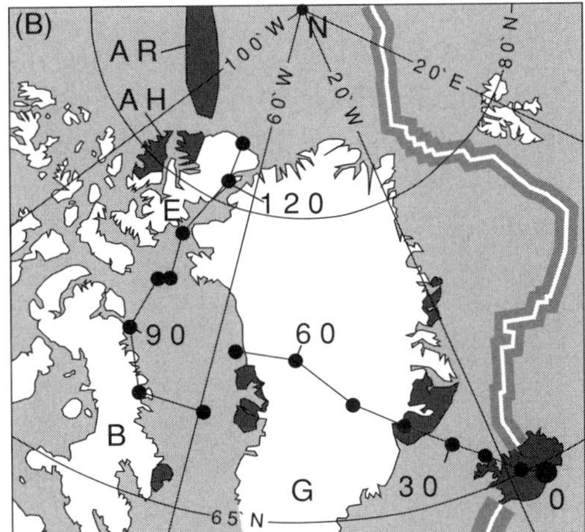

Figure 3.10. The North Atlantic Igneous Province about 63 Ma. A) suggested original distribution of plume-related igneous rocks, and B) trajectory of the Iceland hotspot back to about 130 Ma. Circles in A are interpreted sites of plume head impact with the lithosphere. Hotspot trajectory shows 10-Myr increments, and breaks in the track reflect later rifting. Dark gray denotes hotspot-related volcanics. G = Greenland; B = Baffin Island; E = Ellesmere Island; AH = Axel Heiberg Island; AR = Alpha ridge. From Johnston and Thorkelson (2000), with permission. Copyright © 2000 by Elsevier Science.

the North Atlantic, rifting of Pangea began some 125 Ma, and by 118–80 Ma, rifts propagated northward into the Labrador Sea and the Rockall trough; by about 55 Ma, rifts propagated into the northeastern Atlantic basin. Basalt erupted onto pre-Tertiary continental crust largely composed of sediments of Mesozoic age. Although not well defined before 80 Ma, the Iceland plume appears to have arrived at the lithosphere about 130 Ma and to have left a hotspot track defined by magmatism and uplift to its present position beneath Iceland (Fig. 3.10(b)) (Johnston and Thorkelson 2000).

In West Greenland, a graben system developed during the Mesozoic, and these grabens continued to open during the Tertiary, eventually separating Greenland from Baffin Island. Two major episodes of mafic magmatism are recorded in the North Atlantic Province. The first at 61–58 Ma produced large volumes of basalt and picrite and associated intrusive rocks in Baffin Island, Greenland, and the British Isles. This activity preceded the main breakup and separation of Greenland from northwestern Europe by about 4 Myr. Immature clastic sediments in the North Sea and nearby areas record rapid uplift in this area at 58–55 Ma followed by rapid subsidence (England et al. 1993). Also, the Kangerlussuaq, Skaergaard, and Kap Deichman layered intrusions were emplaced at about 55 Ma. The second major period of volcanism occurred between 50 and 40 Ma as eastern Greenland passed over the hotspot. Hotspot lavas were erupted in three different tectonic settings: on continental crust (these tend to be the oldest), along continental margins forming dipping seismic reflectors, and along the hotspot track defined by the Greenland–Iceland–Faeroe ridge. The slow migration of the lithosphere over the Iceland hotspot between 100 and 50 Ma resulted in gradual thermal erosion and thinning of the lithosphere, as recorded by the rifting history (Johnston and Thorkelson 2000). Separation of Greenland from Europe at about 55 Ma was accompanied by the formation of thick, seaward-dipping reflectors along much of East Greenland and northwest European coasts (Fig. 3.10(a)). These reflectors appear to be chiefly basalt flows with a cumulative volume of almost three times that of present-day Iceland (Larsen and Jakobsdottir 1988). Most lavas were erupted in subaerial or shallow marine environments.

Iceland is the largest exposure of volcanics in the North Atlantic Province and provides a sample of relatively uncontaminated volcanics coming from the Iceland plume. It appears that the Mid-Atlantic Ridge overrode the Iceland plume about 15 Ma and was captured by the plume. Iceland can be divided into two tectonic provinces: an axial rift zone and marginal volcanic plateaus (Saemundsson 1986). New crust is forming in the axial rift zone as Iceland separates along the ocean ridge, and both central and fissure eruptions are found on the island. Most of the lavas are tholeiitic basalts (90%) with small amounts of intermediate to felsic rocks. Volcanics in the marginal plateaus comprise over 65% of the island, and the oldest crust is only 15 Ma. Seismic studies indicate the Iceland crust ranges from 25 to 30 km in thickness, and at least the upper 10 km comprises chiefly lavas (Bjarnasson et al. 1993). Flows dip towards the axial rift system, and dip angle increases with decreasing distance to rifts. In the model of Palmason (1986), excess magmatism along the rift zone results in overloading of the lithosphere, causing subsidence of the surrounding crust. Flows closest to the rift subside the most and are buried by later flows. The deeper the flows are buried, the higher the metamorphic grade. The model predicts that the axial rift is underlain by a zone of partial melting at only 12 km depth. A combination of results from seafloor bathymetry, the chemical composition of basalts erupted along the Mid-Atlantic Ridge south of Iceland, and the thickness of oceanic crust around Iceland suggest that the thermal effects of the Iceland plume extend well beyond the boundaries of Iceland (Saunders et al. 1997).

Ontong Java and Hikurangi Plateaus

General Features

The Ontong Java plateau in the western Pacific is the world's largest oceanic plateau and covers an area the size of Alaska (1.86×10^6 km^2) (Gladczenko et al. 1997; Neal et al. 1997). The plateau surface rises to depths of about 1700 m in the center, but most of it is 2–3 km below sea level. Deep ocean basins bound it on the north and east, and it has collided with the Australian plate on the south (Fig. 3.11). The top of the plateau, although relatively smooth, is capped by several large seamounts and covered in most places by up to 1 km of pelagic sediments (Phinney et al. 1999). As previously discussed, the average thickness of Ontong Java is about 33 km. The volume of magma erupted is probably larger than that of the present plateau because Early Cretaceous oceanic flood basalts in the Nauru basin to the northeast and the East Mariana basin to the north are closely related to Ontong Java basalts both in eruptive age and in composition (Gladczenko et al. 1997).

Because clearly defined linear magnetic anomalies are not found on the Ontong Java plateau (Fig. 3.11), it is probable that lavas were erupted onto at least somewhat older oceanic crust. However, grabens along the eastern margin of the highest part of the plateau and questionable magnetic anomalies across the northern half of the plateau have led some investigators to propose an origin at or near an active ocean ridge (Hussong et al. 1979; Gladczenko et al. 1997). Others have suggested that Ontong Java formed at a migrating triple junction or along a transform fault during ocean ridge jumping accompanied by intense volcanism (Winterer 1976; Neal et al. 1997). Mahoney (1987) proposed that the plateau formed by a ridge-centered or near-ridge plume that may have been the Louisville hotspot (Fig. 2.2). Geochemical data from Ontong Java basalts indicate that the magma source includes a variable contribution from depleted mantle, a finding consistent with a near-ridge plume setting if the depleted mantle source is shallow (Tejada et al. 1996). Recent ^{40}Ar/^{39}Ar isotopic ages from Ontong Java lavas show a strong bimodality at 122 ± 3 and 90 ± 4 Ma (Neal et al. 1997). Data indicate that most of the plateau was erupted during the 122-Ma episode and that only the eastern slope was formed during the younger event.

The island of Malaita in the Solomon Islands (Fig. 3.11) is an obducted fragment of Ontong Java emplaced during collision with the Solomon arc some 25 Ma. The island presents a unique opportunity to study an approximately 4-km section of the basement rocks of Ontong Java. The rocks in this section, which have ^{40}Ar/^{39}Ar ages of about 122 Ma, comprise a monotonous sequence of pillowed tholeiitic basalts and associated sills and dykes (Petterson et al. 1997). Lavas and sills vary from about 1 to 50 m in thickness. There is very little sediment in the section, only a few minor layers of pelagic chert and carbonate, ranging from a few millimeters to a few centimeters thick. Because

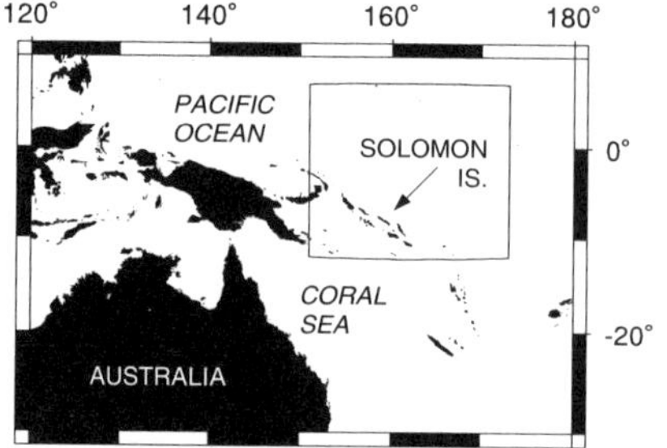

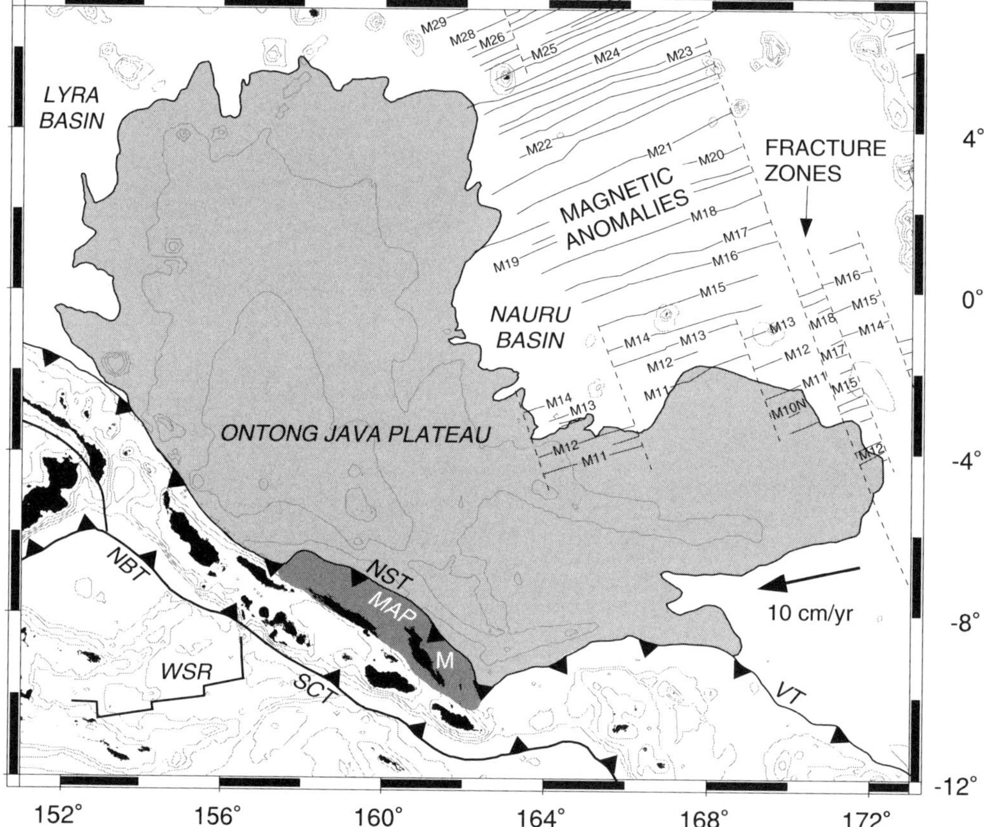

Figure 3.11. Map of the Ontong Java oceanic plateau in the Southwest Pacific. MAP = Malaita accretionary prism; M = Malaita; NST = North Solomon trench; SCT = San Cristobal trench; NBT = New Britain trench; WSR = Woodlark ocean ridge; VT = Vitiaz trench. Modified after Phinney et al. (1999).

chert greatly exceeds carbonate in abundance, most of the basalts were erupted below the carbonate compensation depth.

Tectonic History

One model for the origin of the Ontong Java plateau based on plume–ridge interaction is illustrated in Fig. 3.12. Before 122 Ma, an ocean ridge approached the Ontong Java plume head (a), and by 125–122 Ma, some of the plume head became incorporated in

a. Approaching plume, pre-122 Ma

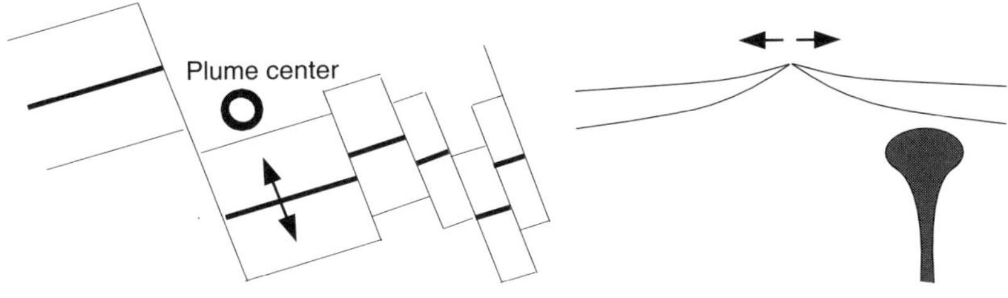

b. Main OJP emplacement, 125–122 Ma

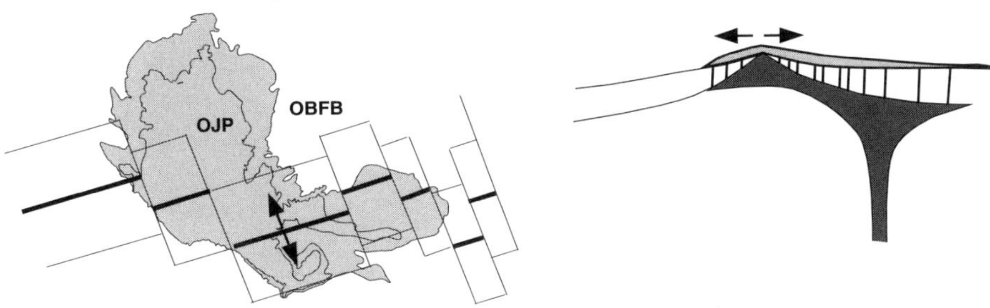

c. Second OJP emplacement, ≈90 Ma

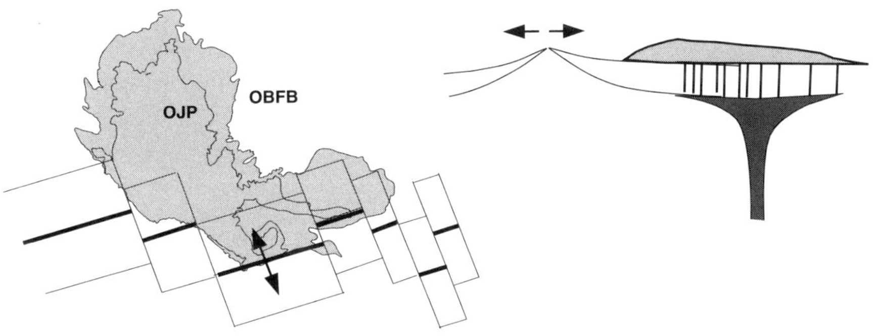

d. Normal seafloor spreading, post-90 Ma

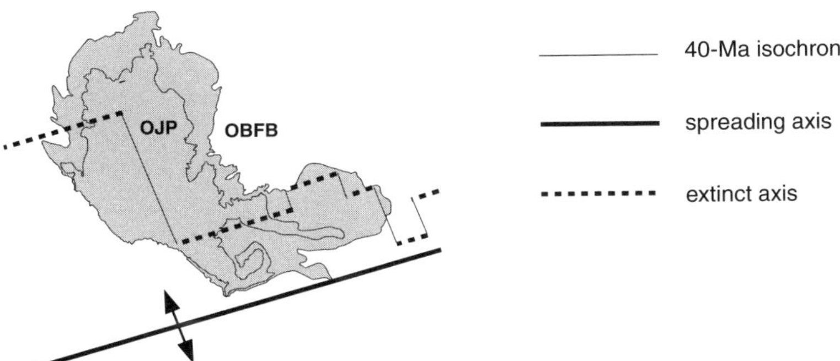

Figure 3.12. Maps showing the interaction of the Ontong Java plume with an ocean ridge at 125-90 Ma. OJP, Ontong Java plateau; OBFB, ocean-basin flood basalts. Modified from Gladczenko et al. (1997), with permission. Copyright © 1997 by the American Geophysical Union.

the spreading center; the bulk of the plateau was then erupted along the flank of the spreading center (b). Magmas came from both plume and mixed plume–ridge sources at this time, thus explaining the geochemical and isotopic compositions of the lavas. At least part of the plateau was erupted on older oceanic crust in the last 40 Myr. At the same time, less voluminous eruptions of oceanic flood basalt also occurred in the nearby basins accompanied by the injection of sills and dykes (Castillo et al. 1994). As plume magmatism waned some 120 Ma, the ocean ridge jumped to a new location south of Ontong Java, and seafloor spreading became established at the new location, which is consistent with linear magnetic anomalies found east of the plateau today (c). At about 90 Ma, either the plume underwent renewed activity or a new plume head developed in response to continued mass transport through the plume tail. At this time the second and smaller magmatic episode occurred with little if any magma input from the ocean ridge.

The Ontong–Australian Plate Collision

The portion of Ontong Java that contains Malaita probably formed along the southern edge of the plateau thousands of kilometers from the present Solomon Islands (Petterson et al. 1997). During the Eocene, the Pacific plate began to subduct beneath the Australian plate, and the Solomon arc was born above the subduction zone (Fig. 3.13(a)). Between 25 and 20 Ma, Ontong Java arrived at the Solomon arc (b), but the lack of deformation in the Malaita rocks indicates that this was a "soft docking." Ontong Java began to clog the subduction zone as it resisted subduction 15–10 Ma (c), and arc volcanism shut down. However, at about 12 Ma it had started again in response the initiation of subduction of the Australian plate beneath the Pacific plate (d). Increased coupling between the Solomon arc and Ontong Java led to deformation of the leading edge of the plateau with the main compressive deformation beginning at 4 Ma. This resulted in the obduction of the upper part of Ontong Java onto the old Solomon arc block (e). The final stage of obduction resulted in the detachment of a flake at a depth of 5–10 km followed by transportation of the flake over the Solomon block between 4 and 2 Ma. During this time the rocks in Malaita were compressed, and high angle thrusts developed with northeast vergence. Obduction continued to the present, elevating Malaita above sea level.

The fate of Ontong Java in the future is problematic. Will it remain on the Pacific plate as an extinct passenger and sometime in the future be accreted to a continental margin? Or will it be subducted beneath some other plate before or during collision with a continent? The studies of Cloos (1993) suggested that oceanic plateaus greater than about 30 km thick cannot be subducted, which implies that Ontong Java, with a thickness of 33 km, will not subduct but someday will collide with, and be accreted to, a continent.

Hikurangi Plateau

Besides the Caribbean plateau, the Hikurangi oceanic plateau east of North Island in New Zealand is the only other oceanic plateau known to be colliding with another plate today (Fig. 3.14). This plateau, which is being subducted into the Hikurangi trench, comprises a thick succession of submarine basalts erupted some

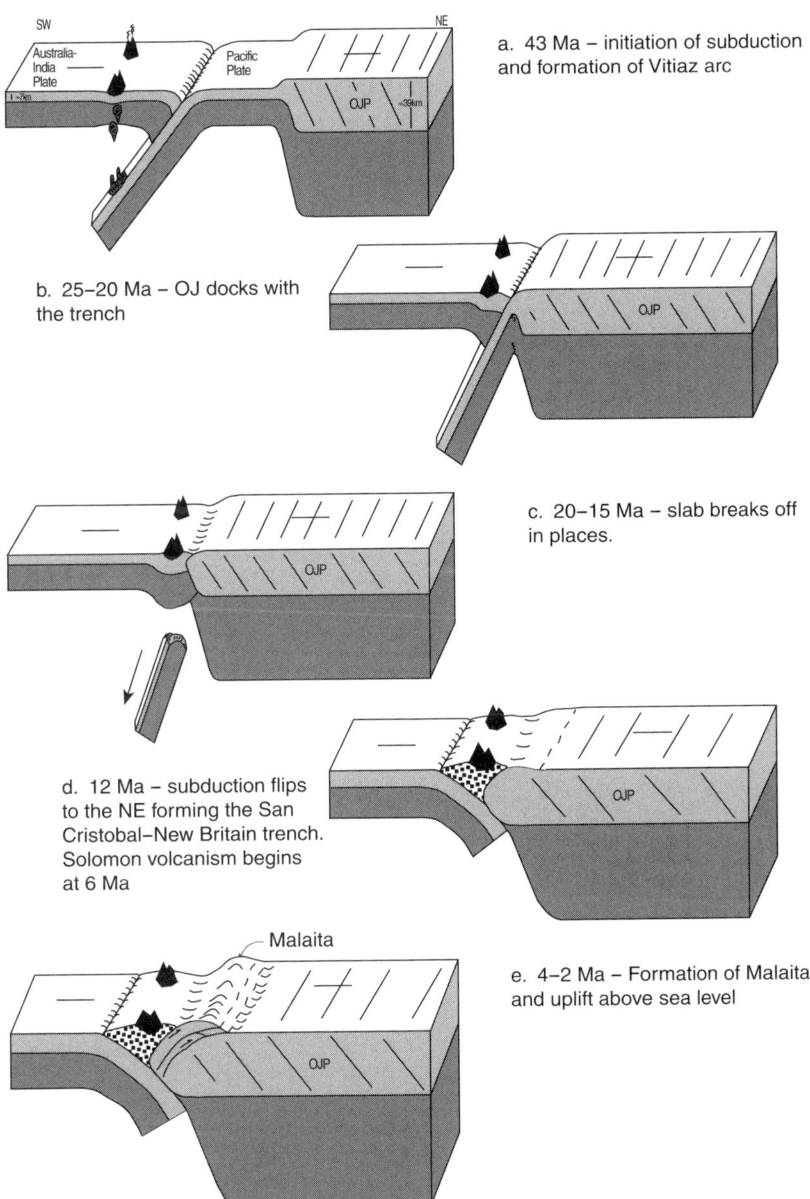

Figure 3.13. Schematic sequence of events showing collision of the Ontong Java plateau with the Australian plate and the subduction zone flip that occurred afterwards. OJP, Ontong Java plateau. Modified from Petterson et al. (1997), with permission. Copyright © 1997 by Elsevier Science.

80 Ma, and it is bounded on the south by the Chatham Rise, which is a block of continental crust (Strong 1994; Mortimer and Parkinson 1996). If the plateau is a product of the Mid-Cretaceous superplume event (see Chapter 8), it could have components as old as 120 Ma. The plateau is 10–15 km thick and thus is subductable from buoyancy considerations (Cloos 1993). Dredged basalts have chemical and Nd–Sr isotope characteristics similar to Ontong Java basalts. Hikurangi plateau is

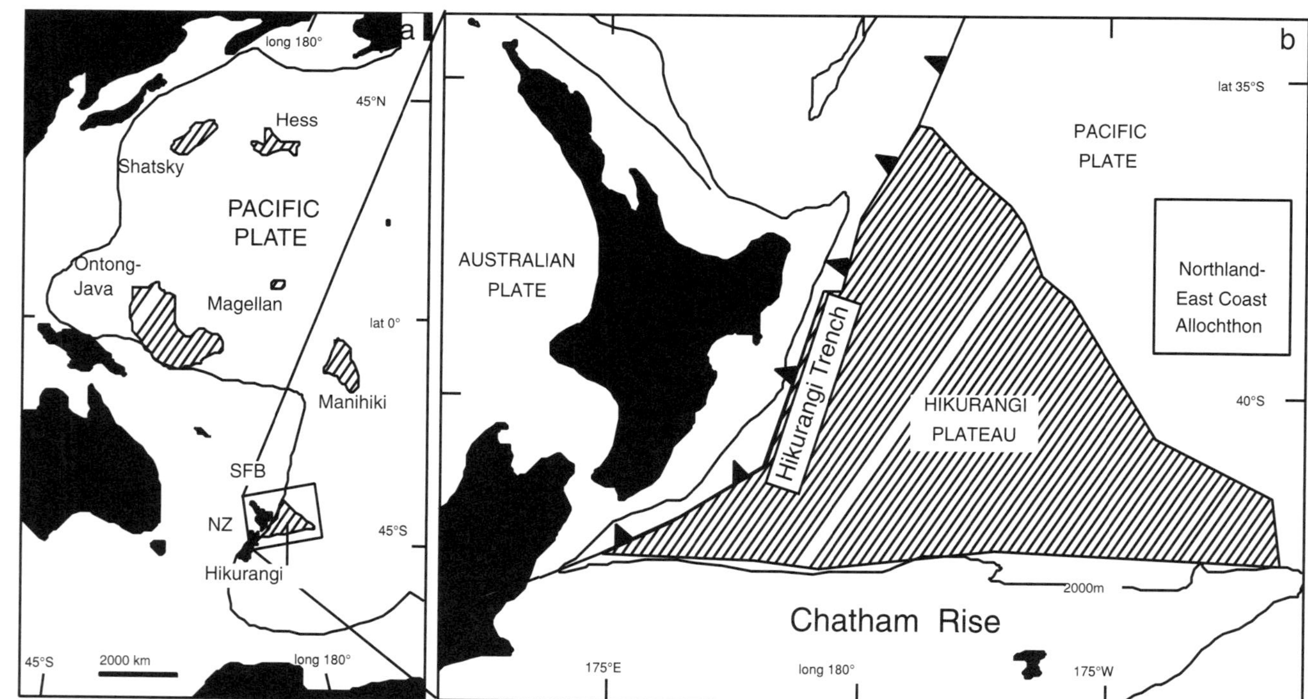

Figure 3.14. Map showing the Hikurangi oceanic plateau, which is being subducted into the Hikurangi trench adjacent to New Zealand. SFB, South Fiji Basin. Courtesy of Nick Mortimer.

only about half the size of Manihiki plateau east of Ontong Java (Fig. 3.14). At the present Pacific–Australian plate convergence rate of about 5 cm/yr, 150–200 km of the plateau have been subducted beneath North Island in the past 20 Myr (Wood and Davy 1994). However, the exact arrival time of the plateau at the Hikurangi trench is unknown.

Two terranes on North Island, the Northland and East Coast terranes, may represent parts of an obducted allochthon of the Hikurangi plateau (Mortimer and Parkinson 1996). This allochthon appears to have been thrust westward over northern New Zealand beginning about 20 Ma (Sporli and Ballance 1988). Similarities in composition of the basalts in the allochthon to those dredged from the Hikurangi plateau support this idea.

Siberian Traps

The Siberian traps are one of the largest flood basalt provinces and were erupted in less than 1 Myr at the Permian–Triassic boundary. The province covers an area of about 4×10^6 km^2 with an average thickness of about 1 km. Basalts were erupted on the northwestern margin of the Siberian platform, which has been a stable craton since the end of the Precambrian (Khain 1985; Zolotukhin and Al'mukhamedov 1988). Individual flows in the province are a few tens of meters thick and extend for tens to hundreds of kilometers. One of the unusual features of the Siberian traps in comparison to other flood basalts is the relative abundance of mafic pyroclastic rocks (Zolotukhin and Al'mukhamedov 1988). Basaltic tuffs and breccias reach a maximum thickness of about 700 m in the Tunguska basin near the center of the field (Fig. 3.15, no. 4) and then thin to the north and west where lavas dominate. In the southern part of the province, mafic pyroclastics greatly exceed lavas in abundance. Dykes and sills are exposed chiefly around the margins of the Tunguska basin, where they intrude

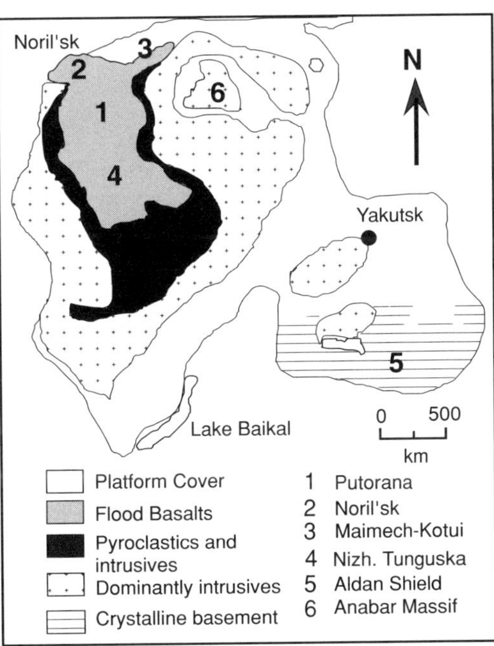

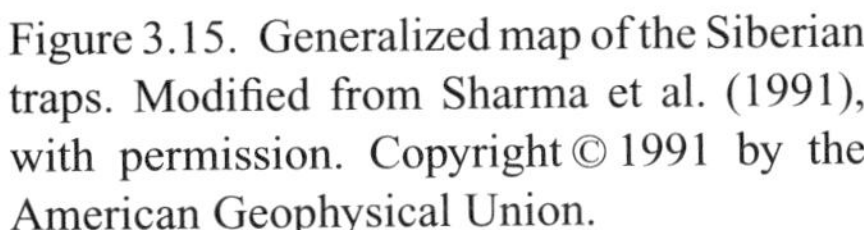
Figure 3.15. Generalized map of the Siberian traps. Modified from Sharma et al. (1991), with permission. Copyright © 1991 by the American Geophysical Union.

surrounding basement rocks. Some are thick-layered intrusions. In some of the surrounding Devonian sediments, sills related to the Siberian eruptions comprise as much as 50% of the succession. Although not exposed, field and geophysical studies indicate that the majority of the flows and sills were erupted from linear fissure systems. The origin of the mafic pyroclastics, however, is not well understood.

In terms of composition, the Siberian traps are divided into four regions (Fig. 3.15) (Lightfoot et al. 1990; Sharma et al. 1991). The Putorana region (1) in the north-central part of the province comprises chiefly tholeiitic basalts. Volcanics of the Noril'sk region in the extreme northwestern part of the province (2) comprise rocks of variable composition ranging from picrites through tholeiites to alkali basalts and basaltic andesites. In the Maimech–Kotui area in the northeast (3), alkali basalts and related undersaturated lavas dominate, whereas in the Nizhnyaya–Tunguska region in the south (4) basaltic tuff and breccia dominate.

The ages of volcanism in the Siberian traps are of considerable interest because volcanic activity in this province may be related to mass extinctions at the end of the Permian. Because some samples are contaminated with radiogenic argon and others have lost argon, the precise ages of eruption have been a subject of disagreement among laboratories (Sharma 1997). Recent high-precision $^{40}Ar/^{39}Ar$ ages of basalts and U/Pb baddeleyite ages seem to be converging on an eruptive age for basalts in the Noril'sk region of 250 Ma with an error on the order of 1 Myr. This is indistinguishable from the precisely dated Permian–Triassic boundary at 250–251 Ma (Renne et al. 1995).

Morgan (1981) suggested that the Siberian traps are related to the initiation of the Jan Mayen hotspot north of Iceland (Fig. 2.1) and that the Lomonosov Ridge is the hotspot track. Subsequent geophysical data, however, suggest this ridge is continental and was possibly rifted from the Siberian craton when the Nansen Ridge (a continuation of the Mid-Atlantic Ridge) formed and the Arctic basin opened (Jokat et al. 1992). Although the lack of uplift associated with the Siberian traps is difficult to explain with a mantle plume model (Czamanske et al. 1998), geochemical and rare gas results that constrain the source composition of the volcanics, the large volume of magma erupted, and the short duration of eruption (≤ 1 Myr) all favor the plume hypothesis. However, because there is no obvious connection of these basalts with the Jan Mayen or any other modern hotspot, where is the plume that was the source of the magmas? The most likely answer is that the Siberian plume has become extinct sometime in the last 250 Myr and that most or all of the older seafloor has been subducted, removing the Siberian hotspot track.

Paraná–Etendeka Flood Basalts

The Paraná flood basalts in southern Brazil and the Etendeka flood basalts in Namibia are remnants of an enormous flood basalt province that appears to have been erupted from the Tristan hotspot in the South Atlantic (Fig. 2.1). The Paraná field alone covers at least 1.2×10^6 km^2 and averages about 0.7 km in thickness (Fig. 3.16) (Peate 1997). In Namibia in southwestern Africa, the Etendeka lavas occur as scattered remnants over an area of about 0.8×10^5 km^2. Large dyke swarms are associated with both fields and suggest that they were probably much larger before erosion. Thick, probably correlative lavas occur in offshore basins along the coast of South America and

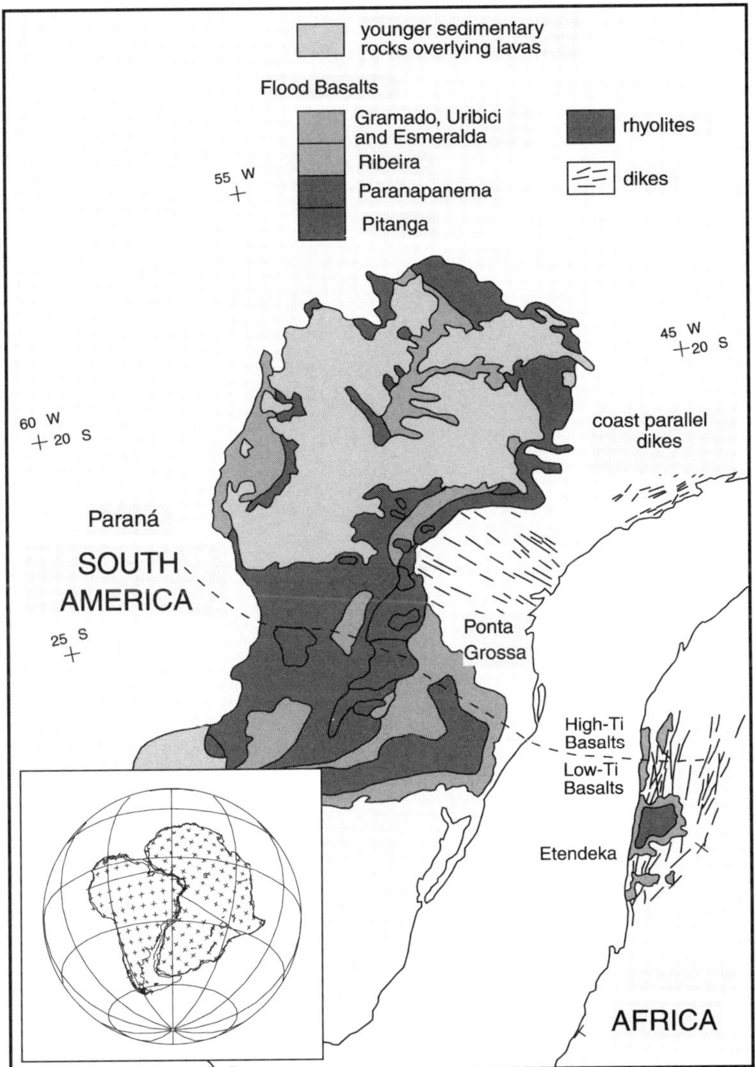

Figure 3.16. Geologic map of the Paraná–Etendeka flood basalt province before the breakup of Africa and South America. Modified after Peate (1997).

Southwest Africa. In terms of composition, the Paraná–Etendeka lavas are strongly bimodal with few if any intermediate compositions. Tholeiitic basalts comprise over 90% of both volcanic fields. In South America, basalts were erupted on sediments of the intracratonic Paraná basin that was established in the Ordovician. High-precision $^{40}Ar/^{39}Ar$ ages from basalts in the Paraná–Etendeka province yield ages ranging from about 138 to 127 Ma, and the greatest volume of lavas erupted between about 135 and 131 Ma (Turner et al. 1994; Stewart et al. 1996; Peate 1997). In this flood basalt province the main-phase volcanism lasted for 4–5 Myr in contrast with others such as the Deccan, Siberian, and Ethiopian for which the bulk of the plateau was emplaced in less than 1 Myr (Table 3.1). In addition to the dyke swarms, several large alkalic complexes were emplaced around the margins of the province, both in South

America and in Africa. These rocks cover a wide compositional range including carbonatite, alkali gabbro, phonolite, syenite, and granite (Milner et al. 1995). Many of the complexes in Namibia are high-level intrusions associated with caldera collapse structures.

Eruption of the Paraná–Etendeka lavas was associated with the opening of the South Atlantic beginning about 150 Ma and progressing from south to north (Hawkesworth et al. 1992; Wilson 1992). Between 135 and 131 Ma, the flood basalts were erupted from the Tristan plume, and, as the South Atlantic continued to open, two hotspot tracks formed: the Walvis Ridge, extending southwest from Namibia, and the Rio Grande Rise, extending southeast from Brazil (Fig. 3.1). Ages on both tracks decrease in going from the continents towards the Tristan hotspot. Also, isotopic and geochemical data suggest a common source for the Paraná–Etendeka and Walvis Ridge basalts. By 126–119 Ma, rifting had progressed to the easternmost tip of Brazil, and subsidiary rifts, such as the Benue trough in Nigeria, propagated into the cratons, becoming aulacogens. The final separation of South America from Africa occurred about 100 Ma. Until 70 Ma, the Tristan hotspot was ridge centered. After this time, the Mid-Atlantic Ridge migrated westward away from the hotspot, and magmatism stopped along the Rio Grande Rise (O'Conner and Duncan 1990).

Deccan Traps

No other flood basalt province has received more attention than the Deccan traps in west-central India (Fig. 3.17). These basalts were erupted very rapidly – probably within 1 Myr – at the Cretaceous–Tertiary boundary 65 Ma when western India lay above the Réunion hotspot now located east of Madagascar (Courtillot et al. 1986; Mahoney 1988) (Fig. 2.1). The flood basalts were erupted on the Archean craton of western India over an area possibly exceeding 1.5×10^6 km^2 (Raja Rao et al. 1978). Offshore drilling and geophysical surveys indicate that the traps extend to the edge of the continental shelf west of Bombay (Fig. 3.17). The Deccan basalts comprise a relatively thick succession of remarkably flat-lying (dips $\leq 1°$) flows that are dominantly tholeiites. Only in the Cambay and Narmada grabens are the flows cut by normal faults and sections tilted from the horizontal. The succession of flows ranges to 2-km thick in the Western Ghats near Bombay and thins to the east. Dykes are associated with the flows in the western and northern parts of the field, and these range in thickness from about 1 m to over 300 m (Agashe and Gupta 1972). Although the Deccan flows are largely tholeiites, a few localized alkali basalt flows occur in the Narmada graben and along the west coast north of Bombay. Because of the interest in Deccan eruptions as a possible cause for mass extinctions at the end of the Cretaceous, a major effort is being made to date the eruptions accurately. The most precise ^{40}Ar/^{39}Ar dates suggest that most of volcanism occurred at 65 Ma with an uncertainty of less than 1 Myr (Courtillot and Cisowski 1987; Courtillot et al. 1999).

Although the source vents for the Deccan basalts have not been recognized, several lines of evidence indicate they lie to the west of the Western Ghats beneath the ocean. Both the total thickness of flows and the greatest number of flows occur along the west coast of India, which is consistent with a western source (Lightfoot 1985). Although

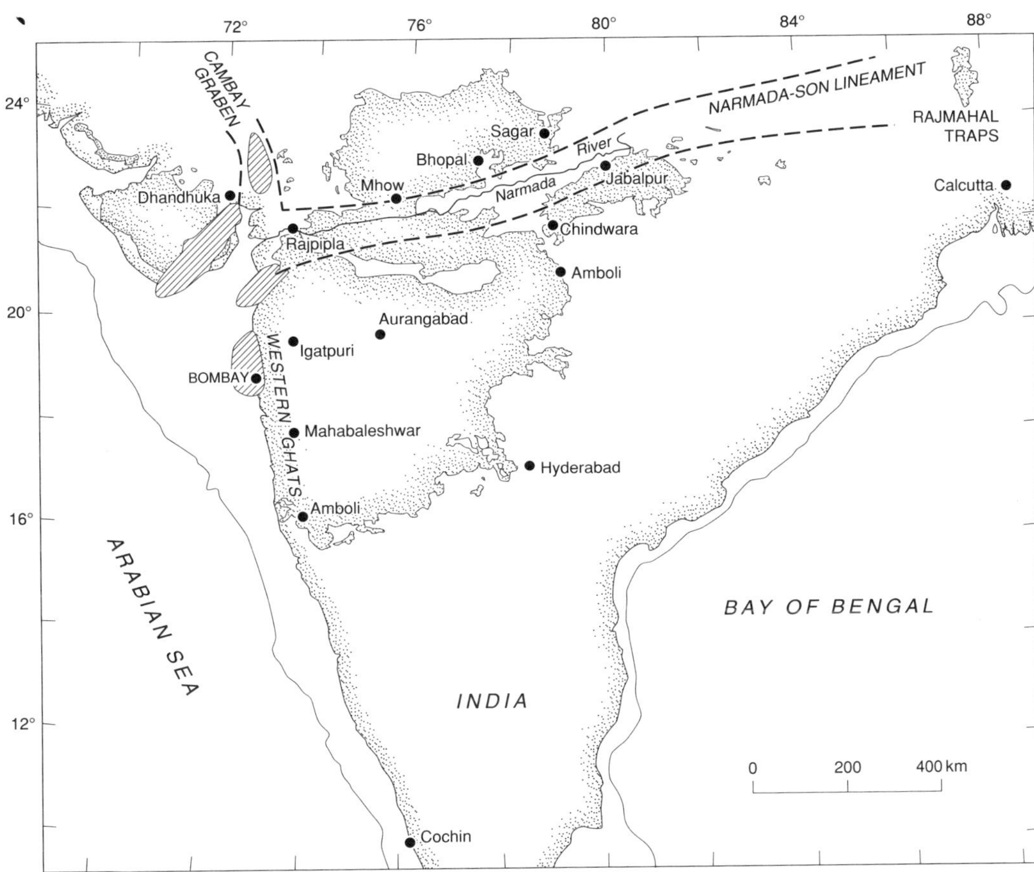

Figure 3.17. Map of India showing the Deccan and Rajmahal traps. Positive gravity anomalies indicated by diagonal line pattern. Courtesy of John Mahoney.

pyroclastics are not abundant in the Deccan, they increase in importance to the west, again supporting a western source. Also, the main rift faults that may have been feeders for magma probably lay on the continental shelf because the breakup of India and the Seychelles block occurred in this area. Positive gravity anomalies west of Bombay (Fig. 3.17) on the continental shelf are also interpreted as large mafic intrusives perhaps associated with vents for the flows.

The separation of the Seychelles block from mainland India coincided precisely with the eruption of the Deccan basalts (White and McKenzie 1989). At the time of separation, the Réunion hotspot lay beneath the western edge of the Deccan, and as the breakup progressed a hotspot track formed as now defined by the Mascarene plateau and the Chagos–Laccadive ridge (Duncan 1981) (Fig. 3.18). The gap in the track is caused by offset due to later spreading along the Carlsberg ridge.

Kerguelen Plateau

One of the largest oceanic plateaus is the Kerguelen plateau in the southern Indian Ocean (Fig. 3.1). Unlike Ontong Java, which is largely below sea level, Kerguelen has numerous islands that provide access to sampling. Together with Broken Ridge

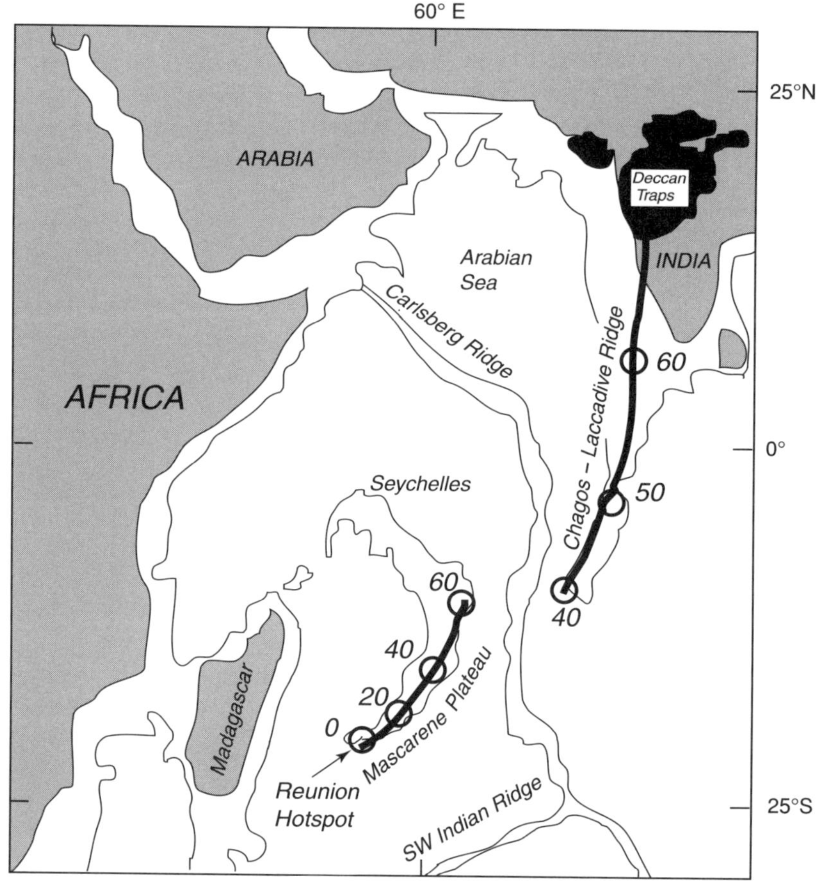

Figure 3.18. Mascarene–Chagos–Laccadive hotspot track in relation to the Reunion hotspot. Note the track is split by development of the Carlsberg ocean ridge. Ages on track are given in Ma. After White and McKenzie (1989).

(Fig. 3.1), which is now rifted away from Kerguelen, this LIP formed in the Cretaceous in response to the Kerguelen plume (Duncan and Storey 1992; Nicolaysen et al. 2000). The plume initially impinged on the Antarctic plate ($\geq$116 Ma) and later, owing to a southward jump of the Indian Ridge 100 Ma, it became anchored beneath the Indian plate (Frey et al. 2000). The Rajmahal flood basalts in northeastern India record the first major plume magmatism at about 116 Ma. Most of the Kerguelen plateau was constructed on the Antarctic plate from the plume head between 110 and 114 Ma. As the India plate moved rapidly northward over the plume head between about 85 and 38 Ma, a long hotspot track, the Ninetyeast Ridge, formed on the seafloor (Fig. 3.19). The frosting of island volcanoes, known as the Kerguelen Archipelago, formed in the last 40 Myr – either from bending of the plume head (Weis et al. 1992) or in response to another ridge jump that again placed the plume beneath the Antarctic plate or centered it on the Indian Ridge (Frey et al. 2000). Crustal production rates were relatively high (3.5 km^3/yr) during formation of the plateau (Table 3.1) and dropped off as the Ninetyeast Ridge (0.5 km^3/yr) and Kerguelen Archipelago (0.42 km^3/yr) formed in the last 80 Myr (Nicolaysen et al. 2000).

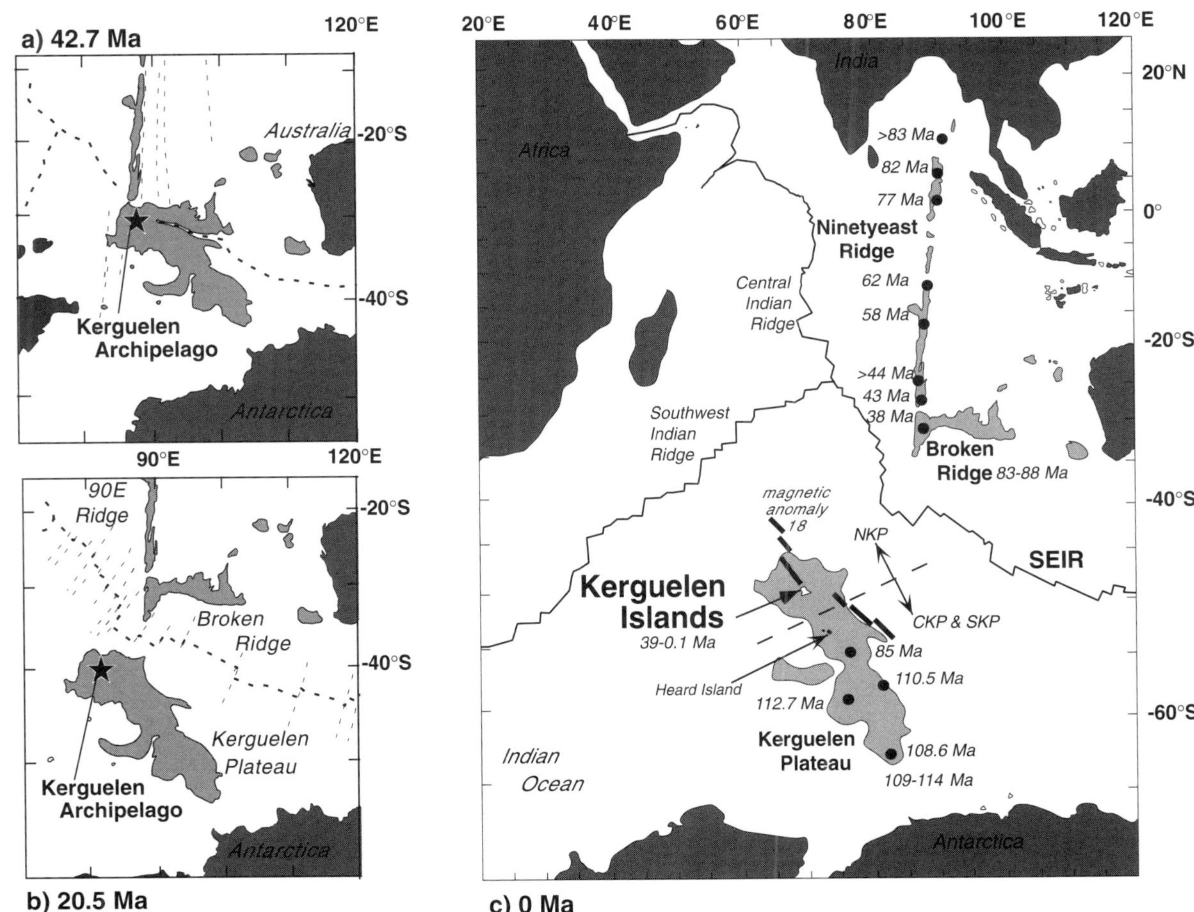

Figure 3.19. Plate reconstructions in the Indian Ocean showing the tectonic history of Kerguelen. (a) propagation of the Indian ocean ridge beneath Kerguelen at 42.7 Ma, (b) rifting of Broken Ridge and the Ninetyeast Ridge from Kerguelen at 20.5 Ma, and (c) current tectonic regime showing ages of basalts from Kerguelen, Broken Ridge, and the Ninetyeast Ridge. NKP, CKP, and SKP denote northern, central, and southern Kerguelen Plateau, respectively. Courtesy of Kirsten Nicolaysen.

The Kerguelen Archipelago is unusual because it contains a wide range of igneous rocks from mafic to felsic in composition with many plutonic rocks (Yang et al. 1998). More than 85% of the plateau, however, comprises extensive tholeiitic flood basalts with a few dykes. Flow characteristics and the presence of wood fragments, pollen, spores, and seeds in shallow water sediments overlying the igneous basement show that parts of the plateau emerged above sea level (Frey et al. 2000). Eruptive ages decrease from northwest to southeast, and younger volcanic structures are built on platforms of older flood basalts. Mafic granulite xenoliths suggest a density for the lower crust in Kerguelen that is lower than typical oceanic crust (Gregoire et al. 1998). This gives Kerguelen isostatic buoyancy and prevents it from sinking or from subduction. Sr, Nd, Pb, and Os isotopic studies of Kerguelen basalts show a dominant mantle plume source with no evidence of mixing of mid-ocean-ridge basalt (MORB) and plateau basalt sources (Yang et al. 1998).

The possibility that part of Kerguelen is a microcontinent has received considerable attention because granitic rocks are generally absent in oceanic plateaus. However, Sr, Nd, and Pb isotope compositions of many Kerguelen granitoids are similar to those of associated basalts derived from the Kerguelen plume, which suggests they do not come from older continental crust but were produced by fractional crystallization of the basaltic magmas (Yang et al. 1998). However, Os and Pb isotopic data from mantle xenoliths and some lavas dredged and drilled from the southeastern part of the plateau and from Broken Ridge have continental affinities (Mahoney et al. 1995), suggesting at least a minor continental component in Kerguelen. Indisputable evidence of continental affinities for at least part of Kerguelen is the presence of clasts of felsic gneiss in fluvial conglomerates from one of the Deep Sea Drilling sites (Frey et al. 2000). Results of a wide-angle seismic survey show a 22-km-thick crust with three crustal layers (Fig. 3.4), including a peculiar 4–6-km-thick zone just above the Moho that is highly reflective (Operto and Charvis 1996). Velocity structure of the crust is consistent with a largely mafic composition. The strongly reflecting zone as well as underlying upper mantle show a strong seismic anisotropy. This lends support to the idea that Kerguelen is an intensely stretched microcontinent overlain by flood basalts and associated intrusives and that it was isolated from Antarctica during the opening of the southern Indian Ocean.

On the basis of new geochemical and isotopic data from recent drill sites at Kerguelen, some investigators question whether Kerguelen is really an oceanic plateau. At four sites drilled on Kerguelen, there is good to indisputable geochemical evidence that the mafic magmas passed through continental crust before reaching the surface (N. Arndt, 2000, personal comunication). It is possible that a large part of the plateau is composed of continental flood basalts. The best analogues are the North Atlantic reflector sequences, which are also composed, in large part, of drowned, subaerial, crustally contaminated basalts. A transition from flood basalt volcanism to submarine plateau volcanism similar to that found in going from the Kerguelen plateau to the Kerguelen Archipelago is also found in the Ethiopian Province in East Africa.

Karoo–Ferrar Province

Remnants of a vast flood basalt province associated with the fragmentation of Gondwana in the Jurassic are found throughout southern Africa, in Droning Maud Land,

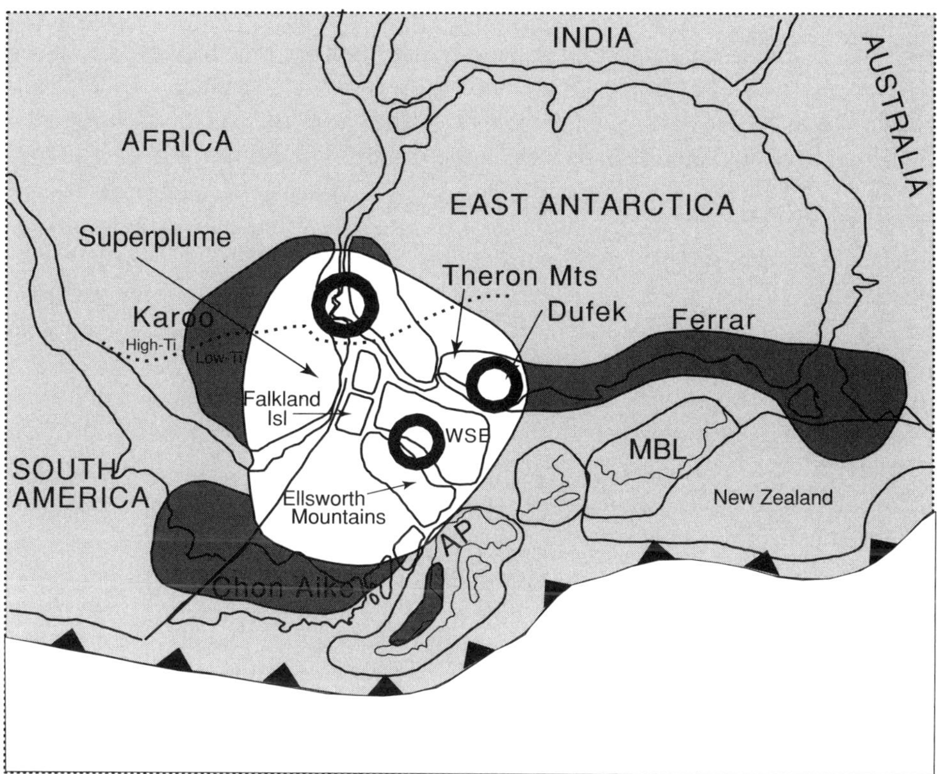

Figure 3.20. Gondwana reconstruction showing location of three plumes (black wheels) associated with a possible superplume. Modified after Storey et al. (2000). AP = Antarctic peninsula; MBL = Marie Byrd Land; WSE = Weddell Sea embayment. Dashed line is the boundary between high- and low-titanium basalts in the Karoo Province. Courtesy of Bryan Storey.

and the Ross orogen in Antarctica (Fig. 3.20) (Hergt et al. 1991; Brewer et al. 1992). The largest and best known basalts in this province are the Karoo volcanics in southern Africa. Exposures are widely separated and suggest an original extent on the order 3×10^6 km^2. Like the Siberian traps, the Karoo Province contains a great range in rock compositions (Cox 1988). Although tholeiitic basalt flows dominate, picrites, rhyolites, and alkaline rocks such as nephelinite are locally important. Intrusive phases, chiefly dykes, are dominantly diabase and gabbro, but granite, syenite, and other alkaline intrusives occur in some regions. Karoo volcanism began about 195 Ma in continental areas covered with terrestrial sediments in the central part of the province and in the southern Lebombo area. The second major pulse of volcanic activity occurred at 180–177 Ma and concentrated mainly in the northern part of the province. In Antarctica, the Ferrar Province consists of chiefly mafic flows, dykes, and sills with a total volume in excess of 10^6 km^3 (Kyle et al. 1981). Geochemically, the Ferrar Province is divided into two subprovinces: one in Droning Maud Land, which has a relatively depleted mantle source, and one in the Ross orogen (Ferrar magmatic province, Fig. 3.20) with more enriched sources (Brewer et al. 1992). The earliest magmatic episode in Droning Maud Land at 193 Ma appears to reflect the initiation of rifting in this area with major magmatism and rifting concentrated at 180–170 Ma.

The mantle plume or plumes responsible for Karoo–Ferrar magmatism appear to have been centered near the Africa–Antarctic connection prior to breakup of Gondwana (White 1997) (Fig. 3.20). Extensional environments related to the plume or plumes, and perhaps also to back arc extension, were concentrated in a belt over 2000 km behind the subduction zone on the west. In particular, the Weddell Sea embayment was the site of extensive extension, whereas the Transantarctic Mountains are characterized by strike slip motions associated with converging and accreting terranes. In the melting model of White and McKenzie (1989), the plume or plumes supply melt over an area on the order of 1000 km radius, but it is not the driving force for the breakup of Africa and Antarctica. These authors concluded that the Ferrar magmatism is unrelated to the Karoo plume, which is consistent with its linear pattern extending for thousands of kilometers along the Transantarctic Mountains. However, recent studies in the Ferrar magmatic province indicate that tholeiitic magmatism occurred in a remarkably short interval of time (<1 Myr) 177 Ma and was associated with an active rift system extending along the axis of the Transantarctic Mountains (Elliot et al. 1999). The Ferrar magmas may have been transported along fault systems associated with the rifting for as much as 3500 km along the Transantarctic Mountains (Fig. 3.20).

Dalziel et al. (2000) have proposed that compressive deformation in the Gondwanide orogen extending from southern South America through the Weddell Sea into eastern Australia (barbed pattern, Fig. 3.20) was affected by the mantle plumes in this region. They suggest that the oceanic plate descending beneath Gondwana was flattened by the impingement of one or more of these plumes on the bottom of the plate. This also caused deformation in the overlying Gondwana plate and led to a temporary cessation in arc volcanism (by eliminating the mantle wedge) until a new subduction zone was established on the west. Finally, the plume or plumes broke through the flattened plate, stretching the continental lithosphere and leading to the voluminous Karoo–Ferrar magmatism.

Considered as a whole, the geochemical data suggest two or three sites of melt generation (Elliot et al. 1999; Storey et al. 2001): one in the northern part of the Karoo Province that gave rise high-Ti tholeiites (discussed later), one centered in the Weddell Sea region that gave rise to low-Ti tholeiites in the Karoo Province and was responsible for the Chon Aike volcanism in Argentina, and one centered beneath the Dufek layered intrusion in East Antarctica that gave rise to the Dufek intrusion and Ferrar Province (Fig. 3.20). If correct, the long distance of transport of flood basalts in Antarctica derived from the Dufek source suggest that LIP magmas may travel great distances before erupting. A similar source for the Dufek intrusion and Ferrar basalts is supported by the remarkably similar chemistry of the magmas. Three principal periods of felsic magmatism are recognized in the Chon Aike Province in Argentina at 184, 170, and 158 Ma and geochemical models indicate a crustal source (Pankhurst and Rapela 1995). Although this province formed adjacent to the Chilean subduction system, the felsic volcanics of the first two episodes are not subduction related. Instead, partial melting of the lower crust may have been initiated by a mantle plume.

Storey and Kyle (1997) and Storey et al. (2001) have suggested that the three centers of melting are part of a large superplume centered near the Falkland Islands (Fig. 3.20). These three centers of melting may be represented today by the Shona, Bouvet, and

Discovery hotspots in the southern Indian Ocean (Fig. 2.1). If this idea survives, the superplume produced one of the largest LIPs known in Earth's history, including the Karoo and Ferrar Provinces and at least part of the Chon Aike Province in southern Argentina. Initial rifting of Gondwana occurred across the top of the thermal anomaly with formation of triple junctions (over each hotspot) beneath the Weddell Sea and beneath the Mozambique basin. Extension along the arms of each triple junction may have controlled the channeling of plume material into the Ferrar and Chon Aike Provinces.

Ethiopian and East African Plateaus

Eruption of the Ethiopian and East African plateaus started some 45 Ma, before formation of the East African rift system, and they are generally attributed to one or two mantle plumes beneath northeastern Africa. The Afar plume, centered in northeastern Ethiopia, eventually resulted in the separation of Arabia from East Africa (Burke 1996). The northeast African flood basalt plateaus are the largest of several mid-Tertiary LIPs in Africa, none of which has hotspot tracks – a feature due to the relatively small motion of the African plate during this time (Fig. 3.1). Nearly 1×10^6 km^3 of mixed tholeiitic and alkaline mafic volcanics were erupted in this area in the last 45 Myr (Table 1.1). The bulk of the Ethiopian and East African basalts, however, were erupted in ≤ 1 Myr at 30 Ma (Hofmann et al. 1997; Pik et al. 1998, 1999), and the remainder of these basalts was erupted chiefly at 17–23 Ma. The volcanics in these plateaus are mainly tholeiitic with some transitional alkaline members. Both plateaus are overlain by large, chiefly alkaline, shield volcanoes with ages ranging from 30 to 15 Ma.

Rogers et al. (2000) presented convincing geochemical and isotopic evidence for the existence of two mantle plumes beneath northeastern Africa. Basalts erupted in the East African plateau in Kenya have distinctly different Nd and Sr isotopic compositions from those associated with the Afar plume in Ethiopia. The Kenyan basalts may be associated with a second mantle plume, the Kenya plume, centered beneath Lake Victoria. Early and mid-Tertiary basalts from southern Ethiopia are more similar in composition to Kenyan basalts than to Afar basalts, suggesting that the Kenya plume has been active for at least 45 Myr. Migration of magmatism southward from southern Ethiopia is consistent with the northeastward movement of the African plate over the Kenya plume head in the last 50 Myr.

As pointed out by Ebinger and Sleep (1998), it is possible that the thick lithosphere in northeast Africa was thermally thinned by Late Cretaceous and Early Tertiary rifting. When the Afar plume head arrived at the lithosphere 45 Ma, relief on the bottom of the lithosphere may have deflected the plume, resulting in ponding of plume material beneath the thinnest lithosphere (Sleep 1997). Decompression melting should be most intense in these plume "ponds" beneath thin lithosphere. It is possible that melts generated from a single plume could flow along channels in the base of the lithosphere, explaining swells associated with small amounts of volcanism in other parts of East Africa (Fig. 3.21).

Xenolith and seismic data indicate the lithosphere in East Africa was more than 150 km thick before arrival of the Afar plume. Volcanism began over the plume head 45 Ma, some 20 Myr before rifting responsible for the East African rift system. The

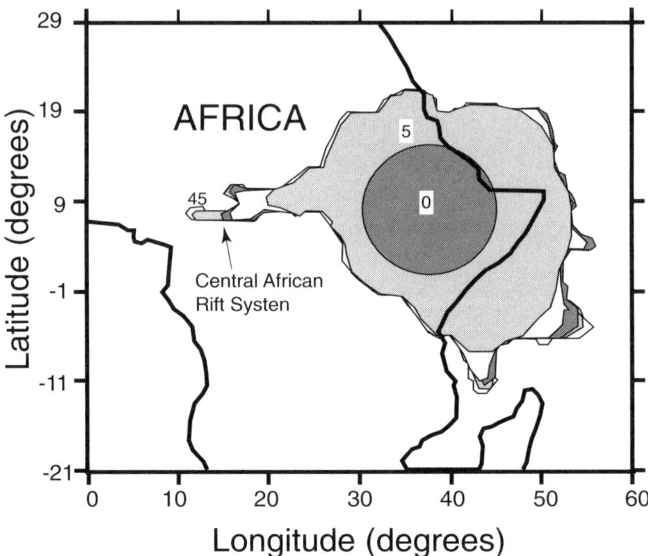

Figure 3.21. Map showing the lateral extent of mantle plume material beneath northeastern Africa during the last 45 Ma. Contour zero is 45 Ma, and the present is represented by contour 45. Notice the channeling of plume material along older rifts with time. After Ebinger and Sleep (1998).

geochemical and isotopic signature of the plume is recognized in volcanics as far away as the Central African rift system and around the Red Sea and Gulf of Aden (Fig. 3.21) (Ebinger and Sleep 1998). Magmatism and rifting seem to radiate from the inferred position of the plume head in southwestern Ethiopia beginning about 40 Ma and continuing to the present. As plume material was channeled by relief along the bottom of the lithosphere, it led to regional uplift, more thermal thinning of the lithosphere, and more decompression melting. In the model of Ebinger and Sleep (1998), the Afar plume head with a diameter of about 750 km impinged on the base of the lithosphere 45 Ma, producing a thick, low density layer that resulted in uplift of the Ethiopian and East African plateaus. By 40 Ma, plume material had traveled 500–800 km south and east to previously thinned lithosphere along the margin of the Indian Ocean and beneath the Arabian peninsula following Mesozoic/Early Tertiary rifts (Fig. 3.21). And finally, between 35 and 10 Ma, a narrow tongue of the plume burrowed westward beneath the Central African rift system, reaching and elevating the Adamawa plateau by 5 Ma. Although this model matches the timing of volcanism in East Africa and part of Central Africa well, it cannot explain the swell-related volcanism in North and West Africa (Hoggar and Tibesti swells) unless undiscovered rifts lie beneath the Sahara.

The model of Ebinger and Sleep (1998) is important because it shows that basalts may not always coincide with the center of a starting plume head and that several magmatic provinces may be produced from a single plume at different times. As we saw previously, a similar mechanism can explain the Ferrar flood basalts in Antarctica, which may have traveled thousands of kilometers from their source plume head. Recently, a similar model has been proposed for the Columbia River basalts (Johnston and Thorkelson 2000). If the lithosphere-channeled plume model proves to be important, continental flood basalt provinces may not necessarily mark the arrival time of a plume head and the initiation of melting. Instead, plume magmatism beneath continents may be episodic, depending on the penetrability of the lithosphere, basal relief on the lithosphere, and deflection of plumes by subducting slabs (Johnston and Thorkelson 2000). Actual flood basalt magmatism may be restricted to thin spots in the lithosphere that allow plumes to rise to shallow levels where they undergo extensive decompressive melting.

Plumes and Sediments

When a mantle plume impinges on the lithosphere, the surface of the Earth should be elevated, and for large plumes (the order of 1500–2000 km across), uplift should be 500–2000 m, depending on the viscosity of the plume head (Griffiths and Campbell 1991; Sleep 1990). This should effect sedimentation patterns and thereby provide an independent, field-based tool for identifying and locating ancient plumes. Examples of expected changes include radial paleocurrent patterns in sediments beneath flood basalts and unconformities over regions of maximum uplift (Cox 1989; Rainbird 1993; Rainbird and Ernst 2000). From studies of several sedimentary basins whose sources may have been elevated by mantle plumes, Rainbird and Ernst (2000) proposed the following idealized uplift-sedimentation history.

1. Depositional shoaling and perhaps regression, both with a radial pattern reflecting the initial rise of a domal source.
2. Thinning of sediments towards the center of the uplift and erosion of sediments from the center of the uplift. Isopachs define a roughly circular pattern around the uplift.
3. Deeply incised canyons leading away from the focal point of the uplift and block faulting on the uplift due in part to collapse of the uplift from magma chamber deflation.
4. Eruption of flood basalts and emplacement of dyke swarms.
5. Continued uplift of the dome reflecting a long-lived plume tail.
6. Final subsidence and marine transgression as the lithosphere cools in response to plume cooling.

One of the best-documented examples of plume-related sedimentation is described by Rainbird (1993) in the Neoproterozoic Shaler Supergroup in northern Canada. Marine platform and terrestrial sediments of the Shaler Supergroup underlie the Natkusiak flood basalts and related Franklin mafic dyke swarm, both of which were emplaced at about 720 Ma (Rainbird et al. 1996). The Shaler sediments beneath the flood basalts display stratigraphic and sedimentological features indicating that uplift preceded eruption, and this uplift is interpreted as a consequence of doming by a mantle plume that generated the Natkusiak magmas (Rainbird 1993). The Shaler sediments comprise two depositional sequences that reflect dominantly fluvial clastic sedimentation followed by shallow marine carbonates (Kilian Formation) (Fig. 3.22). In the upper part of the section, marine carbonates are abruptly overlain by coarse fluvial sediments of the Kuujjua Formation that thin to the northeast and eventually are completely eroded. Here, the flood basalts of the Natkusiak Formation overlie the marine carbonates. These facies relationships imply extensive and rapid uplift towards the northeast, where a mantle plume

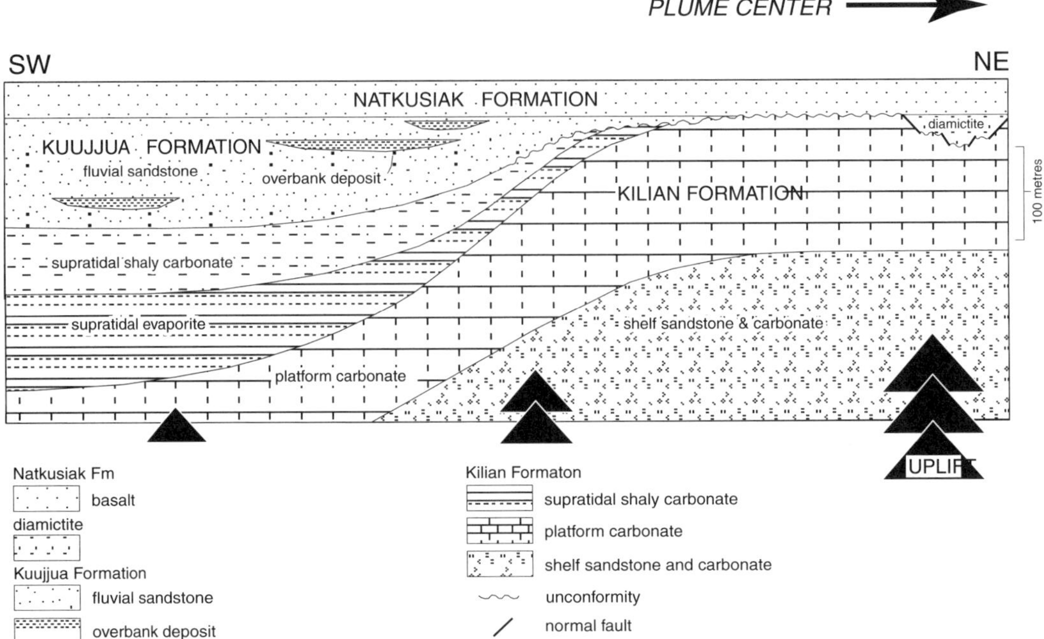

Figure 3.22. Correlation of strata in the uppermost Neoproterozoic Shaler Supergroup of the Minto Inlier, Victoria Island, northern Canada. Note the progressive offlap of marine facies and thinning toward the northeast (NE). Courtesy of Rob Rainbird.

is thought to have been centered. The focal point of the radiating Franklin dyke swarm suggests the plume center was approximately 400 km northwest of the Shaler sediments.

Permian sediments of the Gondwana Supergroup in northeastern India provide another example of possible plume-controlled sedimentation. In this part of India, Permian sediments are immature sandstones and conglomerates deposited in narrow rift basins from sources to the southeast (Kent 1991; Rainbird and Ernst 2000) (Fig. 3.23). Provenance studies of these rift sediments suggest derivation, in part, from Precambrian high-grade terrains in East Antarctica, which, before the breakup of East Gondwana, were adjacent to northeast India. These sedimentation patterns were long-lived and perhaps extended into the Cretaceous with a duration of about 150 Myr. Kent (1991) suggested that these long-lived sedimentary dispersal patterns may have been maintained by uplift over the Kerguelen plume head during passage of East Gondwana.

LIPS on Mars and Venus

Large igneous provinces are also common on other terrestrial planets and offer a unique opportunity to understand LIPs on Earth better. Unlike Earth, Mars and Venus have minor surface erosion and lack plate tectonics; hence, geologic surface features are better preserved and last longer. Also, one-plate planets such as Mars and Venus can help us understand the long-term effects of mantle plumes and their variability under different thermal regimes in space and time (Solomon 1977; Head and Coffin 1997). In addition, the planetary LIP record can be instructive in terms of chronology and

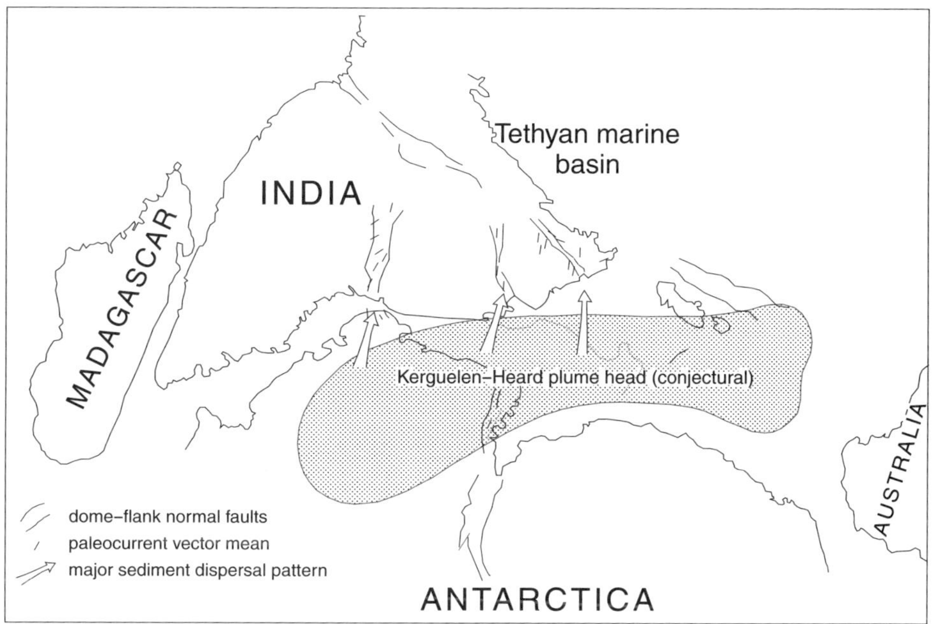

Figure 3.23. Reconstruction of Gondwana 250 Ma showing location of active rifts that channeled sediments off an inferred uplift that may have been produced by a mantle plume. Courtesy of Rob Rainbird.

episodicity of mantle processes – especially those processes operative during the early stages of planetary history (>3 Ga).

Studies of Martian and Venusian plume tectonics complement and expand our knowledge of terrestrial plumes. Common to plume tectonics on all three planets are the following features (Mege and Ernst 2001).

1. Impact of plumes on the base of the lithosphere generally results in uplift followed by thermal subsidence as plumes cool.
2. Rapid eruption of basalts in less than 1 Myr followed by a prolonged period in which evolved magmas are erupted.
3. Emplacement of one or more central intrusions with radiating dyke swarms feeding the volcanism.
4. A strong positive gravity anomaly resulting from underplating by mafic magmas.
5. Radial extension leading to the formation of one or more successful or failed rifts.

Now we will review some examples of mantle plumes on Mars and Venus.

Martian LIPS

Information from Mars comes from spacecraft exploration, Earth-based observations, and from Martian meteorites (Kieffer et al. 1992). Mars has large shield volcanoes similar to smaller examples on Earth and Venus. They exhibit a wide range of rift development, deformation, flank and slope failure, and caldera formation. Martian

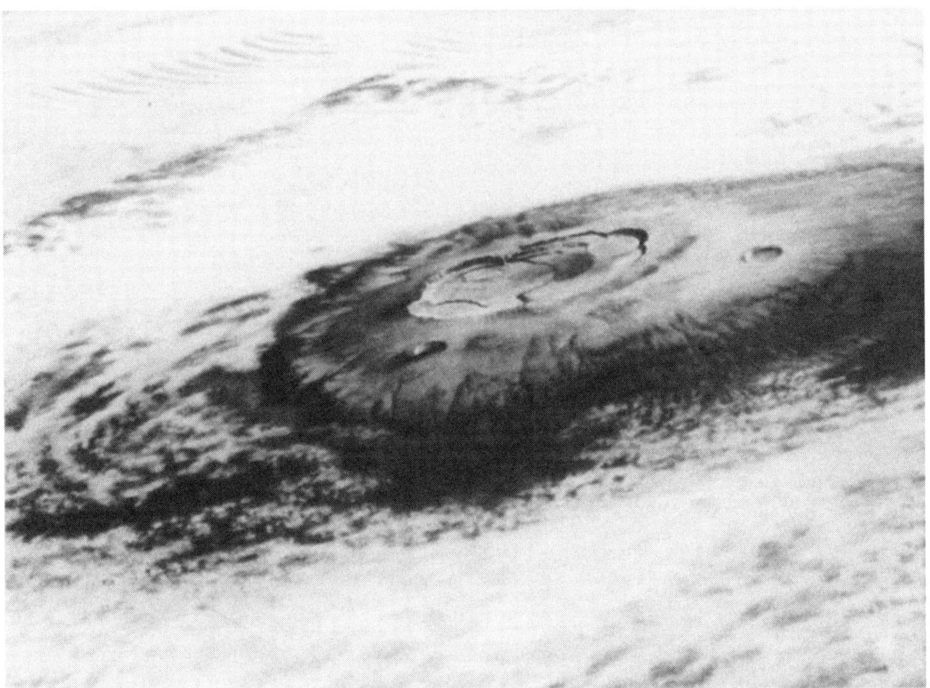

Figure 3.24. The multi-ringed caldera on top of Olympus Mons rises above a cloud bank that obscures the lower two-thirds of the volcano. Painting by Gordon Legg based on images from the Viking Orbiter 1, July 1976. Courtesy of Lunar and Planetary Institute.

shield volcanoes have widths of hundreds of kilometers with heights up to a factor of three greater than Hawaii. Volumes of erupted material are gigantic. Olympus Mons has a volume of about 2×10^6 km^3 compared with 1.1×10^6 km^3 for the entire Hawaiian–Emperor chain (Fig. 3.24). Other single volcanoes on Mars have volumes comparable to those erupted in terrestrial LIPs (Table 3.1). Because the lithosphere has not moved laterally on Mars, mantle plumes concentrate their magmatic products in one spot rather than being spread out in conveyor-belt fashion such as along the Hawaiian–Emperor chain. Hence, magmas stack up vertically, forming huge shields, loading the lithosphere, and causing flexure, deformation, and failure on the flanks (Carr 1973). One factor also contributing to the large height of Martian volcanoes is the thick lithosphere, which is at least 150 km thick beneath Olympus Mons (Comer et al. 1985).

Mars has two especially large LIPs: the Tharsis and Elysium rises. The Tharsis rise is a gigantic LIP covering at least 20% of the Martian surface (Fig. 3.25). This LIP, which greatly exceeds the size of any terrestrial LIPs, comprises extensive volcanic plains spanning a wide range in age and massive, superimposed shield volcanoes. From geologic mapping and crater chronology, it appears that Tharsis volcanic activity ranged over 10^8 to 10^9 years, unlike terrestrial LIPs where most of the action occurs in 10^7 years or less. Development of the Tharsis LIP involved complex episodic tectonic and volcanic events on both local and regional scales (Head and Coffin 1997). Early ideas for the origin of Tharsis focused on convective upwelling in the mantle, producing a gigantic mantle plume that caused the uplift and magmatism (Tanaka et al. 1992).

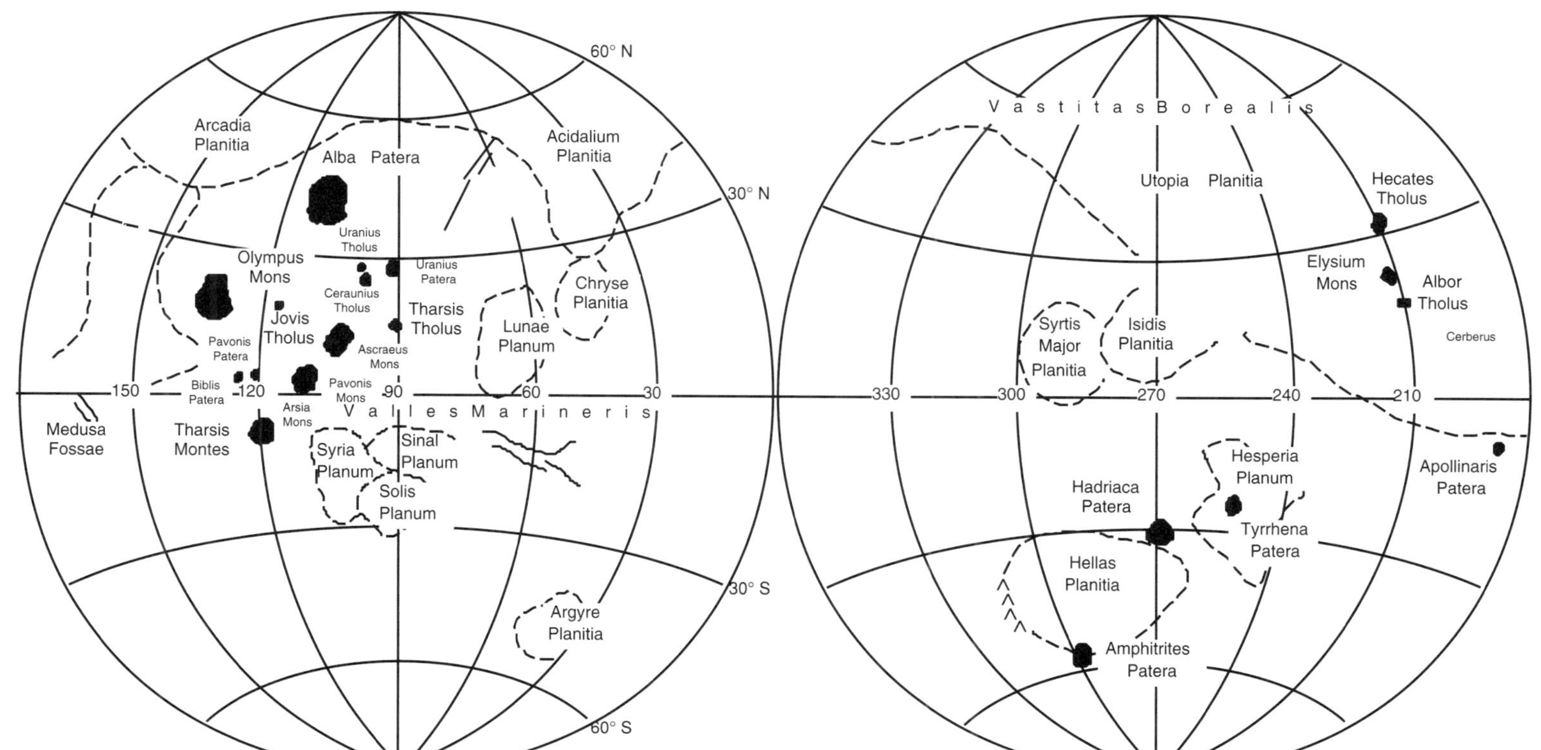

Figure 3.25. Generalized map of Mars showing the two major rises, Tharsis and Elysium, the distribution of impact basins and volcanic provinces (dashed lines), and large shield volcanoes (black spots). Modified after Head and Coffin (1997).

Later studies showed that the uplift cannot be explained by dynamic processes alone and that at least two different events are necessary. One of the major outstanding questions is, What currently supports the Tharsis region some 4 Gyr after its formation? Isostatic compensation will not work because of a large free-air gravity anomaly. Some models suggest that Tharsis is partially supported by the elastic strength of the lithosphere with additional support from low-density crustal roots at depths of 50–100 km (Banerdt et al. 1992). During the early history of Tharsis rise, transient support may have been supplied by a mantle plume. Later, the plume cooled and the support was removed, leaving the load to be supported by the cooling, thickening lithosphere.

One of the intriguing questions about Mars is, Why does it have only two prominent, long-lasting LIPs? Models of the Martian mantle suggest that cylindrical plumes are the probable form of upwelling and that return flow occurs in an interconnected network of planar sheets (Schubert et al. 1992). The number of upwellings is a function of the geometry of heating. Increased bottom heating, which was likely early in Martian history, resulted in a decreased number of upwellings but an increase in their intensity. As the planet cooled, convection style changed, and upwellings increased in number but decreased in strength. If this model is correct, the Tharsis and Elysium LIPs record two of the intense upwellings early in Martian history when the planet was very hot. Similar large LIPs may have been produced on Earth some 4–4.5 Ga, but because of rapid recycling caused by plate tectonics, these have not survived in the geologic record.

Venusian LIPS

Because Venus is about the same size and density as Earth, it offers an important way to test ideas about planetary evolution and, in particular, about the early phases of evolution of silicate planets. Venus, however, differs from Earth in having a very dense CO_2-rich atmosphere, a very slow rotation with a retrograde spin, and the absence of a magnetic field. The high surface temperature (about 475 °C) precludes the presence of liquid water, which results in preservation of landforms in a near pristine state. Approximately 80% of the surface of Venus comprises volcanic plains with a wide variety of volcanic landforms that are probably largely basaltic in composition (Head et al. 1992). Although Venus has folded mountain chains, global rift zones, and large volcanoes, no evidence was found from the Magellan mission to support the existence of plate tectonics. Stratigraphic relations revealed in surface units suggest that volcanism has decreased with time.

The volcanic record on Venus provides examples of LIPs similar to oceanic plateaus, flood basalts, and giant radiating dyke swarms found on Earth. At least 11% of the Venusian surface appears to be covered with flood basalts (Magee and Head 2000). On the volcanic plains, over 150 major shield volcanoes have been identified with diameters exceeding 100 km. The heights of these volcanoes, however, are much less than those of large shield volcanoes on Earth and Mars and typically are less than 2 km above the surrounding plains. Contributing to the formation of relatively low shields is the dense atmosphere and high surface temperature, both of which favor large primary magma reservoirs with numerous conduits that build small, relatively flat shields (Head and Coffin 1997). Also, models indicate that, for a given magma supply rate, magma reservoirs on Venus should stabilize and grow laterally with less magma reaching the surface.

Coronae, features that appear to be unique to Venus, are circular uplifts ranging from 60 to 2000 km across (most are 200–500 km across) with a central plateau, a raised annulus, ridges and grabens, and a surrounding moat (Fig. 3.26). These features are

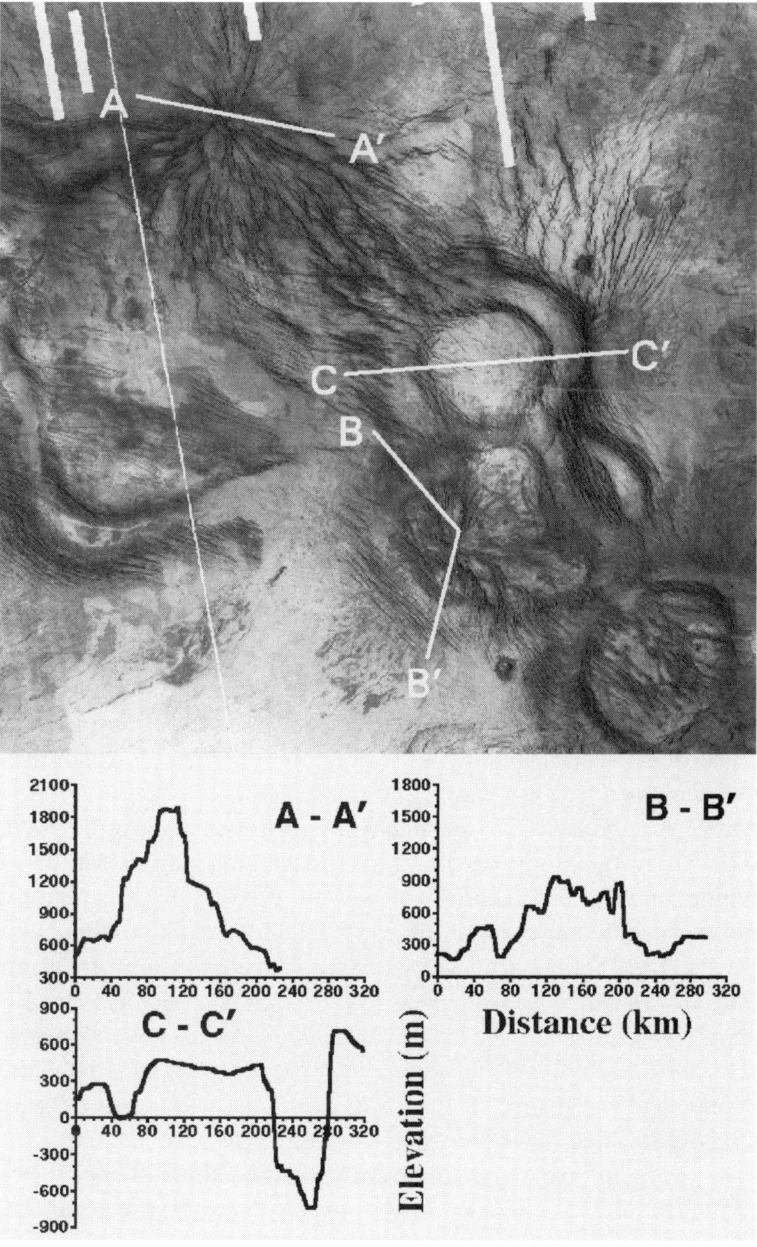

Figure 3.26. Features along Parga Chasma corona on Venus with three topographic profiles. Profiles illustrate stages in the evolution of coronae produced in response to mantle diapirs: A–A′ = doming producing radial fractures; B–B′ and C–C′ = circumferential faulting and collapse. From Herrick (1999), with permission. Copyright © 1999 by the American Geophysical Union.

widely distributed on the Venusian surface and are generally interpreted as the surface manifestation of relatively small mantle plumes that have risen to shallow levels and spread laterally (Janes et al. 1992; Koch and Manga 1996). Coronae also have associated volcanic flow fields and are commonly found in areas of rifting (Plate 3). Many of the Venusian coronae appear to have developed contemporaneously with rifting. The differences in size and structure of coronae are thought to reflect differences in plume size and strength and the thermal strength of the lithosphere. The lithosphere is elevated and deformed by a mantle plume, and as the plume cools the circular uplift subsides, producing a central depression and moat (Fig. 3.26). These features have been likened to swells in Africa, which are thought to form in response to small mantle plumes (Burke 1996; Herrick 1999). Furthermore, Herrick (1999) has suggested the source of these small plumes, or more appropriately mantle diapirs, to be a mantle undergoing hard turbulence – a process discussed in Chapter 4. The wide range in diameters of Venusian coronae is consistent with production by mantle diapirs that lose their tails at various stages of growth. The gravity signatures of coronae also suggest that the diapirs do not have trailing conduits of hot mantle material.

Among the other features found on the volcanic plains of Venus are sinuous channels resembling rilles on the Moon, which are thought to have formed by thermal erosion from lavas erupted at high rates. Many of these channels exceed 300 km in length, and the longest is nearly 7000 km long. Such large lava channels imply very high eruption rates at high temperatures. Flood basalt provinces, some of which are larger than terrestrial analogues, are also common on Venus. Some of these have flow volumes on the order of 2×10^4 km^3 (Magee and Head 2000). Many of the flood basalts appear to have been erupted from coronae, rifts, or long fracture zones. Geologic relationships suggest that lithospheric extension and thinning accompany formation of the majority of the Venusian flood basalts. This is in contrast to many terrestrial flood basalts in which eruption of the main magmatic phase preceded significant extension and crustal thinning. The close association of large flood basalt fields with rifts and fracture zones indicates that extension and thinning of the lithosphere are necessary for the formation of most Venusian flood basalts (Magee and Head 2000).

Many rift zones on Venus are associated with broad rises of the order 2000–3000 km across that tower several kilometers above the surrounding plains (Plate 3). These rifts, which commonly radiate from central volcanoes and have positive gravity anomalies, may reflect large mantle upwellings – much larger than the plumes thought to produce most of the coronae. Preliminary analysis of the Magellan data from Venus suggests that the styles of volcanism have changed with time (Basilevsky and Head 1996). The tessera terrains, which appear to be the oldest terrains exposed on the Venusian surface, are relatively high and complexly deformed. These terrains are partly covered by major volcanic plains, and the plains, in turn, serve as a base for many relatively young volcanoes with associated flows. Recent studies of the tesserae on Venus suggest they are of different ages and may have formed when large mantle plumes interacted with relatively thin lithosphere early in the history of the planet (Hansen et al. 1999).

The size–frequency distribution of impact craters on the Venusian surface suggests that the average surface age is 500–300 Myr and that most of the craters have not been modified by later volcanism. It would appear that Venus underwent a global tectonic and volcanic resurfacing at this time that destroyed the earlier crater record (Magee and Head 2000). Following that event, volcanism was very minor in volume and aerial extent (Strom et al. 1994). Whatever the cause of such an event (plate tectonic overturn, global volcanism in response to widespread plumes), it appears to have occurred over a short time ($\leq$100 Myr).

Giant Dyke Swarms

Introduction

Giant dyke swarms are groups of genetically related dykes that occur over tens to hundreds of thousands of square kilometers (Halls 1982; Ernst et al. 1995). Although generally mafic in composition, they may be any composition and occur in a variety of tectonic settings. Most dykes in giant swarms have widths of 10–50 m with some ranging to 200 m, and cross sections and geophysical profiles of these dykes suggest they narrow downwards. Most large dyke swarms appear to be related to mantle plumes and are often associated with divergent plate margins (Fahrig 1987; Ernst et al. 1995). Smaller swarms may be related to subduction or collisional plate boundaries. Sheeted dyke swarms are an important component of ophiolites having formed at ancient spreading centers (Baragar et al. 1987). More than one dyke swarm can occur in the same region, and precise isotopic dating is necessary to distinguish between swarms. Single dykes in large swarms have been traced for distances up to 1000 km and, on the basis of geophysical methods, many appear to extend for more than 2000 km.

Dykes are generally oriented normal to the minimum stress direction in the crust and trend in the direction of maximum compressive stress, but many have a radial distribution. Studies of dyke emplacement indicate that most dykes are injected laterally and occur as vertical slabs centered at levels of neutral buoyancy in the crust (Ernst et al. 1995b). Some giant dyke swarms have radial patterns (Fig. 3.27), a feature suggesting they were injected from a point source – probably over a mantle plume head. Ernst et al. (1995a) have recognized six basic types of swarms based on geometric relationships. Types I and II exhibit a distinct fanning pattern; Type I swarms show a relatively continuous fan, whereas Type II swarms have gaps between radiating subswarms (Fig. 3.28). The most famous example of a Type I swarm is the Mackenzie swarm in northern Canada, which fans out over an angle exceeding 100° (Fig. 3.27). Type III dykes also occur as subswarms, but they appear to radiate from a point. Types IV–VI consist of subparallel dykes, those that are broadly distributed (IV), those restricted to a narrow zone (V), or those that define a curved pattern (VI). When examined at a larger scale, some Type IV and V swarms may be portions of swarms I, II, or III.

The model of Baragar et al. (1996) has been widely accepted to explain major differences between dyke Types I, II, and III. In their model, magma chambers develop at different places around the center of a plume, and each chamber is responsible for injection of a subswarm of dykes. If chambers are close together, there are no gaps in dyke generation, but when they are not, gaps occur between subswarms. If rifts are present in the crust over a plume head, dykes may parallel the rifts. Fahrig (1987) proposed that dyke swarms associated with the breakup of supercontinents are emplaced both parallel and perpendicular to the rift margin (Fig. 3.29). Examples of parallel dykes are those swarms paralleling the Red Sea and the East Greenland dykes in the North Atlantic.

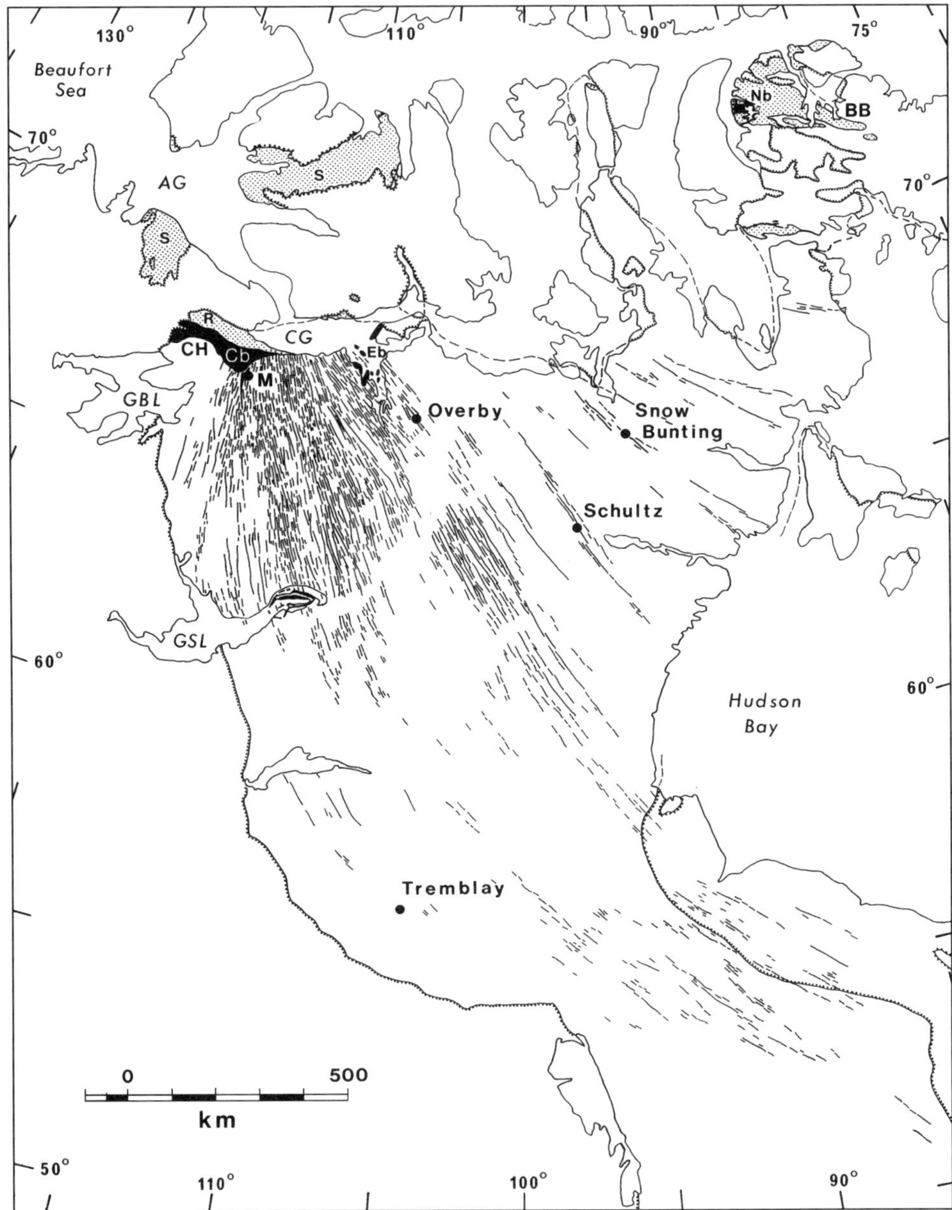

Figure 3.27. Map showing the Mackenzie dyke swarm intruded into the Canadian shield 1267 Ma. The Muskox intrusion (M) and the Coppermine River flood basalts (Cb) are part of the same event. GBL = Great Bear Lake; GSL = Great Slave Lake. Courtesy of A. N. LeCheminant.

Radiating dykes, which are injected into failed rift arms in the Fahrig model, come from the plume head. Examples are the radiating dykes in the Paraná province in Brazil (Fig. 3.16) and the Proterozoic Grenville dyke swarm in Eastern Canada. Most rift-parallel dykes are destroyed or highly deformed during later plate collisions (Fig. 3.29, no. 3), whereas at least portions of radiating swarms may survive. This rather simplistic model explains why many dyke swarms are truncated at one end by younger orogens.

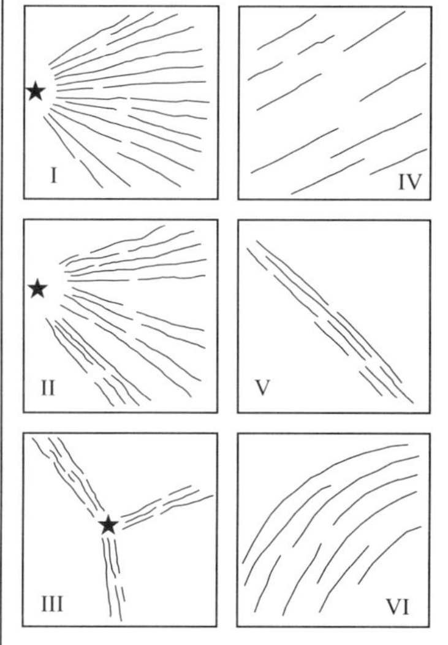

Figure 3.28. Patterns of giant dyke swarms. Stars show probable mantle plume centers. From Ernst et al. (1995), with permission. Copyright © 1995 by the American Geophysical Union.

Relationship of Dyke Swarms to Plumes

Many radiating dyke swarms appear to be remnants of LIPs, which, because of erosion, have lost most or all of the flood basalt component. A classic example of this is the Central Atlantic Volcanic Province (Fig. 3.30). Dykes in this swarm were injected laterally for more than 2800 km from the plume head, which was located near the triple junction of North and South America and Africa prior to breakup of the Atlantic 200 Ma (Ernst and Buchan 1997). Precise $^{40}Ar/^{39}Ar$ dating of dykes and flows in the Central Atlantic Province show that magmatism in this province occurred within an interval of about one million years with peak activity at 200 Ma and the dykes were injected over an area larger than 7×10^6 km^3 (Marzoli et al. 1999; Hames et al. 2000). It appears that the Fernando plume, located now off the tip of Brazil (Fig. 2.1), was responsible for the gigantic magmatic event at 200 Ma. Sedimentation in rift basins in the eastern United States stopped just before the onset of plume volcanism – presumably in response to uplift of the lithosphere caused by the impinging plume head. Despite the great extent of the dyke swarms, only small volumes of associated flood basalt have been recognized (some in the basement of Florida and some in Morocco). Large sill complexes occur in the Guiana shield in South America and in West Africa (Fig. 3.30).

The Mackenzie dyke swarm in northern Canada emplaced at 1267 Ma is one of the largest known dyke swarms (Fig. 3.27). It covers an area of 2.7×10^6 km^2 and extends for more than 2600 km from the probable plume source north of the Coppermine River area (Fahrig 1987). Near the focus of the swarm are coeval flood basalts (the Coppermine River basalts) and a small layered intrusion, the Muskox Complex, which is a surface exposure of a large layered complex to the north. Precise U/Pb isotopic dating has shown that all of these magmatic components were emplaced in less than 5 Myr at 1267 Ma (LeCheminant and Heaman 1991). Both the radiating pattern of dykes and stratigraphic evidence for uplift in the focal region preceding magmatism support a plume source for the flood basalts, layered intrusion, and giant dyke swarm.

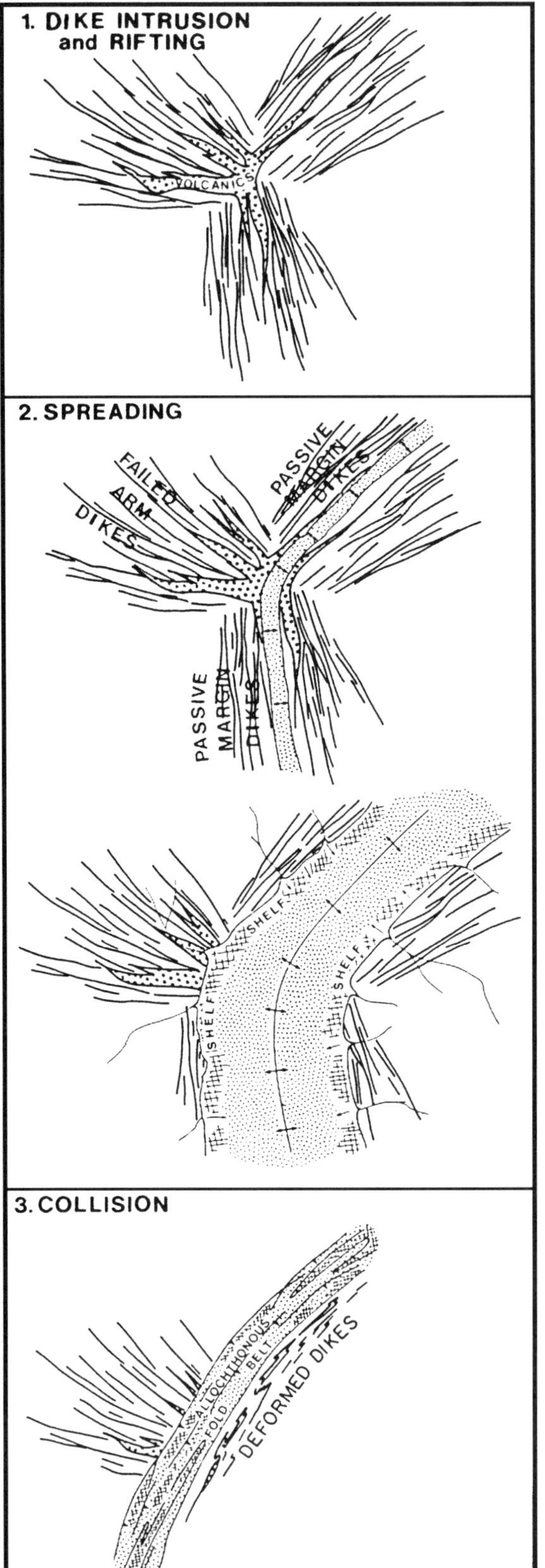

Figure 3.29. Hypothetical model for the relationship of giant dyke swarms to the opening and closing of an ocean basin. Modified after Fahrig (1987).

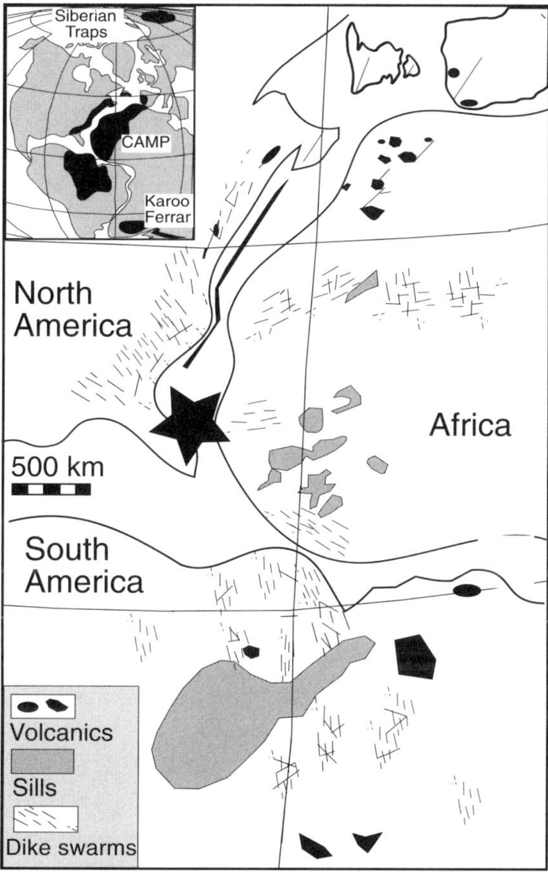

Figure 3.30. Map showing distribution of the Central Atlantic Igneous Province 200 Ma before the breakup of Pangea. Inset also shows the Siberian and Karoo–Ferrar flood basalts. Star is inferred location of mantle plume. Courtesy of Paul Renne.

In addition, studies of the magnetic fabric in the Mackenzie dykes indicate vertical flow in the region over the plume center and lateral flow away from the center (Ernst and Baragar 1992). In the Mackenzie plume model of Baragar et al. (1996), partial melts from the plume ($\pm$ melts from the overlying lithosphere) feed crustal magma chambers and are also injected as dykes for up to a distance of 2600 km. An unknown but probably large volume of magma also underplates the crust near the alleged plume head.

Dyke swarms are related to uplift accompanying plume magmatism. Models suggest that uplift topography varies during plume ascent, but at the time of peak magmatism (about 5 Myr after maximum uplift), the outer boundary of uplift approximately coincides with the edge of the plume head (Griffiths and Campbell 1991). In the case of the Mackenzie swarm, this distance is about 1000 km. In some cases, the convergence point of dyke swarms can be used to locate the point of maximum uplift (Ernst and Buchan 1997). As an example, the calculated convergence point for the Mackenzie swarm based on 10 dykes distributed evenly across the swarm is 70°N, 244°E with a 95% confidence uncertainty of $\pm$2 degrees (Ernst et al. 1995b). Although most plume volcanism occurs near the center of uplift, in some cases magmas appear to have traveled hundreds, if not thousands, of kilometers before major eruption. One example, as previously discussed, is the Ferrar basalts in Antarctica, which may have traveled more than 3000 km from their plume source; at the time of magmatism this source was located off the present-day coast line of South Africa (Ernst and Buchan 1997) (Fig. 3.20). Another example is the Columbia River basalts, which may have traveled northward up to 500 km from the Yellowstone hotspot (Fig. 3.8). Extensive lateral movement of plume magmas may be related to rift systems that channel the magma, or plumes may flow laterally along irregularities in the base of the lithosphere, as was suggested earlier for the Afar plume in eastern Africa.

The pattern of magma flow in giant radiating dyke swarms has been related to melt generation in plume heads. The transition from vertical to horizontal flow with increasing distance from the plume center may mark the outer boundary of melt generation in the plume (Ernst and Baragar 1992). As determined from magnetic fabrics in dykes, this transition occurs about 500 km from the focus in the Mackenzie swarm, and in other swarms the distance varies from 200 to 500 km. As plumes flatten at the base of the lithosphere, they should spread to distances much greater than 500 km, as discussed in Chapter 4. Why are distances greater than 500 km in the change of dyke flow direction not observed in giant dyke swarms? Perhaps the thick lithospheric roots of cratons, and especially of Archean cratons, prevent plume magmas from reaching the surface except over the center of the plume head. This could also account for the highly focused plume magmatism that may have produced the Bushveld Complex in South Africa without appreciably destroying the deep lithospheric root. One of the exciting yet controversial aspects of radiating dyke swarms is their potential in locating and dating the breakup of supercontinents. The main problem is to distinguish between radiating dykes related to supercontinent breakup and dykes in which there was no fragmentation of a supercontinent.

Dyke Swarms on Venus and Mars

Some of the most spectacular surface features on Venus as revealed by the Magellan mission are radiating fracture and dyke systems that radiate from volcanic structures and from coronae (Grosfils and Head 1994) (Fig. 3.31). Unlike Earth, where erosion and plate tectonics have destroyed or partially destroyed many dyke swarms, those on Venus are well preserved and provide a more complete record of dyke swarm development with time (Parfitt and Head 1993; Ernst et al. 1995a). Radiating dyke systems on Venus are characterized by radiating fractures, fissures, grabens, and dykes, some of which have propagated more than 2000 km from their foci (Fig. 3.32). The average length, however, is close to 200 km (Ernst et al. 1995a). The focal regions of Venusian dykes are sometimes surrounded by concentric structures (25–600 km in diameter), some of which appear to be calderas and others are coronae (Fig. 3.32). The role of large volume, long-distance, lateral dyke emplacement is illustrated by the Venusian coronae. These may represent analogues to the now-eroded, giant radiating dyke swarms on Earth and associated but largely eroded flood basalts. In some structures, radial grabens completely surround the central structure. Studies by Grosfils and Head (1994) show that 72% of the structures must have involved subsurface dyke swarm emplacement,

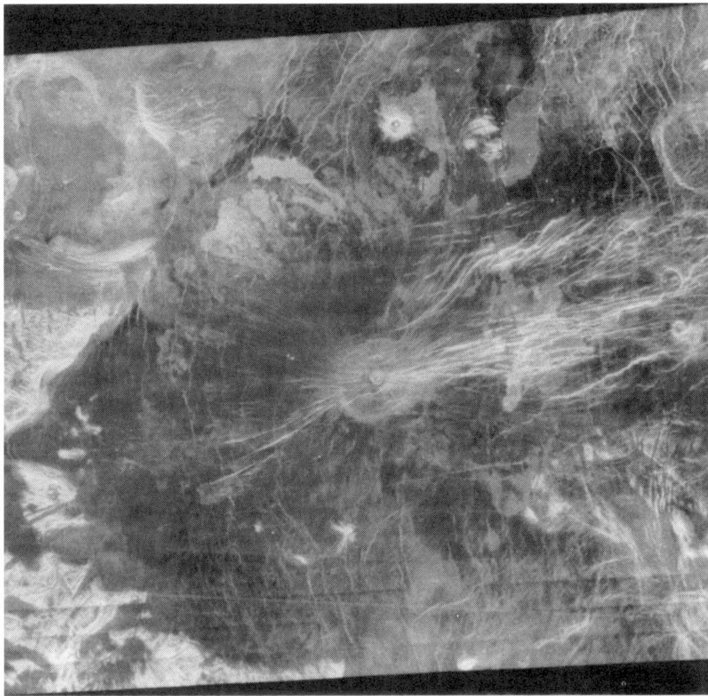

Figure 3.31. Example of giant radiating dyke swarm on Venus. The small caldera at the center of the photo is at the focus of a 200-km radiating system of grabens, fractures, and fissures interpreted to reflect a subsurface dyke system. Magellan photo F-MIDR; distance across the photo is 525 km. Courtesy of Eric Grosfils.

and only 9% appear to have formed chiefly from uplift. Almost all of the radiating dyke systems on Venus are associated with some sort of volcanism. Most common are lobate flows, probably basalts, emerging from single or multiple radial lineaments. Small shield volcanoes are associated with many radial systems.

On Earth, Mars, and Venus, radiating dykes appear to come from a central source and to be injected for great distances (>2000 km). That dyke swarms on Venus are mostly Type I or II supports the idea that many terrestrial swarms, now partially destroyed, radiated in a similar fashion when they were emplaced. This suggests that giant radiating dyke swarms can be used as a restoration tool for ancient plate reconstructions on Earth. In Precambrian shields, dykes were intruded at depths of 5 to 14 km, which may be the level of neutral buoyancy in the shield crust (Percival and Card 1983). In contrast, on Venus, where the host rocks are likely mafic in composition and the atmospheric pressure is much greater than on Earth, the neutral buoyancy level should be less than that on Earth.

Giant radiating dyke swarms on Venus appear to have been emplaced after the formation of the tessera terrain, the wrinkle-ridge topography, and most lava flooding on the regional plains (Grosfils and Head 1996). The cross-cutting relationships documenting this timing are illustrated on a cartoon-like map in Figure 3.33. These relations indicate that most giant dyke swarms on Venus formed during or after the period of widespread volcanic resurfacing of the planet 500–300 Ma. Structures with radial fracture–dyke

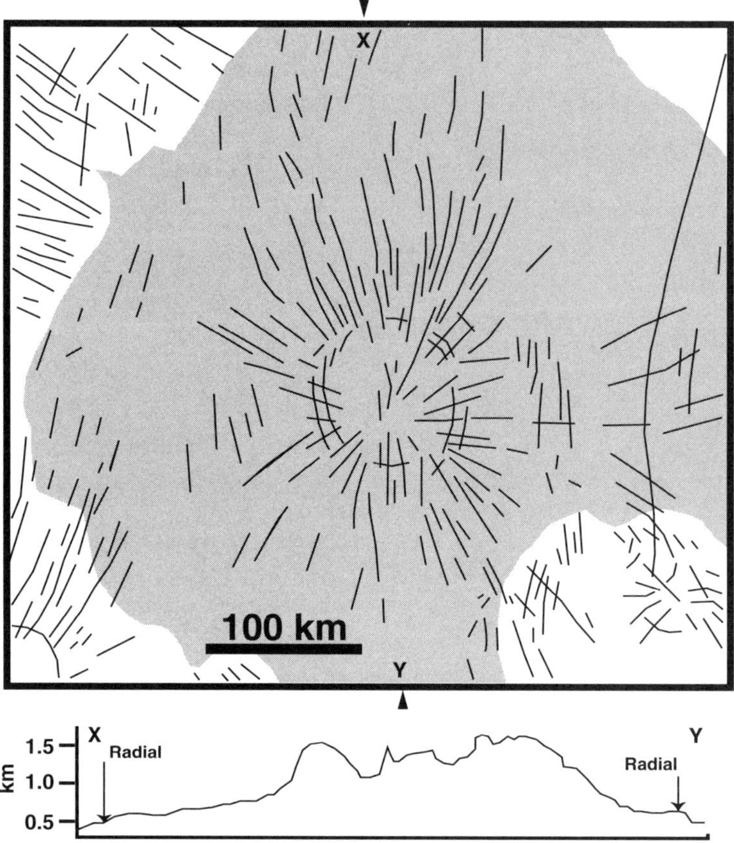

Figure 3.32. Sketch map and cross section of radial lineaments caused by dyke swarms radiating from a corona on Venus. Patterned area is lobate lava flows. From Ernst et al. (1995), with permission. Copyright © 1995 by Elsevier Science.

systems, although widely distributed on Venus, occur chiefly within 40° of the equator, and many are concentrated around longitude 200–300°. This is quite unlike the distribution of hotspots on Earth, which are concentrated in the two large mantle upwellings (Fig. 2.1). Although the significance of the distribution on Venus is not yet clear, it may imply that Venus does not have large mantle upwellings similar to those found on Earth.

The most famous radiating fracture system on Mars is in the Tharsis region (Fig. 3.25), where a vast system of fractures and grabens radiate from the central part of the Tharsis Rise (Tanaka et al. 1991). The regularity of graben widths strongly suggests that a radiating dyke system underlies the grabens. These grabens, which extend radially for thousands of kilometers around the Tharsis uplift, show no displacement gradients along the boundary faults, which is typical of most grabens on Earth. Mege and Masson (1996), using morphological and structural data from the grabens, proposed that most of the grabens formed in response to dyke emplacement. The dykes, in turn, may have been generated by a large mantle plume centered near the maximum uplift. The Mege–Masson model successfully explains the geometric distribution of dykes and grabens, the extensive flood basalts in and around the uplift, and the wrinkle-ridge topography

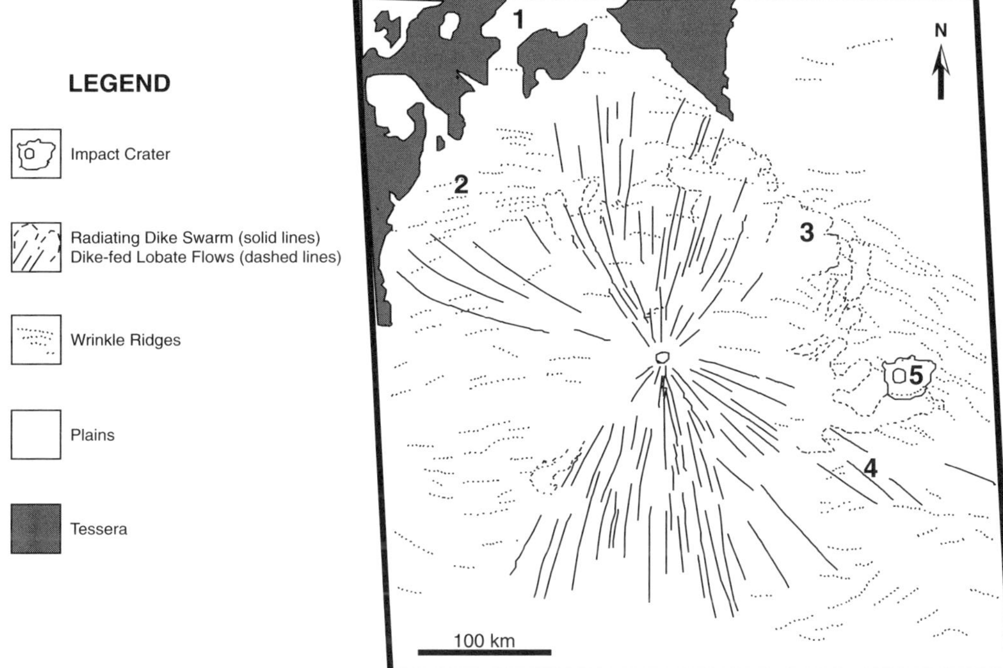

Figure 3.33. Sketch map of a radiating dyke swarm on Venus illustrating relative age relations between events. Tessera, the oldest unit, is embayed by plain deposits (1), and then the plains are deformed into wrinkle ridges (2). At (3) surface eruptions fed by the radiating dykes bury the wrinkle ridges, and some dyke-induced fractures cut other flows and wrinkle ridges (4). A small impact crater is superimposed on the deformed plains and dyke-fed flows at (5). From Grosfils and Head (1996), with permission. Copyright © 1996 by the American Geophysical Union.

on Tharsis, which is interpreted to have formed by subsidence related to the lava load on the crust.

Large Layered Intrusions

Although not exposed at the surface, large mafic intrusions may be an important component of plume magmatism. Some large layered mafic intrusions, such as the Muskox intrusion in northern Canada and the Dufek intrusion in Antarctica, are closely associated with flood basalts, giant dyke swarms, or both, and appear to have formed by intrusion of plume-related mafic magmas. As shown by the uplifted root zones of the Caribbean oceanic plateau described earlier in the chapter, layered intrusions, sills, and dykes are probably also important in the crustal roots of oceanic plateaus. Some of the largest mafic intrusions, such as the Bushveld Complex in South Africa and the Stillwater Complex in Montana, do not have associated flood basalts, and there is considerable debate as to whether they are the products of plume magmatism.

In this section we will review two large mafic intrusions proposed by some investigators to have mantle plume sources: the Muskox intrusion in Canada and the Bushveld Complex in South Africa.

The Muskox Intrusion

The Muskox layered intrusion in northwestern Canada, emplaced at about 1267 Ma, is closely associated with the cogenetic Coppermine River flood basalts and the Mackenzie giant dyke swarm (Fig. 3.27). This plume-related magmatic province is one of the few in which all three components (flood basalts, dyke swarm, and layered intrusion) are exposed (Baragar 1969).

The Muskox intrusion, which is intruded into 1.9-Ga basement, includes two marginal groups: a layered series and a granophyric roof zone (Irvine 1970; Francis 1994). The layered series is about 1.8 km thick and has 42 mappable cumulus layers, 18 different rock types, and 25 cyclic units. The cyclic units are interpreted as injections of new batches of magma into a fractionating magma chamber. The first phase of crystallization involved olivine followed sequentially by clinopyroxene, plagioclase, and orthopyroxene (Irvine 1970). Higher in the series, orthopyroxene appears before clinopyroxene, and chromite occurs as a minor cumulus phase throughout much of the succession. In addition to the layered series, there is a long feeder dyke that shows symmetrical zonation from the margins inwards of various gabbroic and noritic components (Barnes and Francis 1995).

The Muskox intrusion may be an exposed example of many small layered intrusions and thick sills that underplated the crust during plume magmatism 1267 Ma in northern Canada.

The Bushveld Complex

General Features

The Bushveld Complex in South Africa is one of the largest layered intrusions exposed at Earth's surface. It was emplaced into a stable craton at 2060 Ma and is generally considered the intrusive equivalent of flood basalts (Cawthorn and Walraven 1998). Perhaps the Dullstroom basalts in this part of the Kaapvaal craton are remnants of flood basalts associated with the Bushveld event. Both plume and impact origins have been proposed for the Bushveld, although the lack of definitive impact structures and minerals strongly argues against an impact origin (Buchanan and Reimold 1998). The Bushveld Complex comprises four lobes that are probably interconnected at depth, and each lobe contains widespread cumulus-layered mafic and ultramafic rocks (Fig. 3.34). The Lower Zone consists of pyroxenite, harzburgite, and dunite layered on scales of less than 1 m to hundreds of meters. The Critical Zone includes layered pyroxenite with numerous chromite bands and a cyclic upper unit consisting of repetitions of chromitite, pyroxenite, norite, and anorthosite. The Main Zone comprises gabbronorite, anorthosite, and norite, and the Upper Zone consists of gabbronorite and diorite. Progressive changes in mineral composition reflect fractional crystallization, and sudden changes in some horizons (such as at the Merensky reef) represent injection of new batches of magma (Eales and Cawthorn 1996). An east-striking mafic dyke swarm of probable Bushveld age connects the Bushveld Complex with the Molopo Farms Complex, a layered intrusion of Bushveld age in Botswana. This implies crustal stress patterns with north–south extension and east–west compression (Uken and Watkeys 1997).

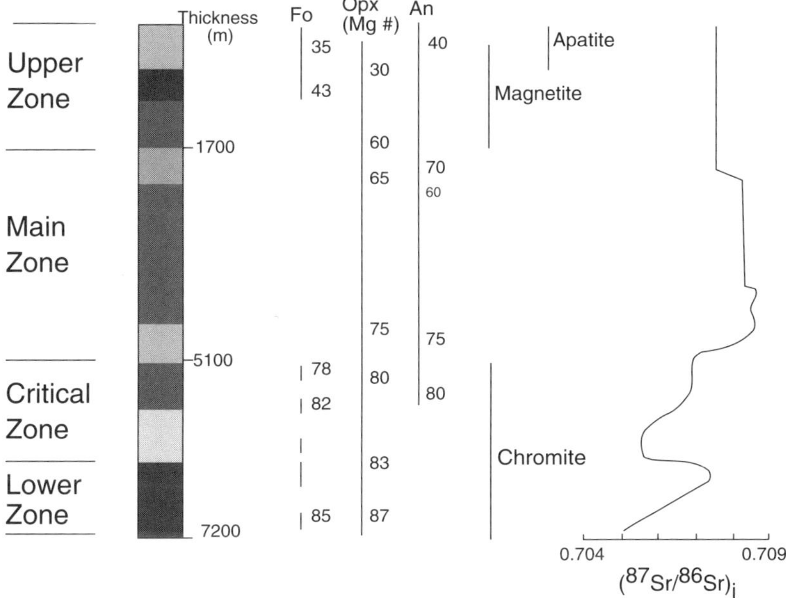

Figure 3.34. Generalized stratigraphy of the Bushveld Complex in South Africa. Shown are the appearance of cumulus minerals, mineral compositions, and intitial $^{87}Sr/^{86}Sr$ ratios. Fo = forsterite; Opx = orthopyroxene; An = anorthite. Modified after Cawthorne and Walraven (1998).

The parent magma of the Bushveld was probably a komatiite (Barnes 1989) contaminated in varying degrees by upper crust. The high and variable $^{87}Sr/^{86}Sr$ ratios indicate variable amounts of crustal contamination, with the greatest contamination above the Merensky reef (5100 m) (Fig. 3.34). Osmium isotopic ratios indicate crustal contamination of the Critical Zone of 5–20% (Schoenberg et al. 1999). Rapid changes in $^{87}Sr/^{86}Sr$ ratios and breaks in cumulus mineral composition indicate injection of new batches of magma.

Crystallization

Calculations of cooling rate trends and mineral differentiation trends suggest that the 8-km-thick layered sequence crystallized in about 150,000 years and that magma addition to the complex ceased after about 60,000 years (Cawthorne and Walraven 1998). The short duration of Bushveld magmatism results in a magma production rate (6 km^3/yr) comparable to that of the Deccan traps (Table 3.1). During stage 1 of crystallization, the magma cooled about 100 °C in only 250 years, crystallizing about 25% of the magma (Fig. 3.35). The entire ultramafic sequence (stages 1–6) crystallized in about 19,000 years, and when plagioclase began to crystallize in stage 7 the cooling rate and crystal accumulation decreased. The Critical and Lower Main Zones crystallized in about 75,000 years, and the Upper Main and Upper Zones in 100,000 years. From mineral composition breaks, it appears that the Bushveld magma was replenished for the first 75,000 years and that thereafter magma addition ceased.

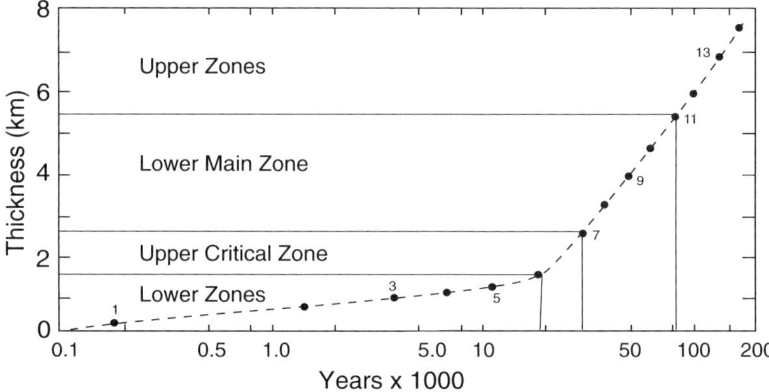

Figure 3.35. Fractional crystallization model for the Bushveld Complex. Numbers refer to times mentioned in text. After Cawthorne and Walraven (1998).

A Plume Origin

Hatton (1995) proposed that the Bushveld magmas were derived from the mantle plume, which ascended through the lithosphere into the lower crust. One of the many problems with this model, however, is how to preserve the deep lithospheric root beneath the Kaapvaal craton. Xenoliths in kimberlite pipes formed much later than the Bushveld Complex indicate the presence of a thick lithospheric root that formed in the Archean. Seismic tomography also supports the existence of thick lithosphere today (James et al. 1999). The Bushveld event is not the only time that plume-generated magmas penetrated the thick Kaapvaal lithosphere. The Ventersdorp flood basalts erupted at 2700 Ma and the Karoo basalts erupted at 190 Ma are examples of other plume-generated LIPs in the Kaapvaal craton. During all three of these events, it would appear that plumes did not rise through the mantle lithosphere and that the lithosphere did not thin appreciably. However, recent Re/Os isotopic studies of mantle xenoliths from the Premier kimberlite, which intrudes the Bushveld, suggest that the subcontinental lithosphere was substantially modified during emplacement of the Bushveld magmas (Carlson et al. 1999).

One way to preserve a thick lithospheric root during a mantle plume event is for the plume-generated magmas to travel hundreds, if not thousands, of kilometers as dyke swarms before final intrusion or eruption. Migmatitic injections and metasomatism associated with such dykes may isotopically alter large volumes of the lithosphere, disturbing the Re/Os isotopes. In such a scenario, the "Bushveld mantle plume" may have been located beneath now subducted oceanic lithosphere adjacent to the Kaapvaal craton.

Kimberlites, Diamonds, and Mantle Plumes

Although kimberlites are not LIPs, some or all kimberlites may be related to mantle plumes, and thus they will be considered in this chapter. Kimberlites are small, alkaline, ultramafic breccias well known for their mantle xenoliths, some of which contain high-pressure mineral assemblages, including in some cases, diamonds. Although prior to the late 1980s, kimberlites were commonly thought be generated in the subcontinental mantle lithosphere, discoveries of diamond xenocrysts containing syngenetic microinclusions of majorite garnet, the high-pressure garnet described in Chapter 1, and of xenoliths that originally contained majorite, indicate a deeper origin for most or all kimberlites (Ringwood et al. 1992; Haggerty 1999). From experimental studies, we know that majorite forms at depths of 400–660 km, and it appears that kimberlitic magmas are produced by very small amounts of melting in this depth range, which is a finding that strongly supports a mantle plume origin (Haggerty 1994). Mg-perovskite and magnesiowustite inclusion assemblages in some kimberlitic diamonds support an even deeper source, at or below the 660-km discontinuity (Suzuki et al. 1995; Haggerty 1999). Although most kimberlites are younger than 600 Ma, this age distribution may reflect erosion and depositional cover of kimberlite pipes. There are a large number of kimberlites with ages of 120–80 Ma, which correlates with a major superplume event in the Pacific (Chapter 8).

One thing that is not clear is why plumes that give rise to kimberlites do not produce LIPs. It is noteworthy that most kimberlite pipes occur in Archean cratons where the lithosphere is anomalously thick (>200 km) (Haggerty 1994, 1999). Perhaps plumes do not ascend to levels shallow enough for appreciable decompression melting to occur beneath such thick lithosphere. Instead, only small amounts of melting occur, producing CO_2-rich kimberlitic melts. As these gas-rich melts rise through the lithosphere, they may react with metasomatized components in the lithosphere, increasing their alkali content (Mitchell 1986). However, because of the thick, cool lithosphere, only small volumes of kimberlitic melt rise as far as the upper crust. In contrast, in areas where the lithosphere is thinner, plumes rise to shallower levels and undergo large amounts of decompression melting, leading to the production of LIPs. Yet another equally plausible explanation for the lack of LIPs associated with kimberlites is that kimberlites are not generated in mantle plumes but are produced in the asthenosphere by very localized melting triggered by high contents of CO_2, methane, or both.

LIP Magma Composition

As with ocean-island volcanism, erupted LIP magmas are largely tholeiitic basalts, and although they show the effects of both fractional crystallization and partial melting, the former greatly dominates. In addition, flood basalts may be contaminated with continental crust. A large proportion of LIP tholeiites have low MgO (<8%) and Ni (<100 ppm), which indicates considerable amounts of fractional crystallization – probably in crustal magma chambers. Even most of the less-evolved rocks have compositions displaced from parent liquids, indicating removal of olivine (Peate 1997). LIP tholeiites and tholeiites from giant dyke swarms commonly plot on or near the 1-atm cotectic in the CIPW normative diagram, which implies low-pressure, near surface fractional crystallization just before eruption (Thompson et al. 1983; Neal et al. 1997). Both major and trace element distributions show the effects of fractional crystallization involving the removal of olivine, pyroxenes (dominantly clinopyroxene), plagioclase, and Fe–Ti oxides, as illustrated, for instance, by the fractional crystallization vectors in Figure 3.36 (Neal et al. 1997). On the TiO_2–MgO diagram, high-Ti and low-Ti basalts define two arrays subparallel to the fractional crystallization vector (Fig. 3.36(a)). On

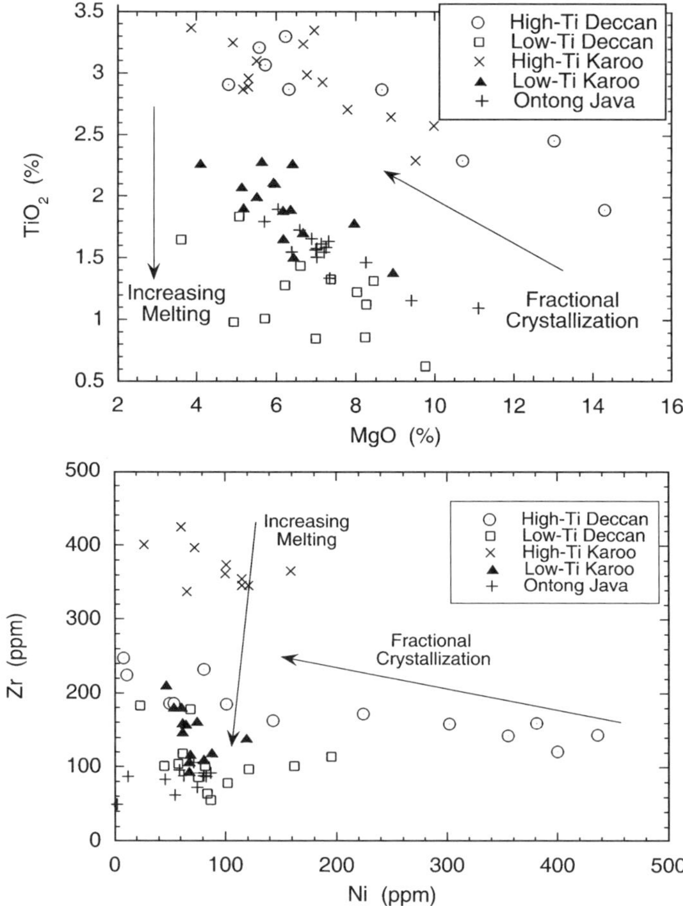

Figure 3.36. Batch melting and fractional crystallization trajectories compared with data from LIP basalts. (a) TiO_2–MgO graph, and (b) Zr–Ni graph. Data from Sweeney et al. (1994), Peng et al. (1994), Tejada et al. (1996), and Yang et al. (1998).

a Zr–Ni plot, the high-Ti Deccan basalts define a fractional crystallization trend (Fig. 3.36(b)). Considerable variation in flow composition is recognized both stratigraphically and geographically within flood basalt provinces (Sweeney et al. 1994; Hooper 1997). The variations occur both in incompatible element contents that reflect varying degrees of fractional crystallization, melting, and contamination and in incompatible element ratios and isotopic ratios, which reflect different magma sources and magma contamination. Some units have undergone up to 50% fractional crystallization.

Another characteristic feature of LIP basalts is the relatively high Ni content of the most primitive lavas ($\geq$300 ppm) (Fig. 3.36(b)). High-temperature melting in the mantle results in strong partitioning of Ni into the melt, whereas at lower temperatures, such as temperatures characteristic of mantle wedges above subduction zones, more of the Ni is retained in mantle olivine (Arndt 1991). Thus, at a given MgO content, plume-generated magmas should contain more Ni than subduction-generated magmas (Fig. 3.37).

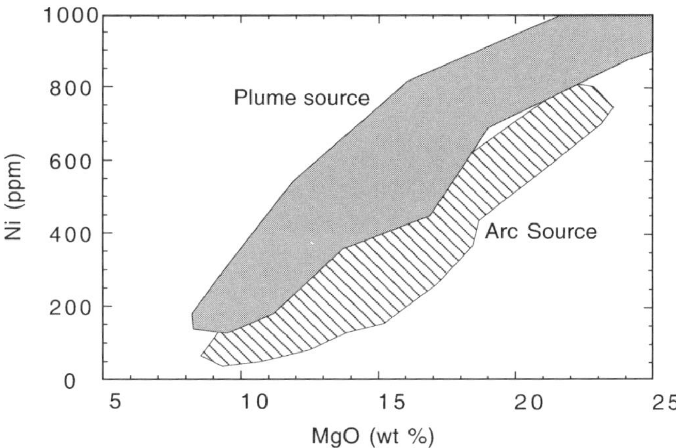

Figure 3.37. Ni–MgO diagram showing distribution of basalts derived from mantle plume and arc sources. After Arndt (1991).

One of the intriguing and controversial features of flood basalts is the presence of high- and low-Ti groups within the same province. Although both groups show a fractional crystallization relation with MgO, they show a distinct break in TiO_2 at about 2% and in Zr at about 200 ppm (Fig. 3.36). Often the two groups have a distinct geographic distribution, as, for instance, in the Paraná–Etendeka and Karoo provinces (Figs. 3.16 and 3.20). Some investigators have called upon variable degrees of melting of plume heads to produce the two groups. As an example, Fodor (1987) argued that both groups in the Paraná province can be produced from a plume head with high-degree (low-Ti) melts generated over a hot plume axis and low-degree (high-Ti) melts from a plume periphery. Some investigators have called upon mantle lithosphere sources as an explanation (Hawkesworth et al. 1988, 2000) and still others on a combination of different extents of partial melting combined with contamination of low-Ti basalts by continental crust (Wooden et al. 1993). Arndt et al. (1993) suggested that lithosphere thickness is an important factor, for it controls the extent of partial melting and renders basaltic magma more (low-Ti) or less (high-Ti) susceptible to contamination. In most instances, differences in radiogenic isotopic ratios between the two groups rule out a model involving only variable degrees of melting of the same source (Peate 1997; Sharma 1997; Lassiter and DePaolo 1997). Unlike flood basalts, oceanic plateau basalts are usually low-Ti tholeiites, reflecting relatively high degrees of melting in plume heads. This is illustrated by Ontong Java basalts in Figure 3.36. Unlike continental flood basalts, oceanic plateau basalts have relatively flat REE distributions and reflect sources only slightly less depleted than MORB sources.

Giant dyke swarms also record shallow fractional crystallization, and, in the case of the Mackenzie swarm in Canada, the degree of fractional crystallization tends to increase as a function of distance from the focal point of the dykes (Baragar et al. 1996). This is shown on a TiO_2–MgO diagram in which dykes at 400–500 km from the focal point have low TiO_2 values and relatively high MgO values (least-evolved samples), whereas those at 1000–1500 km have high TiO_2 and low MgO (Fig. 3.38).

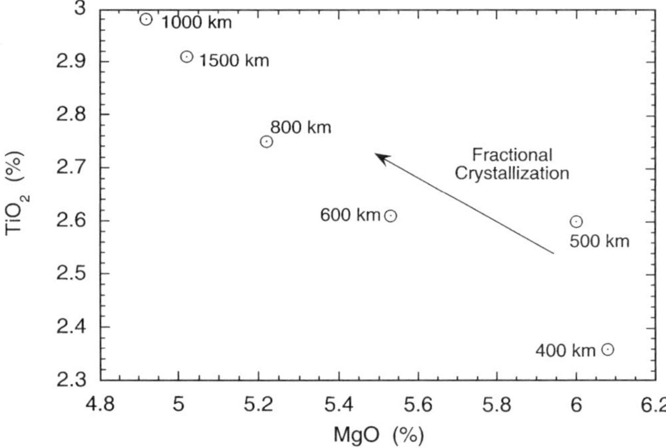

Figure 3.38. TiO_2–MgO diagram showing fractionation trend of dykes from the Mackenzie swarm as a function of distance from the focus. Data from Baragar et al. (1996).

Baragar et al. (1996) suggested two possible mechanisms to explain this trend.

1. Crystal settling in a dyke as it propagates, and thus the upper part of the dyke is always more fractionated. A problem with this mechanism is the lack of phenocrysts in the dykes and the improbability that all phenocrysts could have settled below the current levels of exposure.
2. Fractional crystallization takes place in large magma chambers near the focus. In this case magma sources rise as the plume head rises, increasing the potential energy of the source magma chambers with time. This allows evolving magmas to be injected to increasing distances with time. Thus, if they were close to a plume head, dykes would record all stages of fractionation, whereas at greater distances dykes receive only the more evolved fractions. Supporting this interpretation is the fact that trends in dyke composition with distance mirror trends in the Coppermine River flood basalts with decreasing stratigraphic height near the focal point.

Some flood basalt provinces, such as the Paraná in Brazil and the Karoo and Ethiopian in Africa, contain significant volumes of rhyolite. The rhyolites in these provinces also can be divided into high- and low-Ti groups (Harris et al. 1990). High-Ti rhyolites are typically rich in plagioclase phenocrysts and have high incompatible element contents and similar Nd, Sr, and Pb isotopic ratios to high-Ti basalts (Peate 1997). In contrast, low-Ti rhyolites are nearly aphyric, have low incompatible element contents, and are similar isotopically to low-Ti basalts. Calculated eruptive temperature for both groups is relatively high for felsic magmas (950–1100 °C; Garland et al. 1995). In both the Paraná and Karoo provinces, rhyolites are concentrated around modern rifted continental margins, which indicates a close link to the breakup of Pangea. This fact, together with a compositional gap between basalts and rhyolites, suggests to some investigators an origin by partial melting of continental crust (Hawkesworth et al. 1988). Melting of underplated basalt, which originally ponded at the base of the crust, could occur during the final stages of continental breakup as the lithosphere thins and decompresses.

However, the coincident geographic occurrence and similar isotopic composition between high-Ti basalt and rhyolite and between low-Ti basalt and rhyolite suggest that the mafic and felsic components in each group are related by fractional crystallization (Garland et al. 1995). Isotopic evidence for crustal contamination in some low-Ti rhyolites (such as the Palmas rhyolites in the Paraná) can be explained by AFC (assimilation-fractional crystallization) with variable amounts of assimilation of upper crustal rocks.

LIP Mineral Deposits

Although LIPs are not known for their economic mineral deposits, there are some notable exceptions. These are the Ni, Cu, platinum-group element (PGE), and chromite deposits associated with some LIPs. The best examples are the Ni–Cu–PGE deposits found in the Noril'sk–Talnakh area in the Siberian flood basalts and the vast PGE and chromite deposits in the Bushveld Complex. Less important deposits of these types occur in the Duluth Complex in Minnesota related to the Keweenawan flood basalts (1 Ga), the Insizwa Complex in the Karoo Province in southern Africa (190 Ma), and the Ungava orogen in northeastern Canada (2 Ga). In addition, chromite deposits occur in many layered intrusions.

The Ni–Cu deposits at Noril'sk are the third largest Ni deposits in the world, and they are second only to the Bushveld Complex in their reserves of PGE (Naldrett 1992; Czamanske et al. 1995). The ores at Noril'sk consist of three main types: (1) disseminated ore in picrite and gabbro, (2) massive ore associated with mafic intrusions, and (3) copper ores in sediments and collapse breccias. Sulfur is enriched in the heavy sulfur isotope (^{34}S), which is a feature generally interpreted to mean that the sulfur comes from adjacent evaporites or sulfur gases from older sediments. The important characteristics of the Noril'sk deposit are as follows:

1. They are related to the mantle plume that produced the Siberian flood basalts.
2. They are associated with widespread rifting.
3. Extrusive and intrusive magmatism are controlled by major faults.
4. As previously discussed, at least some of the associated basalts are contaminated with continental crust.
5. The contaminated basalts are depleted in Cu, Ni, and other chalcophile elements.
6. Sedimentary evaporites underlie the basalts.

Naldrett (1992) proposed a model for the Noril'sk deposits that is dependent upon crustal contamination at the top of a vertically extensive, fault-controlled magma chamber and on separation and sinking of an immiscible sulfide melt. As the sulfide melt sinks, it depletes the basaltic magma in chalcophile elements, and it finally comes to rest at the base of the magma chamber, forming the Cu–Ni–PGE ore body. The enrichment of PGE in the Noril'sk basalts supports a source for the Siberian mantle plume near the core–mantle boundary because PGE should be concentrated in the core.

Ni-Cu and PGE sulfides also occur in the Bushveld Complex in South Africa and in the Muskox intrusion in Canada. In the Bushveld Complex, most of the PGE are concentrated in a thin layer (10–25 m thick), the Merensky Reef, located near the top of the Critical Zone (Fig. 3.34). Platinum-group elements are found in many finely disseminated minerals in this layer, which has a cumulus origin (Hunter 1978; Eales and Cawthorn 1996). Cumulus silicate minerals in the Merensky Reef are unusually large (2–3 cm) compared with other layers, and the basal contact is very irregular. The origin of the Merensky Reef still remains a topic of heated debate, for some models calling on mixing of new magma (rich in PGE) and others on in situ fractional crystallization (Eales and Cawthorn 1996).

Some of the largest occurrences of chromite occur in layered intrusions as primary cumulus layers (Eales and Cawthorn 1996; Roach et al. 1998). Most of the cumulus chromite in the Bushveld Complex occurs in the Lower and Critical Zones, whereas in the Muskox intrusion it occurs at higher stratigraphic levels. Models for precipitation of the chromite include gravitational sorting, increases in oxygen fugacity, pressure changes in the magma, and magma mixing. Sharp contacts and remarkable lateral continuity of cumulus layers demand that, whatever the processes, they must have occurred at the same time over the entire magma chamber. Because only about 1000 ppm of Cr is possible in the Bushveld and Muskox magmas (Barnes 1986) the large concentrations of chromite in these bodies demands lateral flow of Cr-poor liquids and crystallization in some other part of the magma chamber or escape from the magma chamber. At present, no model fully explains geochemical, mineralogical, and physical constraints for the origin of cumulus chromite in large layered intrusions.

LIPS in Perspective

Although mantle plumes and LIPs seem to go hand-in-hand, several interesting and puzzling questions have appeared on the horizon in the last few years that must be addressed if this relationship is to survive. For instance, how far can plume material propagate laterally along the base of the lithosphere, and how far can plume-derived magmas travel in the crust in giant dyke swarms? In long-lived plumes that pass beneath continental lithosphere, such as Iceland and Yellowstone, magmatism is episodic, which is a feature that may be related to the thickness of the lithosphere (Johnston and Thorkelson 2000). Could the episodic nature of plume magmatism in these cases reflect migration of "thinspots" in the lithosphere over a plume? The thinned lithosphere would permit the plume head to rise to shallower levels and undergo decompressive melting. Likewise, when thick Archean lithosphere overrides a plume, magma production should drop because melt production is restricted to greater depths, thereby reducing the extent of partial melting and the volume of melt produced. A related question of current interest is, Just how far can plume head material propagate laterally by following structural irregularities at the base of the lithosphere? As discussed above, both the Ferrar and Afar provinces are examples of LIPs for which plume head material may have traveled thousands of kilometers laterally from its source controlled by sublithospheric channels caused by rifting. An alternative mechanism to account for LIPs far from their alleged plume source is that magma is transported by radiating dyke swarms.

Another question beginning to emerge is: Are LIPs, and especially oceanic plateaus, episodic in time? If so, over what time intervals, and what is the significance of this episodicity? What is the significance of multiple magma episodes spread over 30–40 Myr in oceanic plateaus such as Ontong Java and Caribbean? Does this reflect more than one plume arriving beneath an oceanic plateau at different times but from the same source, as suggested by the laboratory models of Bercovici and Mahoney (1994)? Alternatively, did a single plume have several different pulses of magmatism separated by 10–15 Myr, and, if so, why? Could some of the magmatic episodes be caused by lithosphere extension rather than by the mantle plume?

Did LIP events in the past result in mass extinctions and climate change? Are oceanic plateaus eventually subducted, and, if not, what happens when they collide with continental margins? Is accretion of oceanic plateaus to continental margins an important process by which continents have grown? These and related questions will be returned to in later chapters.

The very large and long-lasting plume-related features on Mars strongly suggest that similar features may have formed on Earth during the early stages of its history. LIPs, larger than those found on Earth today, may have been widespread on Earth before the onset of plate tectonics by 4 Ga. Better knowledge of the Martian and Venusian examples should not only enhance our understanding of the early stages (not preserved) of Earth history but also help explain why the evolution of Mars and Venus has been so very different from that of the Earth.

And finally, although the mantle plume model for hotspots and LIPs has been widely accepted among the geological community, it has not gone unchallenged. Some of the problems that have been suggested with the plume model are as follows (Anderson et al. 1992; Anderson 1994; King and Anderson 1995). As discussed in other parts of this chapter and briefly summarized at the end of each alleged problem, each of these problems has been refuted to the satisfaction of many, if not most, investigators.

1. Some hotspot volcanism has occurred over long periods of time, which may be difficult to reconcile with a plume head model where the action should be short-lived. However, long-lived hotspot volcanism is consistent with the long-lived nature of plume tails.
2. A plume head should stretch and elevate the lithosphere prior to, or concurrent with, volcanism – a feature that is not recognized in some LIPs. Some flood basalts have erupted into rifts that contain thick, prevolcanic sedimentary successions, and thus they may have been basins prior to volcanism (Czamanske et al. 1998; Sheth 1999). In these cases, however, uplift may have occurred after basin sedimentation ceased.
3. A large plume head should be resolvable with seismic waves, even when flattened. However, no seismic evidence for plume heads or damaged lithosphere has been reported beneath LIPs. This is no longer the case, and several plume heads (namely Iceland, Hawaii, Tanzania, and Ontong Java) have been resolved with seismic data.
4. Anderson and colleagues suggest that almost all flood basalts are erupted near the edge of Archean cratons, which is a feature not readily explained by a plume source. However, there are many examples of flood basalts erupted beneath cratons or beneath orogens. In addition, many cratonic boundaries, like the western edge of the Karnataka craton in India where the Deccan traps were erupted, were not cratonic boundaries during flood basalt eruption. It was not

until Pangea broke up that these cratons became "marginal" cratons. Before the breakup of Pangea, the Karnataka craton extended into Madagascar and Antarctica.

5. Some LIPs, such as the Siberian traps and Ontong Java plateau, do not have hotspot tracks leading to their parent plumes. Also, the timing and composition of some LIPs are not consistent with alleged parental hotspots. For instance, some Greenland basalts, often related to the Iceland plume head, postdate the hotspot track (Lawver and Muller 1994). The absence of hotspot tracks can readily be explained without abandoning the plume model by either, (a) subduction of the oceanic crust with the hotspot track, such as is probably the case with the Siberian traps or (b) limited motion of the plate overlying a plume that results in no hotspot track being produced, as is the case for the Ethiopian plume. Multiple periods of volcanism related to the same plume may be caused by pulsating magma production in a plume.
6. Some hotspot tracks, such as the Hawaiian–Emperor chain, have no associated LIPs. This seems to be more of a problem of the rate of decompressive melting in plume heads or the probability that many oceanic plateaus have been subducted. To address these difficulties, fictitious or real as the case may be, Anderson and colleagues have proposed a nonplume model in which LIPs are derived from an enriched upper mantle source (King and Anderson 1995). The model provides an explanation for the correlation of some LIPs with craton boundaries (however, see no. 4 above). In the model, a difference in heat flux between thicker, older cratonic lithosphere and thinner, younger lithosphere drives small-scale convection, which moves mantle from beneath thick lithosphere to boundaries between thick and thin lithosphere. Here, partial melting occurs, and magma rises freely without stretching the lithosphere. Instead, melting occurs as mantle moves upward between lithospheric blocks that split, producing rifts at the surface into which sediments accumulate prior to volcanism (Sheth 1999). Because these regions are often sites of earlier subduction, the shallow mantle already may be enriched in incompatible elements and thus may not require deep mantle sources or enriched mantle lithosphere sources to explain the enriched geochemical signatures of some LIP magmas.

4

Mantle Plume Generation and Melting

Introduction

Before describing models of plume behavior, it is important to review the fluid characteristics of mantle plumes. Whereas a fluid is a substance that can undergo an unlimited amount of deformation, a solid will undergo only limited deformation before it breaks. Another distinction is that many solids will deform a certain amount for a given force and then return to their original shape when the force is removed. In contrast, a fluid will keep deforming as long as the force is applied and will stop deforming when the force is removed, but it will not return to its original shape. A Newtonian fluid is a material whose rate of deformation is proportional to the applied force. In the mantle, flowage occurs in response to stresses, which result in strain of the mantle rocks. The proportionality between stress and strain rate is expressed as the viscosity of a fluid. For a viscous fluid undergoing very slow flow, as in the mantle, driving forces are in balance with viscous resisting forces, and the mantle can usually be considered to be an incompressible fluid. Whether all or only some of the mantle behaves as a Newtonian fluid is a subject of considerable uncertainty and debate.

Plume Characteristics

Experimental Models

It is well known that when a hot fluid is injected into a cooler fluid in a laboratory tank, the hot fluid will rise as a plume. The experiments of Griffiths and Campbell (1990) showed that when plume viscosity and density are lower than the surrounding media, the plume will divide into two components: a nearly spherical *head* at the top and a relatively thin conduit or *tail* connecting to the plume source (Fig. 4.1). This is the type of plume thought to form in the mantle. In the experiments of Griffiths and Campbell, the viscosity of the injected fluid determines the diameter of the conduit required to carry a given plume flux, whereas the viscosity of the surroundings determines the rate of rise and size of the plume head. When a plume head reaches the top, it is nearly symmetric and spreads horizontally beneath the surface layer (analogous to the

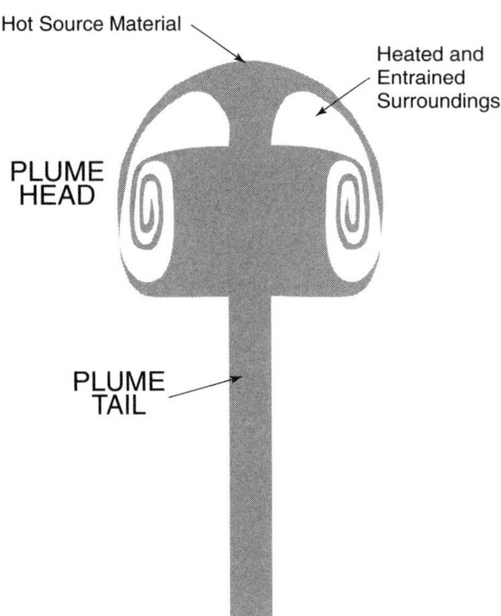

Figure 4.1. A laboratory plume showing plume components. After Campbell and Griffiths (1990).

lithosphere in Earth). After the head stops its vertical motion, the tail continues to supply hot source fluid, which, in the absence of horizontal motions near the surface, accumulates in a "pool" near the axis of the plume head. When motion of the surface plate is added to the experiment, the head may be deformed and carried away, whereas the plume tail is only deflected from the vertical. The size of plume tails critically depends on plume viscosity. The tail is thinner for lower viscosities, reflecting the fact that lower viscosity material requires only a thin conduit for a similar flow rate.

Experiments also show that plumes with the largest volume flux and the largest temperature anomaly (temperature difference between plume and surroundings) ascend most rapidly, accumulate the greatest volume, and are the least diluted by entrainment of surroundings (Griffiths and Campbell 1990; Campbell and Griffiths 1990). With a large buoyancy flux, plumes can penetrate the entire mantle today in as little as 50 Myr or the entire Archean mantle, which was hotter, in less than 10 Myr. Buoyancy flux (b) is related to average plume velocity (v), plume radius (r), and the difference in density between plume and surrounding mantle ($\Delta\rho$) by, $b = [g\Delta\rho] \times [\pi r_2 v]$ (Davies 1999). Mantle plumes with high-buoyancy fluxes ($\geq 10^5$ Ns^{-1}) are not significantly deflected from the vertical by mantle convection and remain connected to their deep mantle source. Weaker plumes with buoyancy fluxes less than 10^4 Ns^{-1} may not make it to the surface, and their plume heads may detach, forming small diapirs. These diapirs may become incorporated in the convecting mantle, or, if they make it to the base of the lithosphere, they may become part of the lithosphere with little or no melting and magmatism.

Experiments also demonstrate that plume heads grow large before "lift off" from the source. The most important variable controlling head size is mantle viscosity, which today is the order of 10^{22} Pa s. Modern mantle plumes must reach 400–600 km in diameter before beginning to rise (Fig. 4.2). When they have ascended 2800 km, their diameter has doubled (800–1200 km across) owing chiefly to entrainment of surrounding mantle on the way up. As the heads flatten against the lithosphere, the diameter

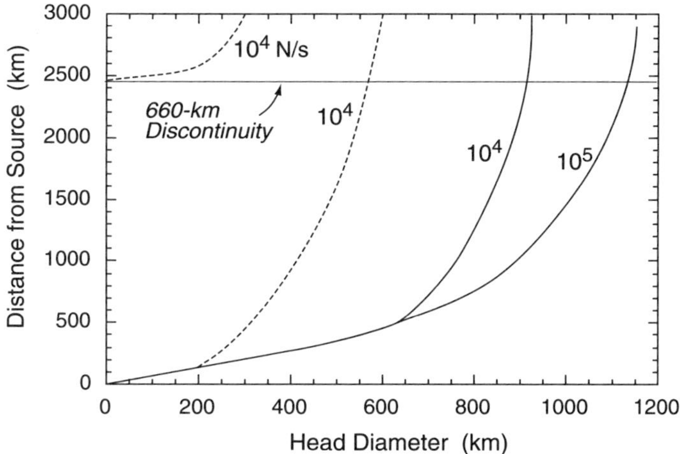

Figure 4.2. Growth of plume heads as a function of distance above the source (the core–mantle boundary). Model is for a source temperature anomaly of 400 °C and a viscosity of 10^{22} Pa s (present mantle, solid lines) or 10^{21} Pa s (Archean mantle, dashed lines). The two Archean examples are for plumes coming from the D″ layer and the 660-km discontinuity, respectively. Buoyancy fluxes given in N/s. After Griffiths and Campbell (1990).

can double again, producing a "pancake" thermal anomaly on the order of 2000 km or more across. In the Archean, when the mantle temperature was higher and mantle viscosity was lower ($\approx 10^{21}$ Pa s), plume heads would have been smaller at the time of lift off (200–400 km across; Fig. 4.2). These plumes would be only 400–600 km across when they reached the lithosphere. The models of Griffiths and Campbell also indicate that plumes originating at the 660-km discontinuity will reach diameters of only about 300 km when they intersect the lithosphere with little or no entrainment (Fig. 4.2). Plume diameters are almost independent of buoyancy flux and show only a moderate variation with temperature anomaly (Fig. 4.3). Regardless of temperature anomaly, in the modern mantle, many plume diameters, on reaching the base of the lithosphere,

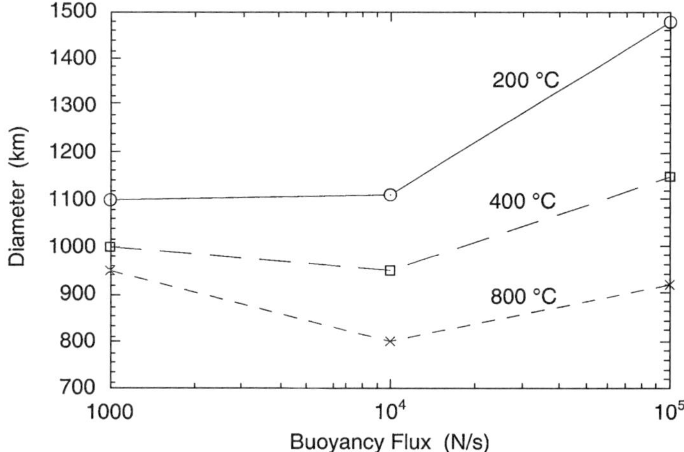

Figure 4.3. Predicted diameters of plumes coming from the D″ layer just before impinging on the base of the lithosphere. Temperature anomalies given are differences between plume and ambient mantle temperature. After Griffiths and Campbell (1990).

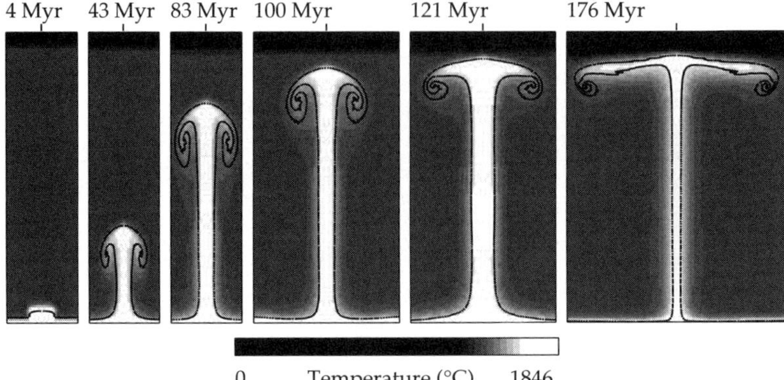

Figure 4.4. Numerical model showing rise and growth of a mantle plume from the D″ thermal boundary layer. Viscosity is a strong function of temperature, and ambient viscosity is 10^{22} Pa s. The bottom boundary temperature is 430 °C above the interior plume temperature, and plume viscosity here is about 1% of the plume interior. From Davies (1999), with permission. Courtesy of Geoff Davies.

are 1000 km or more, and up to 1500 km after flattening, which is consistent with the sizes of small-flood basalt provinces. Superplumes exceed 1500 km in diameter and may reach 3000 km, which is a size similar to that of large flood basalt provinces like the Karoo–Ferrar Province in southern Africa and Antarctica.

Numerical Models

Numerical models of Davies (1999) and others of temperature-dependent viscosity plumes confirm and expand upon the laboratory experiments. A sequence of snapshots of a mantle plume as it ascends and spreads over about 175-Myr period is shown in Figure 4.4. The plume ascends rapidly in the first 100 Myr and then more slowly afterwards as it begins to flatten against the lithosphere. As the hot fluid in the plume tail reaches the top of the head, it flattens against the lithosphere, and the head becomes very thin and increases significantly in radius (176 Ma). Heat liberated from the plume partly escapes and partly heats material entrained in the plume head, and thus the head has a temperature intermediate between the tail and the surrounding mantle. Such features were also recognized in the experimental models of Griffiths and Campbell (1990). As the head continues to grow and cool, it rises more slowly than the tail, which continues to feed the head.

Uplift, Deformation, and Subsidence

General Features

An important question related to the plume model for LIPs is, Do uplift and extensional deformation precede magmatism as predicted by numerical models? The amount of uplift is controlled by plume buoyancy and by lateral variations in the thickness and rheology of the lithosphere. In the simplest case, plume heads with high viscosity, a maximum uplift of 500–1000 m should be attained when the plume is at depths of 100–200 km (Griffiths and Campbell 1991). The uplift should begin 25 Myr before the time of maximum uplift and precede major basaltic volcanism by 3 to 30 Myr. Hence, unless the lithosphere is already thinned from forces in the plates, melting in the top of the plume should not occur until uplift has ceased. Uplift is followed by subsidence, which results from some combination of the following three factors: (1) As the plume head thins and spreads laterally, it loses buoyancy and begins to sink over the plume axis; (2) removal of magma from the top of the plume results in deflation, causing subsidence of overlying lithosphere; (3) gradual heat losses from the surface over hundreds of millions of years also results in some subsidence.

Numerical plume models and field observations, however, have shown the situation is more complicated and results are dependent on poorly known model variables. For example, Monnereau et al. (1993) showed that a large number of plume models are acceptable with considerable variation in the timing of uplift and deformation relative to magmatism. The time scale of uplift is critically dependent on the viscosity at the top of the mantle. In the melting model of White and McKenzie (1989) discussed later in this chapter, the main magmatism and deformation are contemporaneous. Uplift, deformation, and subsidence are affected by lithospheric thinning, magmatic underplating from the plume, and reduction in density of the restite mantle after melt extraction.

Laboratory Models

Experiments of Griffiths and Campbell (1991) showed the nature of uplift of the lithosphere followed by collapse as a mantle plume cools. It is interesting that surface uplift begins while the plume is still in the lower mantle and continues over a period of about 20 Myr. The maximum rate of uplift, however, is attained in the last 4 Myr when the plume head enters the low viscosity and low seismic-wave velocity zone at the base of the lithosphere (Fig. 4.5). The maximum elevation of uplift of 600 m is calculated on the assumption of an average temperature anomaly of only 100 °C in the plume head. This maximum, for a rigid, nonslip upper boundary, is reached when the top of the plume is at a depth of about 200 km (0 Ma) and has a diameter on the order of 1300 km. It is important to remember that the times calculated for uplift in this model are critically dependent upon plume viscosity, which has an uncertainty of a factor of three or more.

After maximum uplift, the surface begins to subside as the plume continues to ascend and spread; after 5 Myr, the thermal anomaly is only 150 km from the surface, and the plume has a diameter of 1800 km (5 Myr in Fig. 4.5). As discussed more fully later in this chapter, a plume must ascend to depths on the order of 40–60 km to produce significant quantities of melt necessary for the formation of LIPs, and this requires one or more of the following: (1) penetration of the lithosphere by the plume, (2) stretching and displacement of the lower part of the lithosphere over the plume head, or (3) rifting of previously thinned lithosphere.

Field and Dating Evidence

Unambiguous field and geochronologic data from LIPs relative to the timing of uplift, deformation, and volcanism are few. Modern examples of domal uplifts produced by

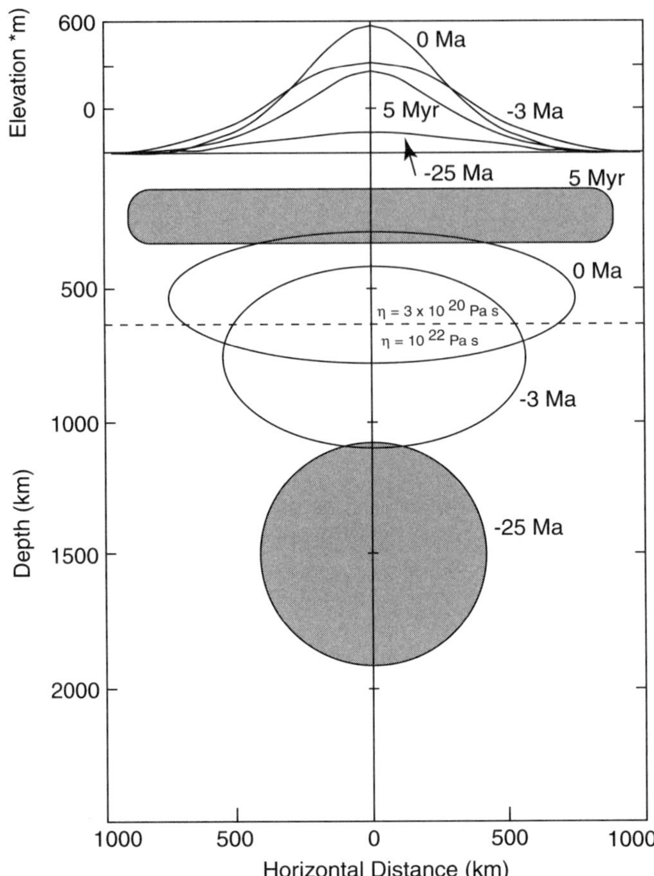

Figure 4.5. Relationship of surface uplift to plume size and rise time based on the laboratory experiments of Griffiths and Campbell (1991). Time in Ma is relative to maximum surface elevation, which is set at $t = 0$. Times in the future are expressed Myr. Model is for a plume derived from the D″ layer with a buoyancy flux of 3×10^4 N/s and a source temperature anomaly of 300 °C.

plumes are the Afar and Kenyan domes in the East African rift system (Sengor 2001). Mapping in the Deccan, Siberian, Rajmahal, Paraná, and Karoo flood basalts suggests that uplift precedes volcanism in these LIPs (Kent et al. 1992), and thus, they are broadly consistent with the experimental models of Griffiths and Campbell (1991) described above. Other investigators, however, have suggested that Deccan and Siberian flood basalts were erupted in nonextensional settings with little or no prevolcanic uplift (Hill et al. 1992; Czamanske et al. 1998). In the Deccan and Columbia River basalts, eruption of the main tholeiitic phase preceded significant extension and lithospheric thinning (Hooper 1990). In contrast, studies of magnetic susceptibility from the Paraná–Etendeka flood basalts imply that rifting preceded volcanism – at least in that part of the field within 100 km of the Mid-Atlantic Ridge (Glen et al. 1997). The lack of evidence for early uplift in many LIPs may be caused by later overprinting due to extension and subsidence. In any case, carefully integrated field and isotopic dating studies are needed to sort out the timing of deformation and volcanism as well as the role of premagmatic uplift in LIPs.

Evidence for syn- to postmagmatic subsidence in LIPs is preserved in many flood basalts. Most flood basalts occur in broad, posteruptive basins. In the Karoo flood basalts in southern Africa, subsidence began before volcanism, as reflected by the thick sedimentary succession beneath the basalts (Campbell and Griffiths 1990). Subsidence leads to the formation of rift valleys in some flood basalt provinces. Examples are the Cambay basin and Narmada graben in the Deccan traps and the Tuli and Nuanetsi valleys in the northern Lebombo area of the Karoo. In the Deccan, major subsidence coincides with the onset of volcanism, which perhaps reflects withdrawal of magma from the plume head. Supporting this interpretation is the occurrence of rift valleys in the Deccan near the inferred plume axis, where maximum melt should be withdrawn. In young flood basalts, such as the Ethiopian and East African Provinces erupted 30 Ma, subsidence has not yet begun.

Using the Afar and Kenyan domal uplifts in East Africa as examples, Sengor (2001) concludes that mantle plumes lead to rifting of the lithosphere, by increasing the potential energy of part of the lithosphere, causing it to disintegrate along one or more rifts. In his model, structural rather than topographic uplift is the chief criterion for the identification of a former plume. Structural uplifts can be identified if the palaeogeography preceding uplift can be reconstructed in detail, if uplift erosion products are preserved (especially in a marine environment to enable precise dating), and if the structural expression of the uplift has not been destroyed by erosion. There is no known process on Earth other than mantle plumes that can form lithospheric domes 1000 km or more in radius and 1 to 2 km high within several million years.

Wrinkle Ridges

One of the important deformational features found in the lithosphere above large mantle plumes is wrinkle ridges formed in response to contractional forces associated with plume emplacement and collapse (Mege and Ernst 2001). Wrinkle ridges are commonly associated with LIPS on Mars and Venus (see Fig. 3.33) and also have been described in terrestrial examples, the best documented of which is the Yakima fold belt in the Columbia River Basalt Province. Typically, wrinkle ridges are on the order of 100 km long, 10–15 km wide, and a few hundred meters high (Mege 2001). An extensive set of wrinkle ridges along the eastern margin of the Tharsis Rise on Mars began to develop as flood basalts were erupted, but this development stopped before eruptions terminated. In general, wrinkle ridges develop normal to grabens and dyke swarms that radiate from a plume center (Fig. 3.33). Wrinkle ridges of the Yakima fold belt in the Columbia River basalts and the wrinkle ridges associated with the Tharsis Rise on Mars have numerous similarities, which suggests a similar mode of origin. In common are the following (Mege 2001):

1. Ridge size.
2. Concentric distribution about a plume center.
3. Development only in stratified basalts.
4. Association with flood basalts erupted on thin, brittle crust underlain by a ductile layer.

5. Ridge growth rate that parallels lava flow eruption rate.
6. Periodic ridge spacing.
7. Ridge formation by combination of folding and thrust faulting.

Models of wrinkle ridge development show that most ridges are related to thermal subsidence as a plume head begins to collapse and that they are enhanced by isostatic adjustment of the load of flood basalts on a thermally thinned brittle crust (Mege and Ernst 2001).

How Fast Do Plumes Rise?

In a Newtonian mantle, plumes can rise from the D″ layer at the base of the mantle to the lithosphere in about 50 Myr (Olson et al. 1987). For non-Newtonian fluids, high-resolution calculations show that mantle plumes rise even faster – on the order of meters per year (Larsen and Yuen 1997). Even with similar Rayleigh numbers for the mantle, non-Newtonian plumes rise to the base of the lithosphere in a few million years, which is an order of magnitude faster than Newtonian plumes. In the models of Larsen and Yuen (1997), plume formation is triggered when a cold downwelling hits the lower boundary layer (D″), resulting in strongly localized flow. In one model, the maximum ascent rate is 36 m/yr, and upon impact with the lithosphere, it drops to less than 1 m/yr. In the presence of both viscous and adiabatic heating, non-Newtonian plumes thin the lithosphere more effectively than plumes with Newtonian properties. Unlike Newtonian plume heads, which are broad and smooth, non-Newtonian plume heads have a complex internal structure and may rise to within a few kilometers of the surface. Non-Newtonian plume tails are relatively narrow and focused with sharp thermal and velocity gradients in comparison to the broad Newtonian tails. These differences between Newtonian and non-Newtonian plumes are due to the different rates of entrainment and mixing with surrounding mantle. Non-Newtonian plumes are important because they provide a means of bringing very hot mantle material to the surface in very short periods of time, which is a feature consistent with producing high-Mg komatiitic melts in plume heads during the Archean.

How Long Do Plumes Survive?

When one tracks the lifetime of mantle plumes using their volcanic records, it is clear that there are both long-lived and short-lived plumes. The Kerguelen plume in the southern Indian Ocean survived for at least 116 Myr (Frey et al. 2000). Its first appearance is recorded by the Rajmahal flood basalts in northeastern India at about 116 Ma. This was followed by rapid growth of Kerguelen and Broken Ridge oceanic plateaus at 116–110 Ma, the Ninetyeast Ridge at 85–38 Ma, and the Kerguelen Archipeligo in the last 40 Myr. The Tristan plume in the South Atlantic began activity with the Paraná–Etendeka flood basalts 137–125 Ma followed by the Walvis and Rio Grande Ridges (135–131 Ma) as the South Atlantic opened and finally by Tristan volcanic island in the recent past (Peate 1997). The Réunion plume, with a lifetime of 65 Myr, generated the Deccan traps, the Maldives–Chagos and Mascarene Ridges, and the modern Réunion volcanic islands (Fig. 2.6) (Courtillot et al. 1999). The St. Helena plume, which began activity 145 Ma, controlled the development of the Helena seamount chain and is still active today (Wilson 1992). The Iceland plume is responsible for the North Atlantic Volcanic Province (130–50 Ma), the Faeroe–Greenland volcanic Ridge, and the recent volcanic activity in Iceland (Saunders et al. 1997). The Hawaiian plume has been active for at least 75 Myr as recorded by volcanic activity along the Hawaiian–Emperor chain (Clague and Dalrymple 1989).

The Siberian traps (246–255 Ma) were erupted from a short-lived plume that survived for at most 9 Myr (Sharma 1997). Most of the flood basalt activity was concentrated in a 1-Myr period at 251–250 Ma. Other examples include the Emeishan flood basalts in China (258 Ma) and the Taymyr, Kuznetsk, and Verkhoyansk–Vilyuy flood basalts in Siberia (about 370 Ma) (Courtillot et al. 1999; Nikishin et al. 1996). In all of these cases, flood basalt eruption was concentrated during a period of 1 Myr or less, and the total duration of magmatism and associated rifting was 5–10 Myr. The Mid-Cretaceous superplume event in the South Pacific, which gave rise to Ontong Java and related oceanic plateaus, was also short-lived with most action occurring at 120–100 Ma (Larson 1991a). In contrast to long-lived plumes, none of the short-lived plumes was followed by fragmentation of continental lithosphere nor long-lived hotspot tracks such as the Hawaiian–Emperor chain.

Long-lived plumes, which can survive for 100–150 Myr, require a continuous supply of hot mantle via a plume tail, which taps a continuously replenished source in the deep mantle. In contrast, short-lived plumes appear to lack a stable plume tail and rise from a source that is rapidly depleted. Perhaps only long-lived plumes are generated in the D″ layer above the core–mantle interface, where a continuous supply of heat is possible. Short-lived plumes may come from shallower mantle depths, perhaps some from the 660-km discontinuity.

Entrainment in Plumes

Both experimental and numerical modeling of plumes show that they entrain material from the surrounding mantle as they rise (van Keken 1997; Davies 1999). This is because hot, buoyant plumes transfer some of their heat to the surrounding ambient mantle, which increases its buoyancy and lowers its viscosity, allowing it to be incorporated into plumes (Richards and Griffiths 1988; Hauri et al. 1994). Hence, plumes may sample not only the source material in the D″ layer but also other mantle geochemical domains as they rise to the base of the lithosphere. Investigators have disagreed over whether plume heads or tails show the most entrainment of surrounding mantle. The disagreement results from the different assumptions made about mantle rheology in various plume models and from disparate results of laboratory and numerical models. Geochemical and isotopic results from basalts generated in plume heads and in plume tails may eventually help resolve this problem.

In a model of Kellogg and King (1997) of a plume consisting of material in which the viscosity is strongly temperature dependent, the plume head grows as it becomes mushroom-shaped. As the plume rises, it entrains surrounding mantle in the head, but very little mixing occurs in the plume tail, as shown by the contours in the left side

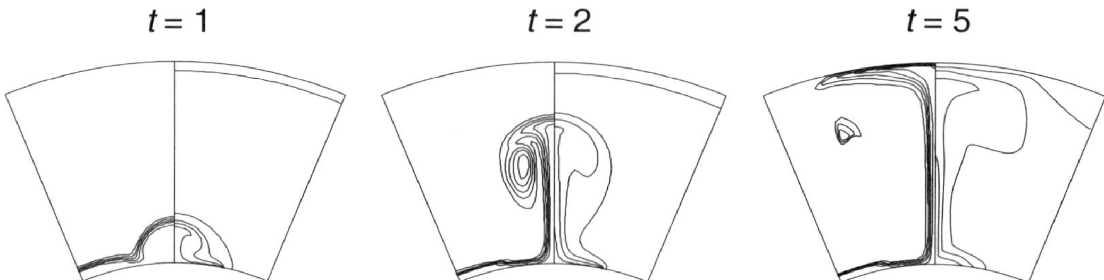

Figure 4.6. Numerical model of Kellogg and King (1997) showing rise and growth of a mantle plume with a strong temperature-dependent viscosity. Time after plume escapes from source given in arbitrary units (t). Contours are temperature (right half of each figure) and concentration of a passive dye (left).

of each frame in Figure 4.6. This indicates that the geochemical signatures of melts coming from the head may be contaminated with ambient mantle, whereas the plume tail should carry a source signature. Thus, flood basalts should reflect mixed mantle sources and oceanic islands, relatively pure samples from plume sources. Also, in the Kellogg and King model, not all of the plume mass arrives at the hotspot, but some of it, like the blob in frame 3 of Figure 4.6 ($t = 5$), breaks off and is mixed back into the upper mantle. Such blobs could produce chemical anomalies in the convecting asthenosphere, which may later appear in ocean-ridge or seamount basalts (Allegre et al. 1984). Farnetani and Richards (1995) suggest that mantle plumes derived from the deep mantle entrain only a very small fraction of surrounding mantle into the region of the plume that undergoes partial melting. Their models, which are for isoviscous plumes with only mildly temperature-dependent rheology, also agree with the earlier numerical and experimental results of Griffiths (1986) and Griffiths and Campbell (1990) in indicating that the original source material in plumes is confined chiefly to plume tails.

More recent numerical models, however, that include strong temperature-dependent rheology, which is likely in the mantle, show that most of the source material is contained in a plume head and that the tail comprises largely or partly entrained material (van Keken 1997; Davies 1999). Davies (1999) has shown that the amount of entrainment is critically dependent on the ratio of plume viscosity to ambient mantle viscosity. As expected, the closer the viscosity of the plume is to that of surrounding mantle, the greater the degree of entrainment. This is illustrated in Figure 4.7 by flow lines in plumes with three different viscosity contrasts. Most of the entrained material comes from the lowest 10–20% of the mantle. Plume tails may also become contaminated with surrounding mantle if they deviate appreciably from the vertical. Vertical plume tails with a strong viscosity contrast can entrain only a very small percentage of surrounding mantle (Stacey and Loper 1983; Davies 1999). However, if the plume tail is inclined to the vertical, as it may be in the upper mantle owing to advection, the amount of entrainment can increase substantially.

Recent geochemical studies of Galápagos hotspot volcanics suggest that the Galápagos plume is chemically zoned and that the zonation has survived for at least 14 Myr (Hoernie et al. 2000). In map view, the zonation is horseshoe shaped with enriched material around the edge of the plume and depleted material in the center. The

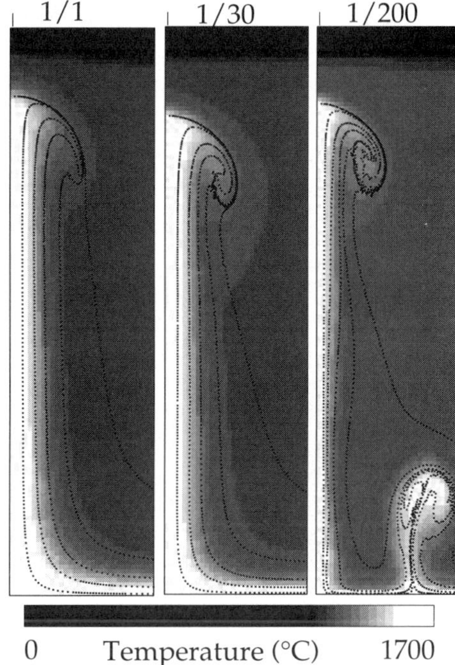

Figure 4.7. Numerical models showing plumes with three different ratios of plume viscosity to ambient mantle viscosity (given at the top). Dotted lines show contamination of the plume as a function of height. Note that the plume tail thins as viscosity contrast decreases. A secondary instability has developed in the right-hand model. From Davies (1999), with permission. Courtesy of Geoff Davies.

main evidence for this is that the compositional zonation continues along the Cocos Ridge hotspot track, one of two hotspot tracks leading away from the Galápagos Islands (Fig. 2.1). The chemical zonation may reflect a plume tail that is draining two or three chemically distinct regions in the D″ source layer. Alternatively, entrainment of enriched mantle in the plume tail could account for the zonation. If entrainment is the source of the zonation, it must occur in the deep mantle because Galápagos lavas have high $^3He/^4He$ ratios indicative of a deep, primitive, or depleted source. In either case, very little mixing is allowed during the ascent of the plume to preserve the zonation.

Plume Roots

Seismic Evidence

Seismic tomography offers a unique method both to identify mantle plumes and to trace the tails of plumes into the mantle. Continued improvement in the resolution of tomographic results now makes it possible for the first time to begin three-dimensional mapping of mantle plumes. An analysis of compressional-to-shear converted seismic phases shows a zone of very low S-wave velocities (<4 km/s) beginning at 130–140 km beneath the island of Hawaii and extending downward into the mantle (Li et al. 2000). The same data indicate that the region between the 410- and 660-km seismic discontinuities is thinned by 40–50 km south and southwest of Hawaii. These observations are interpreted to record the Hawaiian plume tail, which must be 250–300 °C hotter than surrounding mantle and less than 400 km in diameter between the two mantle phase boundaries. This temperature is at least 100 °C higher than the Iceland

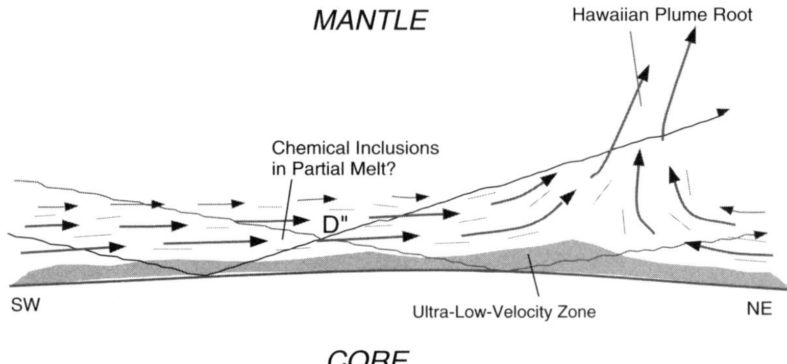

Figure 4.8. Hypothetical sketch showing the source of the Hawaiian plume in the D″ layer above the core. Arrows show flow direction and anisotropy inferred from S-waves. After Russell et al. (1998).

plume, which is in agreement with petrological observations and dynamic models for both plumes.

Seismic tomography has also shown that the tails of the Iceland and Hawaiian plumes can be tracked into the deep mantle – probably into the D″ layer just above the core (Helmberger et al. 1998; Russell et al. 1998). Analyzing ScS and S phases (ScS is a set shear waves that bounce off the core–mantle boundary), Russell et al. (1998) reported a region of lateral gradients in velocity and anisotropy in the lowermost mantle beneath Hawaii, which they suggested represents the source region of the Hawaiian plume. The ScS anisotropy these authors described is unusual. Previous measurements of SH waves in the D″ layer showed horizontal polarization and arrived ahead of vertically polarized waves (SV). This generally has been interpreted to reflect horizontally layered structures or elongated melt inclusions in D″ (Kendall et al. 1996). Only in the Central Pacific, Russell et al. found that this pattern changes to SV arriving ahead of SH, which they attributed to production of a vertical fabric by plume upwelling (Fig. 4.8).

Helmberger et al. (1998) identified an ultralow velocity zone at the core–mantle interface beneath Iceland, and they effectively argued that this anomalous region may reflect hot, partially molten source material feeding the Iceland plume. The model indicates velocity reduction of 10% for P-waves and 30% for S-waves, which is consistent with small amounts of partial melting. Bijwaard and Spakman (1999) described tomographic images of a narrow, low-velocity anomaly beneath Iceland, which extends from the core–mantle boundary to the surface (Plate 4). Results suggest that the plume, which is 500 km or less in diameter, rises from a broad root zone (>1000 km across) and culminates in a plume head about 1200 km in diameter. Also from this image, it would appear that the plume tail is bent near a depth of 700 km. Perhaps resistance at the 660-km discontinuity is responsible for this bending. The "waviness" of the Iceland plume inferred from the seismic data, if real, may suggest that the plume is not stationary but is advected by upper mantle currents. Another model that includes surface and body waves as well as free oscillation data suggests the low-velocity anomaly beneath Iceland essentially bottoms out at the 660-km discontinuity (Ritsema et al. 1999). These data probably map the plume head but may not be sensitive enough to identify the plume tail.

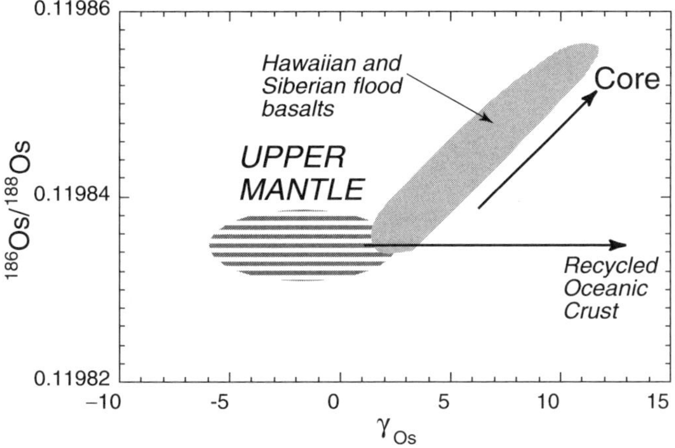

Figure 4.9. Osmium isotope relationships of Hawaiian and Siberian LIP basalts. $\gamma_{Os} = (\{[^{187}Os/^{188}Os]_{sample}\}/\{[^{187}Os/^{188}Os]_{CHUR}\}) \times 100$, where CHUR = chondritic uniform reservoir. Note that the Hawaiian and Siberian flood basalts seem to contain a core component. Data from Brandon et al. (1999).

Osmium Isotope Evidence

As reflected by oceanic island basalts, mantle plumes appear to be characterized by a rather narrow range in Os isotopic composition (Widom and Shirey 1996; Walker et al. 1995). In general, the isotopic compositions are more radiogenic than in MORB or mantle xenoliths and thus require an additional source of radiogenic Os. There are two isotopes of radiogenic Os: ^{186}Os, which is produced from the decay of ^{190}Pt ($t_{1/2} = 450 \times 10^9$ yr) and ^{187}Os, which is produced by the decay of ^{187}Re ($t_{1/2} = 42 \times 10^9$ yr). All three of these platinum group elements strongly follow iron, and thus in the Earth they are thought to be largely concentrated in the core. Hence, even small additions of core material to the mantle source of plumes should dominate the isotopic composition of Os. An excess of ^{187}Os in oceanic basalts could reflect recycled oceanic crust in the plume sources because Re is also concentrated in the crust (Brandon et al. 1998). However, the presence of unusually large amounts of both ^{186}Os and ^{187}Os in some plume-derived basalts in oceanic areas strongly suggests a core contribution (0.5–1.0%) to their sources (Brandon et al. 1999) (Fig. 4.9). This observation is important because it supports a source for at least some mantle plumes in the D″ layer above the core–mantle interface. It also may provide a means to distinguish D″-derived plumes from plumes produced at shallower levels in the mantle.

Plume Families and Head–Tail Detachments

Some hotspots appear to be grouped into families, whereas others occur in isolation (Sleep 1990). Families of hotspots reflect plumes of approximately the same age in the same geographic region. For instance, in the South Atlantic, Bouvet, Marion, Discovery, and Tristan may be a hotspot family, and in the North Atlantic, Cape Verde, Great Meteor, and Canary may be a family (Fig. 2.1). Hawaii, Louisville, and Réunion are examples of isolated hotspots (Fig. 2.2). Families of hotspots have been used as an argument against a mantle plume origin for these hotspots because there seems to be no way to maintain closely spaced plume conduits all the way from the base of the mantle (McHone et al. 1987). Another explanation for hotspot families, which needs to be more fully explored, is that they come from a single superplume that breaks into a group of smaller plumes at shallow depth in the mantle. Storey et al. (2000) proposed such a model for the Shona, Bouvet, and Discovery hotspots in the South Atlantic (as discussed in Chapter 2).

One of the outstanding problems of the relationship between plumes and hotspots is that the great majority of hotspots appear to represent plume tails. Only a relatively small proportion of hotspots (i.e., those that gave rise to large flood basalt fields or oceanic plateaus) are associated with plume heads. Does this mean that most plumes are produced without heads and, if so, how is this possible? Experimental and computer models of plume generation, as discussed above, have not been able to produce plumes without heads. More likely, this phenomenon means that plume tails outlive plume heads and thus that the remnants of older plume heads have not been identified and matched to their surviving tails. So what is the fate of plume heads? Possibilities include incorporation into the lithosphere or mixing with the convecting asthenosphere. Geochemical and isotopic data from plume basalts, as discussed in Chapter 5, render the latter possibility unlikely because plume heads mixed with depleted upper mantle would chemically contaminate this mantle, a feature that is not observed in ocean ridge basalts derived from the upper mantle. This being the case, extinct plume heads may be an important component in the mantle lithosphere.

Plume Temperatures

It is commonly assumed that most or all large mantle plumes arise from the D″ layer at the bottom of the mantle (Stacey and Loper 1983; Olson et al. 1987). On the basis of the mismatch of calculated adiabatic temperatures in the mantle and core, Jeanloz and Morris (1986) estimated a temperature difference across the D″ layer of 800–1000 °C. This temperature difference is considerably higher than that of 200–300 °C in mantle plumes at lithospheric depths as estimated from experimental petrologic constraints (Schilling 1991; Farnetani and Richards 1994). Although entrainment on the way up can lower the temperature of a plume by 10–20%, it cannot be a major factor in explaining the discrepancy between plume excess temperatures at shallow depth and the high temperature gradient across the alleged source just above the core–mantle interface (Farnetani and Richards 1995).

Using a finite element model, Farnetani (1997) evaluated the role that chemically denser material in the D″ layer may have in buffering the excess potential temperature of plume heads. Figure 4.10 shows the time sequence of a mantle plume produced in D″ from a 30-km-thick chemical layer that is 5% more dense than overlying mantle. As the plume rises, a secondary circulation develops along the axis, and material near the base of the plume is entrained downward by counter-rotating flow, as observed also in the model of Kellogg and King (1993). After 75 Myr, the plume head has an excess temperature of 400 °C, and the final excess temperature, upon reaching the base of the lithosphere, is only 300–350 °C, which is similar to that predicted from petrologic data. The role of the chemically dense layer in D″ is that of preventing the lower part

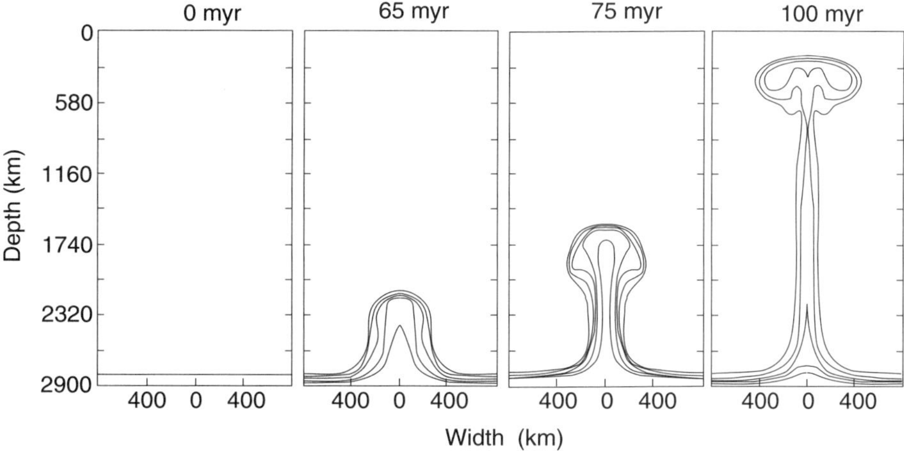

Figure 4.10. Growth of a mantle plume that began in D″ with a temperature gradient of 800 °C and a corresponding density difference of 5%. Contours of excess potential temperature shown in 100 °C increments. After Farnetani (1997).

of the D″ layer from becoming part of a rising plume. Thus, the high-temperature part of D″ remains in the lowermost mantle above the core–mantle interface. These results are important because they show that dense zones or layers within D″ may have an important role in governing plume temperature anomalies.

Phase Transitions and Plumes

As discussed in Chapter 1, there are two major seismic discontinuities in the mantle transition zone caused by phase changes: a transition with a positive Clapeyron slope at 410 km where the most abundant upper-mantle phase, olivine, transforms to Mg–spinel, and a transition with a negative Clapeyron slope at 660 km where spinel breaks down to perovskite and magnesiowustite. Recall from Chapter 1, the Clapeyron slope is the slope of a chemical reaction in P–T space. Transitions with positive Clapeyron slopes enhance mantle convection, whereas those with negative Clapeyron slopes hinder convection; depending on the magnitude of the slope and on the mantle temperature, they may result in an impenetrable barrier (Christensen and Yuen 1985; Davies 1995). Tomographic seismic data suggest that, besides the two large discontinuities, at least one weak discontinuity exists between a depth of 900 and about 1100 km (Niu and Kawakatsu 1997; Cserepes et al. 2000). Some subducted slabs, after penetrating the 660-km discontinuity, appear to be deflected and distorted and perhaps become trapped near depths of 1000 km support the existence of a second barrier at this depth. From high-pressure mineral physics, the weak 900–1000-km discontinuity may be caused by either another endothermic phase change, this time involving garnet, the second most abundant mineral in the upper mantle, or a change in chemical composition (Cserepes et al. 2000).

Whether a phase change has a major effect on rising mantle plumes depends in part on the Clapeyron slope (Davies 1995, 1999). As an example, for a moderate slope of −2 MPa/°C, a rising plume easily penetrates the 660-km discontinuity, although the ascent rate slows in the vicinity of the discontinuity. This is shown in frame 1 of

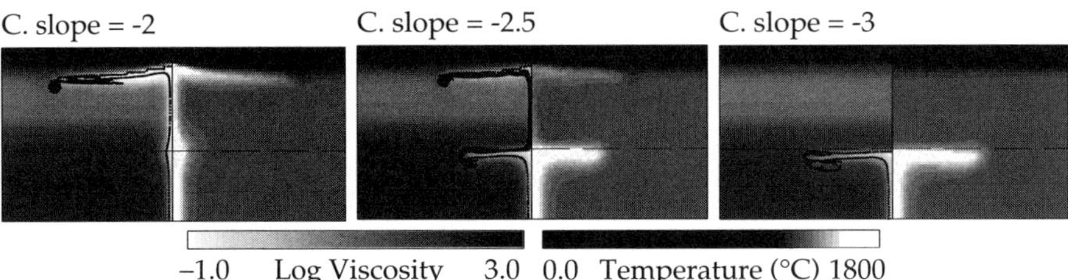

Figure 4.11. The effect of the 660-km discontinuity on mantle plumes. C. slope, is the Clapeyron slope of the spinel to perovskite + magnesiowustite reaction. From Davies (1999), with permission. Courtesy of Geoff Davies.

Figure 4.11, where the plume tail bulges at the discontinuity and then narrows again after it passes through the discontinuity and enters the low-viscosity upper mantle. For an intermediate slope of −2.5 MPa/°C, the plume head penetrates the discontinuity, but most of the tail accumulates at the discontinuity, giving rise to a rootless plume head in the upper mantle (frame 2). For a Clapeyron slope of −3 MPa/°C, the plume is unable to penetrate the discontinuity and spreads laterally at this point (frame 3). Because plumes clearly make it to the base of the lithosphere today, it would appear that the Clapeyron slope of the 660-km spinel transition is close to −2 MPa/°C.

Bercovici and Mahoney (1994) showed from the results of laboratory models that two successive plume heads are generated if there is a significant reduction in mantle viscosity across the 660-km discontinuity. This is one way to explain the episodic nature of oceanic plateau volcanism discussed in Chapter 3. For instance, the 122-Ma event at Ontong Java could reflect the first plume head, and the 90-Ma event, the second plume head. Laboratory results of Kumagai and Kurita (2000) suggest that experimental plumes entrain surrounding mantle and that the density of the plume head increases with time. As previously mentioned, the degree of entrainment depends the viscosity ratio of plume head to surrounding mantle. If the plume head density is smaller than surrounding mantle, the plume passes through the phase transition with very little entrainment. In contrast, if the plume density exceeds the surrounding mantle, the plume head "sticks" at the transition zone, and a new plume emerges from the plume tail and rises through the upper layer.

Numerical modeling by Yuen et al. (1998) suggests that a low-viscosity zone may exist beneath the 660-km phase transition, and, if so, this may allow the formation of small plumes with sources near the discontinuity. This process requires decoupling between upper and lower mantle circulation, which results in the formation of a thermal boundary layer at the 660-km discontinuity. In an upper mantle with relatively high Rayleigh numbers (and thus high temperatures), but with increasing viscosity with depth, as may have been the case in the Archean, large plumes coming from the base of the mantle may be stopped or deflected at the 660-km discontinuity (Fig. 4.11, frame 3). This situation, which results in layered convection in the mantle, could initiate many secondary plumes rising from the 660-km discontinuity.

Numerical models by Marquart and Schmeling (2000) and Marquart et al. (2000) indicate that the wide range in trace element contents of oceanic basalts cannot be explained totally by different temperature and pressure conditions in a mantle plume source. The authors propose a model whereby these chemical differences are produced by interaction of plumes with the 660-km discontinuity. Plumes detaching from the core–mantle boundary are assumed to be of variable size and excess temperature. When plumes arrive at the 660-km discontinuity, they are detained from further rise as a function of their size and excess temperature. Volumes with a radius exceeding about 150 km and an excess temperature of more than 100 °C rapidly cross the phase transition, whereas those with radii smaller than 80 km stagnate. The amount of material entrained in a plume derived from near the phase transition depends on the stagnation time and the thermal growth of the plume head while sitting at the 660-km discontinuity. Plume heads that reach the base of the lithosphere after a stagnation time of longer than 50 Myr may entrain large amounts of material from the transition zone. For a 50-Myr stagnation time, up to 15% of a plume head may come from the phase transition zone.

Numerical plume models of Brunet and Yuen (2000), which include both basal and internal heat sources and viscosity gradients, suggest that some mantle plumes actually may be trapped in the upper mantle between the 410- and 660-km phase boundaries. These investigators recognize three types of plumes in their models. The first is a stable plume originating in the D″ layer that can be stationary for more than 200 Myr. These plumes can produce volcanism with an age progression as found in island chains such as the Hawaiian–Emperor chain. The second type of plume, which also comes from D″, is flexible and can be bent in the convecting mantle; in some cases the plume head becomes detached from the tail. Several plumes of this type can be ejected from a trapped plume head in the transition zone over a prolonged period and rapidly ascend. The spreading of these fast plumes is controlled by advection in the upper mantle. Such irregular plume activity can explain chaotic volcanism in island–seamount chains such as the Society chain. The third type of plume recognized in the models of Brunet and Yuen (2000) is the superplume, which also comes coming from D″. Unlike the smaller plumes, superplumes generate a thick thermal boundary layer as they interact with the 660-km phase transition, and this layer can serve as a site for the launching of new, smaller plumes. An example of this type of behavior has been suggested for the Karoo–Ferrar volcanism discussed in Chapter 3.

Hard Turbulence and Plumes

During the Archean, when the temperature of the mantle was greater than today, the Raleigh number was also higher, and thus the convection regime in the mantle may have been quite different than at present. Experimental and numerical studies have shown the existence of two convection regimes: soft turbulence at Raleigh numbers less than 4×10^7 and hard turbulence for larger Raleigh numbers (Castaing et al. 1989; Hansen et al. 1990). Hard turbulence is characterized by the appearance of numerous, relatively small disconnected plumes (or diapirs) as well as chaotic convection (Yuen et al. 1993). It is unlikely that mushroom-shaped plumes can survive in a turbulent regime. Hard turbulent regimes at high Raleigh numbers also favor layered mantle convection because deep plumes probably will not penetrate endothermic phase transitions. This characteristic has important consequences for Earth's thermal history, because, as the mantle cooled with time, convection may have changed from a strongly layered system to one that is intermittently layered today. Furthermore, the change may have been catastrophic, involving the production of superplumes at several times in the past, as discussed more fully in Chapter 8. The hard turbulent regime, although probably not important in the mantle today, may have characterized the Archean mantle, and, if so, mantle plume regimes in the Archean must have been quite different from post-Archean plume regimes.

Effect of Planetary Rotation on Plume Distribution

It has long been recognized that hotspots are not random on the Earth's surface but are concentrated at low latitudes and especially associated with the Pacific and African mantle upwellings (Fig. 2.1). Similarly, on Venus, hotspot volcanoes occur chiefly within 40° north or south of the equator, and a large concentration is found between 180° and 300° longitude (Crumpler et al. 1993). On Earth, hotspots have an approximately symmetrical distribution with maxima occurring between 20° and 30° of the equator and a curious peak at about 80°S latitude (Fig. 4.12(a)). Because identification

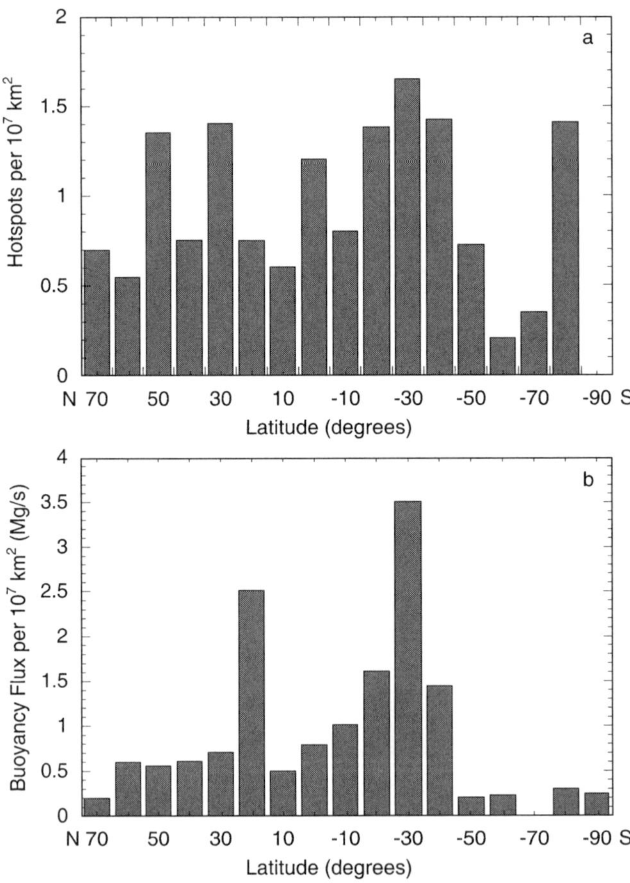

Figure 4.12. Global distribution of hotspots. (a) areal distribution, and (b) buoyancy flux distribution. Data from Oliver and Ghent (2000).

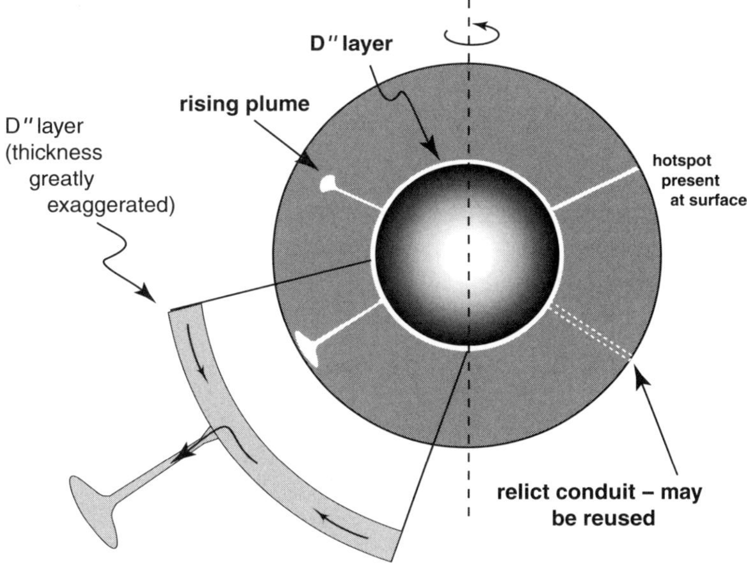

Figure 4.13. Rotational model for mantle plume distribution. Low-viscosity material migrates from high to low latitudes within D″ owing to the tangential component of centrifugal force. After Oliver and Ghent (2000).

of weak hotspots is uncertain, buoyancy flux offers a more rigorous method to evaluate hotspot distribution (Oliver and Ghent 2000). Using 47 hotspots, a strong bimodal distribution is apparent at 20–30° on both sides of the equator (Fig. 4.12(b)). The difference in magnitude of hotspot fluxes between the northern and southern maxima may reflect the greater amount of continental lithosphere in the Northern Hemisphere. Plumes impinging on thinner oceanic lithosphere are more likely to have a strong surface manifestation than those arising beneath thick, continental lithosphere.

To explain this symmetrical distribution of hotspots, Oliver and Ghent (2000) proposed that the hot, low-viscosity material in the D″ layer above the core flows laterally owing to the Earth's rotational forces. Their model suggests that centrifugal and differential rotational forces reinforce each other to move the D″ material from high latitudes to about 30°, where it accumulates (Fig. 4.13). Material stagnates here until it reaches the critical size for a mantle plume to form, or it drains upward into existing plumes. Just how mantle upwellings are related to this model is not yet clear because their centers lie close to the equator, where centrifugal force is at a maximum. It is interesting that this same model may also explain the symmetrical distribution of hotspot volcanoes on Venus, and, if so, this would imply that Venus has, or more likely had in the past, an active D″ layer.

Melting in Mantle Plumes

Introduction

One of the major questions about the production of flood basalts and oceanic plateaus is, How can such large volumes of magma be erupted in such short periods of time?

Four main models have been proposed in the literature, and they all have one thing in common: mantle upwelling beneath the lithosphere.

1. White and McKenzie (1989) proposed that short-lived, relatively voluminous eruptions of basalt are the result of lithospheric rifting of thermally thinned lithosphere.
2. A similar theory proposes that basalts result from melting in a mantle plume head that arrives beneath normal lithosphere (Campbell and Griffiths 1990; White and McKenzie 1995).
3. A recent variant of this idea involves an eclogite component in the plume head that contributes to the melting (Cordery et al. 1997).
4. A small but unrelenting group of investigators have proposed that flood basalts are produced by melting of hydrated subcontinental mantle lithosphere from heat supplied by a mantle plume (Gallagher and Hawkesworth 1992). We will now review each of these mechanisms of magma production.

Rift-Related Melting

If the lithosphere is tectonically or thermally thinned, the asthenosphere rises to fill the space, and, as the asthenosphere decompresses, it begins to melt (McKenzie 1984). Assuming that all of this melt escapes upward, White and McKenzie (1989) calculated the amount of decompression melt produced as a function of temperature and pressure using a parameterized model. They showed that if buoyant asthenosphere rises beneath thinned lithosphere, relatively small increases in temperature (100–200 °C) can lead to the production of enormous volumes of magma by decompression. An increase of only 100 °C above normal doubles the amount of melt produced, whereas a 200 °C increase may quadruple it.

As continental lithosphere is thinned by extension, the surface subsides to maintain isostatic equilibrium. In addition to extension, the following three effects, in order of decreasing importance, contribute to the subsidence: the amount of igneous rock added to the crust, the dynamic support of the mantle plume, and reduction in density of the residual mantle after melt extraction (White and McKenzie 1989). With all of these factors corrected for, the amount of subsidence at the time of rifting as a function of degree of stretching is shown in Figure 4.14. For normal mantle the amount of subsidence for a stretching factor of 5 is about 2.3 km (1280 °C), whereas for a mantle plume with a potential temperature 100 °C hotter than normal (1380 °C), only 500–1000 m of subsidence is predicted. If the potential temperature is 200 °C hotter than normal (1480 °C), the lithosphere is actually elevated by 500–1000 m. For lower amounts of stretching, these effects are reduced. Thus, it appears that a very small temperature increase in upwelling mantle (plume or asthenosphere) has a dramatic effect on the subsidence of rifted basins and continental margins. For typical lithospheric thicknesses of 100 km or less, temperature increases in upwelling mantle of 100–150 °C result in the surface remaining near sea level during rifting.

The model of White and McKenzie (1989) predicts the following sequence of events as asthenosphere rises beneath continental or oceanic lithosphere:

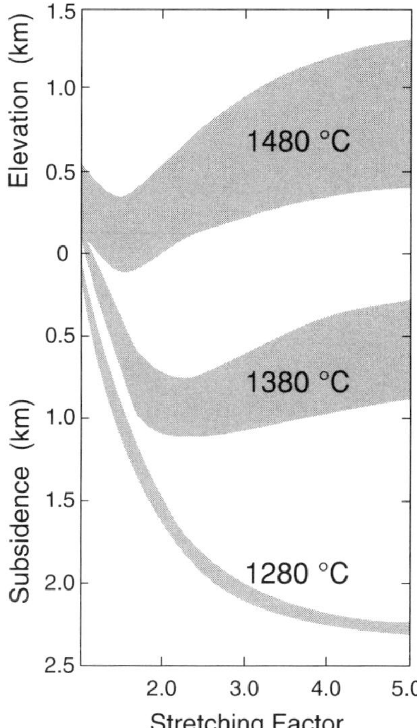

Figure 4.14. Amount of subsidence and surface elevation at the time of rifting as a function of the degree of stretching of the lithosphere. Curves are calculated for an initial lithosphere thickness of 118 km and for the ambient temperatures given. Data from White and McKenzie (1989).

1. The mantle upwelling produces a thermal anomaly 1000–2000 km in diameter with a potential temperature of 100–200 °C and a dynamic uplift on the order of 1–2 km.
2. If the lithosphere rifts across the top of the upwelling, the upwelling continues to rise and decompresses, creating a large volume of melt, which is in part intruded into the crust and in part erupted at the surface.
3. As melt is added to the crust and the mantle upwelling becomes depleted and more buoyant, surface uplift is sustained to maintain isostatic equilibrium.
4. In the case of continental lithosphere, after it has been stretched by a factor of about 5, it breaks, a new ocean spreading center forms, and large volumes of basalt are erupted as the upwelling mantle continues of decompress. The Rio Grande and the Walvis Ridges in the South Atlantic were formed in this manner.

It is important to see how much and at what rate magma can be produced in the White and McKenzie rifting model. Cordery et al. (1997) presented a numerical example of this model in which the asthenosphere is drawn upwards as the lithosphere progressively thins by rifting. In their model, buoyancy forces in the rising mantle and the motion of the overriding plate cause the asthenosphere to elongate and thin in the direction in which the rift is opening. Melting begins very abruptly at about 11 Myr after rifting begins and increases rapidly as the melting region broadens and thickens (Fig. 4.15(a)). A peak in melt production rate is reached in about 5 Myr followed by a gradual drop-off as the plume cools by thermal conduction. At the maximum magma production rate, about 35 km of new crust is produced. Although the volume of magma produced

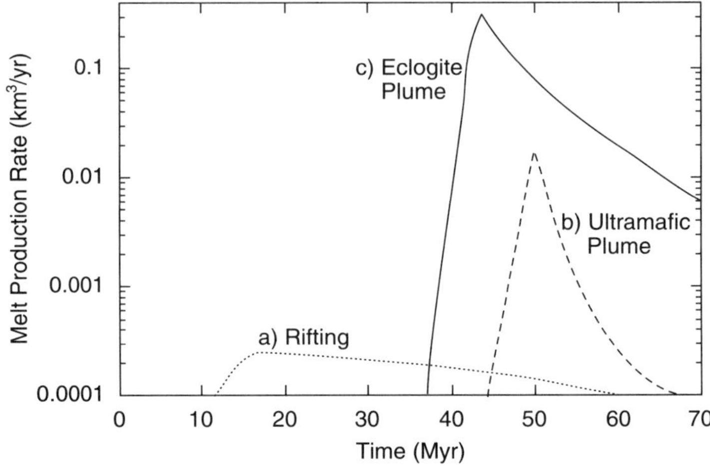

Figure 4.15. Melt production rate versus time for three mantle plume melting scenarios. Time zero is when the plume intersects the lithosphere. Note the large volume of magma produced rapidly in the plume models compared with the rifting model. After Cordery et al. (1997).

(4.5×10^6 km^3) is comparable to that found in LIPs, the total duration of magmatism is 40–50 Myr – much longer than the 2 Myr or less observed in most continental flood basalts. Hence, although the model does not favor rift-related melting for flood basalts, it may be important in understanding one or more of the phases of oceanic plateau volcanism, which commonly occurs over prolonged periods of 30–40 Myr. Leitch et al. (1998) have extended the rifting model to include rifting of lithosphere thinned over a plume head. Their results show that the thickness and width of crust produced from decompression melting in a plume head are very sensitive to the thermal structure beneath the rift. Their model explains magma production in the North Atlantic Igneous Province if the Iceland plume head is thin and strongly flattened. If there is a significant viscosity jump at the 660-km discontinuity, flattening of a plume head may occur when the plume passes through the discontinuity.

Hawkesworth et al. (2000) proposed a melting model in which large volumes of basaltic magma are produced by decompression melting of the mantle lithosphere as it is rifted apart. For a lithosphere stretching factor of four, a duration of melting of 10 Myr, and a lithosphere temperature of 1450 °C, 2 to 4 km of basaltic magma is produced from the lithosphere as it decompresses during rifting. This model is successfully applied to explain basalts that produced the Rio Grande Rise and Walvis Ridge during opening of the South Atlantic 135 Ma.

Melting in a Mantle Plume

Using a finite element analysis, Farnetani and Richards (1994) estimated melt volumes and compositions produced in a mantle plume that rises beneath normal (unrifted) lithosphere. In their models, significant volumes of melt are produced only in very high temperature plumes (1600 °C or less). As a plume rises, it looses heat around its edges, resulting in more diffuse boundary zones after 20 Myr. The zero velocity of the plume at the source tends to retard the plume's upward motion, resulting in the lower part of the plume's becoming elongated with time. As the plume passes into the uppermost mantle, it begins to neck because of the decrease in upper mantle viscosity, and after 40 Myr it hits the lithosphere and begins to spread and thin. After about 44 Myr, the plume begins to melt, reaching a maximum melt production rate at 50 Myr (Fig. 4.15(b)). After the peak, melt production rate falls off less rapidly than the onset of melting, leading to a long period of volcanism ($\approx$22 Myr), most of which occurs in the first 10 Myr. Despite the high plume temperature and long period of magma production, the total melt volume produced (0.14×10^6 km^3) is less than the volume of many flood basalts (Table 3.1). If the lithosphere were less stiff than assumed in this melting model, or if it fractured, the volume of magma reaching the surface could be greater.

Plumes with Eclogite

If oceanic lithosphere is recycled into the deep mantle, as suggested by seismic and isotopic data, a significant fraction of plume sources could be composed of eclogite, which was originally formed by inversion of mafic oceanic crust into eclogite in subduction zones. Cordery et al. (1997) and Leitch and Davies (2000) have shown that a mantle plume of the same potential temperature containing about 15% eclogite can produce a considerably larger volume of magma than in the previous case in which only ultramafic rocks melt. Because the solidus of eclogite is some 150–200 °C below that of ultramafic rock, this melting can also produce flood basalts without invoking unrealistically high plume temperatures. In the eclogite model, melting in the plume head extends to 250 km deep and 30–35% of the eclogite component melts. Unlike the ultramafic melting model, the mixed eclogite–ultramafic source generates about 5×10^6 km^3 of magma, which is comparable to the volume found in many flood basalts (Table 3.1). As with the ultramafic model, the onset of plume melting is rather rapid, beginning at about 38 Myr after it intersects the lithosphere and reaching peak melt production in 4–6 Myr (Fig. 4.15(c)). The duration of peak melting is 1–3 Myr, and the volumes of melt produced are similar to those observed in continental flood basalts (Leitch and Davies 2000). This is followed by a gradual decline over about 30 Myr, with the peak melt production rate (0.34 km^3/yr) more than a factor of 20 greater than in the ultramafic model. As with the ultramafic plume model, however, the eruptive history is too long compared with flood basalts and oceanic plateau volcanism. Reducing the upper mantle viscosity by a factor of 10, however, reduces the melting history to about 3 Myr and increases the melt production rate peak to 3.4 km^3/yr, bringing both values into the range observed in flood basalts (Cordery et al. 1997).

Other attractive features of the combined eclogite–ultramafic melting model are as follows:

1. It explains why flood basalts are iron- and silica-rich relative to ocean ridge basalts because of the eclogite contribution to the melt.
2. Large volumes of tholeiite can be explained without calling upon extensive fractional crystallization of picrites, which is required in the ultramafic-only model. Because they are a minor component in flood basalts and oceanic plateau basalts, picrites are not favored as parental to the voluminous tholeiites.
3. The model is consistent with enrichment in light-REE (rare earth elements) in flood basalts without calling upon unrealistically low amounts of melting.

This is because garnet (which concentrates heavy REE) is left in the restite for moderate amounts of melting of an eclogite-bearing source, but garnet is not in the restite during melting of a garnet lherzolite source.

4. The highly variable range in Nd, Pb, and Sr isotopic composition of flood basalts can be explained with variable melt contributions from eclogite in plume sources.

Lithosphere–Plume Interactions

Plume Erosion of the Lithosphere

Thinning of the lithosphere by mantle plumes has been a topic of much discussion and debate. Plumes heat the lithosphere in three ways: (1) by conduction, (2) by intrusion of magma, and (3) by horizontal extension (Davies 1994). As a mantle plume impinges on the base of the lithosphere, the bottom of the lithosphere will be heated and weakened and may be removed by shear stresses caused by spreading of the plume. The same thing may happen by advection of the asthenosphere. The coupled process of thermal softening and mechanical removal is referred to as thermomechanical erosion. Numerical models in which the plume–lithosphere boundary represents rapid temperature change and models of plume flow in the asthenosphere show that conduction is too slow to cause significant thinning of the lithosphere (Sleep 1994). However, both numerical and laboratory studies demonstrate that small-scale convective instabilities in plume heads are capable of thinning the lithosphere (Yuen and Fleitout 1985; Griffiths and Campbell 1991). Moore et al. (1999) have presented time-dependent, three-dimensional numerical models of mantle plumes impinging on the lithosphere in a temperature-dependent mantle. In these models, lithospheric thinning depends on the formation of convective instabilities in the plume–lithosphere boundary layer with horizontal sizes of the instabilities of a few tens of kilometers. The instabilities form when a plume's viscosity is about an order of magnitude less than the ambient mantle. With excess temperatures of 100–200 °C, efficient thinning of the lithosphere may occur from convective erosion of the instabilities.

Numerical models by Davies (1994) show that substantial thinning of the lithosphere can occur in a few million years. Plume tails can cause localized rapid thinning of the lithosphere because of their small diameter ($\leq$300 km) and their high temperatures (1500–1600 °C). For continental lithosphere that was originally less than 150 km thick, the base of the crust may be heated to the melting temperature within about 25 Myr over a plume tail. Although thinning rates over plume heads are much slower (on the order of 50–100 Myr), the occurrence of large volumes of felsic igneous rock associated with plume heads (such as the Yellowstone hotspot) implies that substantial thinning may also occur over plume heads – particularly if plate motion is slow. During the Archean, when the mantle was hotter, penetration of the lithosphere by plumes was much faster. With lithosphere less than 150-km thick, crustal melting will begin within 15 Myr of plume impingement. Even with thick lithosphere, such as that beneath Archean cratons today, significant thermal erosion of the lithosphere occurs after 10 Myr and crustal melting may begin after 30 Myr.

Dehydration Melting of the Lithosphere

Geochemical and isotopic data from flood basalts have been interpreted by some investigators to reflect sources in the lithosphere (Gallagher and Hawkesworth 1992). Such source regions must be old in order to develop enriched radiogenic isotope ratios, which is an observation used to support a lithospheric rather a mantle plume source. One of the major obstacles of a lithosphere source model for flood basalts is how to produce large volumes of magma in short periods of time in a dry lithosphere. As pointed out by Gallagher and Hawkesworth (1992), however, the addition of small amounts of water to ultramafic rocks appreciably lowers their solidus temperatures and promotes formation of such metamorphic phases as phlogopite and amphibole (Green 1973; Mysen and Boettcher 1975; Bodinier et al. 1990). Fertile mantle can accommodate up to about 0.4% water produced by the hydration of clinopyroxene to form amphibole. If 10–20% of the original phases are converted to amphibole and mica, considerable volumes of melt can be produced at relatively low temperatures and pressures. Also, at low water contents, orthopyroxene melts incongruently, producing residual olivine and a silica-saturated melt. The net result of such dehydration melting is the possibility of producing large volumes of silica-saturated tholeiite.

As an example, experimental data show that the hydrous lherzolite solidus is lowered by up to 500 °C compared with the dry solidus, depending on the depth of melting (Fig. 4.16). For a potential temperature of 1480 °C (200 °C hotter than normal) in the lower part of the lithosphere, which is not unrealistic if it is underlain by a mantle plume,

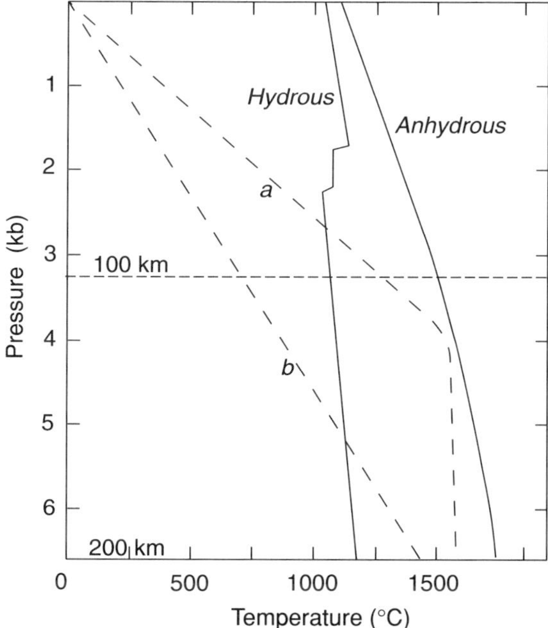

Figure 4.16. The solidus of anhydrous and hydrous mantle (solid lines) shown for two lithospheric thicknesses: (a) 100 km, and (b) 200 km. A potential temperature of 1480 °C is assumed for both geotherms (dashed lines). Note that the hydrous solidus is intersected by the geotherms in the lower part of the lithosphere in both cases. After Gallagher and Hawkesworth (1992).

30–50% melting should occur within 20–40 Myr (Gallagher and Hawkesworth 1992). The actual rate of magma extraction from the lithosphere depends on the degree of extension. In the melting model of Gallagher and Hawkesworth (1992), up to 5 km of melt is predicted for a lithosphere 200-km thick, and a maximum of 25% melting at the base of the lithosphere is anticipated. For more normal lithosphere thicknesses of 100–150 km, the model predicts 1 to 3 km of melt. If the melting is concentrated in a circular area over a plume head of 550 km, for a 2-km-thick melt, 2×10^6 km^3 of melt is produced. The greater the amount of extension of the lithosphere, the more plume-derived magma that enters the system. If extension factors exceed 1.2, plume-derived magmas rapidly dominate the system for a lithosphere 100 km or less thick. For lithosphere 200 km or more thick, however, lithosphere-derived magmas dominate the system until more than 200% thinning has been attained.

Depth of Melting

The relative depth and extent of melting in plumes can be estimated from geochemical data. For instance, with increasing depth of source in the mantle, FeOT (total Fe as FeO) increases and CaO/Al_2O_3 decreases, and with increasing degree of melting at a given pressure, both FeOT and CaO/Al_2O_3 increase (Fig. 4.17) (Lassiter and De Paolo 1997). Before this geochemical screen can be used to constrain depths of melting, the effects of fractional crystallization must be removed. To do this, major element contents are corrected to a given MgO content (say 8% MgO) using a linear regression on a plot of the major element oxide or ratio versus MgO. Figure 4.17 shows the distribution of hotspot and LIP basalts (average for each example) on a $(CaO/Al_2O_3)_8$ versus $FeOT_8$

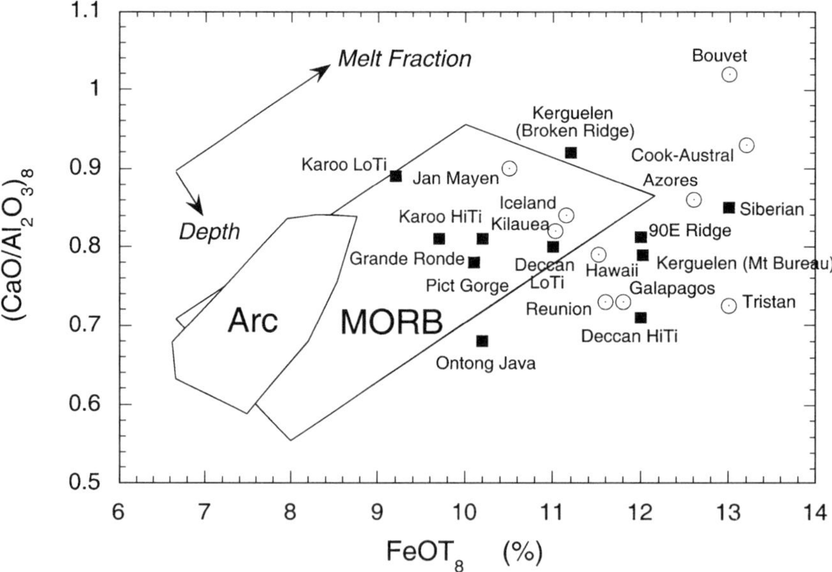

Figure 4.17. $(CaO/Al_2O_3)_8$ versus $(FeOT)_8$ graph showing the average compositions of basalts from oceanic islands (circles) and oceanic plateaus (solid squares). Note the relatively small melt fractions of arc basalts compared with MORB and plume-related basalts. The $(CaO/Al_2O_3)_8$ and $(FeOT)_8$ values are at MgO = 8% in each basaltic suite. In part after Lassiter and DePaolo (1997).

plot, where the subscript refers to correction at 8% MgO. Also shown is the field for the distribution of ocean-ridge (MORB) and arc basalts. MORB provide a reference point corresponding to basalts produced where the lithosphere is very thin (about zero), and thus the depth of melting is very shallow. Arc basalts differ from MORB in recording relatively small degrees of melting of the mantle source. If intraplate plume-derived basalts come from deeper sources than MORB, they should plot off the MORB array such that smaller melt fractions (lower $[CaO/Al_2O_3]_8$) are associated with greater depths and higher potential temperatures (higher $[FeOT]_8$).

Many hotspot islands and oceanic LIPs have average basalt compositions with higher $FeOT_8$ than MORB, thus supporting a plume source for their magmas (higher degree of melting than MORB) (Fig. 4.17). Others overlap the MORB field. Although LIP and island basalts overlap in CaO/Al_2O_3 –FeOT space, LIP basalts range to lower melt fractions than island basalts. This may reflect the relatively cooler plume-head sources for some LIPs compared with the hot plume-tail sources of most island hotspot basalts. Lavas from Bouvet and the Austral–Cook chain show the highest melt fractions yet relatively shallow depths of melting, and lavas from Réunion, Galápagos, Tristan, and Ontong Java show variable melt fractions and large melting depths. Basalts from Iceland, Hawaii, and LIPs such as Karoo and Columbia River overlap depths and melt fractions of MORB with relatively high degrees of melting. It is also noteworthy that low-Ti basalts from the same LIP plot at shallower depths than the associated high-Ti basalts, as evidenced by the Deccan and Karoo data (Fig. 4.17).

Stratigraphic changes in chemical composition of many flood basalts are consistent with increasing melt fraction and decreasing depth of melting. This is shown, for instance, by increases in CaO/Al_2O_3 and decreases in La/Yb ratios with decreasing stratigraphic height in the Siberian traps and in basalts from the North Atlantic Province (Kerr 1994; Lassiter and DePaolo 1997). The La/Yb ratio is particularly sensitive to the amount of garnet left in the restite, which in turn reflects the degree of melting of a garnet lherzolite source. High La/Yb ratios, as commonly found in early eruptive phases of flood basalts, reflect residual garnet (Yb is housed in garnet) and relatively small degrees of melting, whereas later eruptions have lower La/Yb ratios in which garnet was removed from the source by higher degrees of melting. Most investigators have interpreted these trends to reflect thinning of the lithosphere with time by thermal erosion caused by a mantle plume. This thinning permits a plume to rise higher and thus to undergo more decompression melting (Lassiter and DePaolo 1997). It may also result in partial melting of the subcontinental lithosphere. However, the inferred lithospheric thinning occurs over time scales too short to be explained by conductive heating of the lithosphere. It is likely that the thinning is caused by partial melting of hydrous domains in the lithosphere, thereby reducing its viscosity and enhancing the effects of thermal erosion by the plume head.

Magma Composition and Plume Melting

It is now widely agreed that melting in mantle plumes results from adiabatic decompression during their rise near the base of the lithosphere. A numerical plume melting model by Ribe and Christensen (1999) suggests that melting occurs at two depth levels

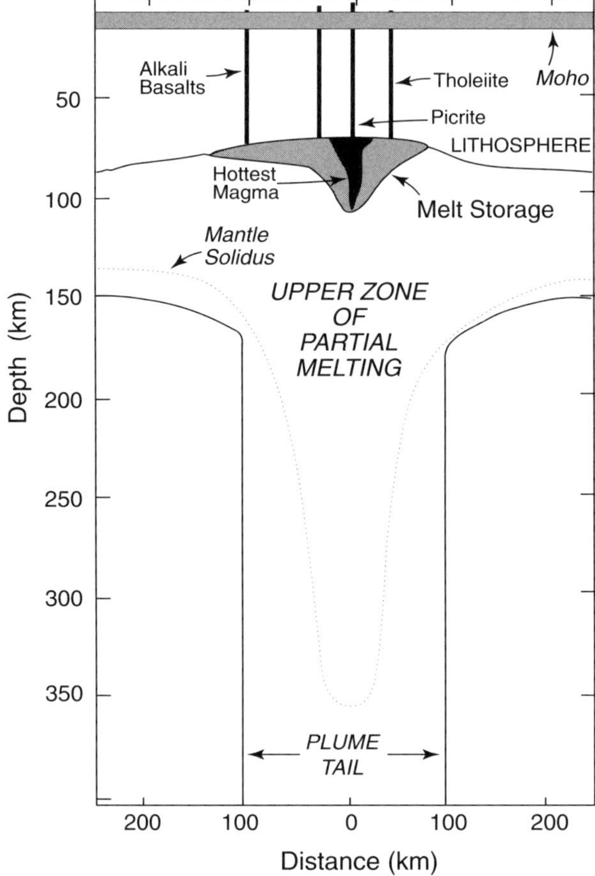

Figure 4.18. Sketch of magma production in an ascending mantle plume.

in a mantle plume: major melting occurs in a primary melting zone in the plume head or at a shallow level in the plume tail, and secondary melting occurs in the plume tail at a depth of 300–500 km with the two melting zones separated by a region of no melting. Melting begins in the hottest, central part of the plume, and the temperature continues to increase at a given depth as the plume continues to rise. These high temperatures lead to large degrees of melting, producing picritic and high-Mg tholeiite magmas (Fig. 4.18) (Wyllie 1988). Some of the picrite and tholeiite magmas pond at shallow depths, where they undergo fractional crystallization and produce more evolved tholeiites such as those found erupting at Kilauea today. Outward from the plume's center in the upper melting zone, the temperature falls off leading to a smaller degree of melting, and it is here that alkaline magmas are produced. At even smaller degrees of melting around the perimeter of a plume head, nephelinite magmas may be produced.

Most of the magma will collect in the central part of a plume and, upon eruption, will form shield volcanoes (small volumes of magma) or oceanic plateaus and flood basalts (large volumes of magma). In addition, if the lower part of the mantle lithosphere is sufficiently hydrated, it will also undergo partial melting. If the top of a plume is deflected by lithosphere motion, the late alkali volcanism can persist even after an island has migrated away from the parental hotspot, thus accounting for post-erosional nephelinites on some volcanic islands. Although this model is useful for many hotspot islands, it cannot be applied to all. For instance, the shield volcanic stage in some islands, such as Tahiti Nui and Raiatea in the Society chain, are composed of alkaline volcanics which suggests that these shield volcanoes formed after the islands moved off the center of the hotspot (Le Dez 1996).

The absence of thin, rifted lithosphere in most areas where major LIPs are formed does not favor a rift-related plume hypothesis for the production of LIPs. This leaves the mantle plume hypothesis as the most promising model for LIP production. The major remaining question is, How much, if any, eclogite is required in the plume source to avoid an unreasonable potential temperature of the plume? High-Mg picrites are common near the base of many flood basalts, and these magmas cannot be produced by melting of eclogites with lower MgO contents. Thus, picrites appear to be derived by partial melting of high-temperature axial zones within plumes (plume tails), which comprise hot mantle material carried from near the core–mantle interface. As illustrated in Figure 4.18, tholeiites come from the region surrounding the plume axis, and alkaline magmas, which record very small amounts of melting, come from the periphery of the plume. As shown by Cordery et al. (1997), a plume head containing about 15% eclogite can produce flood basalts (tholeiites) without calling upon unrealistically high plume temperatures, rifted lithosphere, or hydrated–metasomatized mantle lithosphere.

Do We Need More Plume Modeling?

The answer to this question is yes, we do. Although the quick and easy models have been introduced, as we learn more about the physical and chemical properties of the mantle, plume modeling must become more sophisticated and incorporate more of the results from geophysics and geochemistry. Although most investigators agree that superplumes come from great depth, perhaps the D″ layer at the base of mantle, what about smaller plumes and diapirs? Is there a sufficient thermal gradient at the 660-km discontinuity to generate plumes here? Clearly, we need more precise seismic tomography focused on the 660-km boundary to resolve this question. When the mantle was hotter in the Archean, did most mantle plumes come from the 660-km discontinuity? This has important consequences for cooling of the Earth in the Archean and in the relative volumes of LIPs erupted at that time. The effect of phase transitions on mantle plumes through time needs to be examined in greater detail for various mantle cooling scenarios. If plumes were more important in the mantle during the Archean, as some investigators propose, plumes obviously needed to penetrate the 660-km discontinuity or to transfer their heat somehow to a second set of plumes in the upper mantle. As discussed in Chapter 5, combined results from seismic tomography and isotopic geochemistry introduce a new idea: namely, that most mantle plumes come from intermediate depths (1000–1600 km), perhaps from a boundary zone or zones that dip at appreciable angles to the horizontal. This possibility needs to be tested rigorously by numerical modeling.

And finally, is there anything left to learn about decompression melting in plumes? Again, the answer is yes. We are still a long way from understanding if and when the lithosphere participates in melt generation in mantle plume heads. How rapidly are melts really extracted from plumes during decompression melting? Although geochronology suggests that some LIPs are erupted in less than 10^6 years, how much less? This question is important in terms of whether plume magma eruptions can appreciably affect the Earth's atmosphere and biosphere. Is it necessary to have eclogite in mantle plumes to explain large volumes of melting in short periods of time? This is an exciting possibility because geochemical mass balance models strongly suggest there is a hidden eclogite reservoir in the mantle that sequesters high-field-strength elements like Ta and Nb, which is a subject we will come back to in Chapter 5.

5

Plumes as Tracers of Mantle Processes

Introduction

Magmas derived from mantle plumes provide a powerful method to recognize and map compositional domains in the mantle, which in turn constrain mantle processes. The most definitive tracers are isotopic ratios of daughter elements of radioactive nuclides or concentration ratios of incompatible elements. Incompatible elements are transferred almost entirely to magmas upon melting in the mantle. For moderate or large degrees of melting, both isotopic and incompatible element ratios transfer from a mantle source to a derivative magma, providing a geochemical "signature" of the source.

Incompatible element distributions for basalts derived from four different mantle sources are shown in Figure 5.1 normalized to primitive mantle composition. Elements are arranged from most incompatible (during lherzolite melting) on the left (Rb) to least incompatible on the right (Y). Because incompatible elements are largely transferred from source to liquid, the element distributions in each basalt should reflect the element distributions in the respective mantle sources. Ocean-ridge basalts (MORB) record a depleted mantle source in which the most incompatible elements are most depleted. This source, which appears to reside in the upper mantle, is one of the restite reservoirs formed as continental crust has been extracted from the mantle over time (e.g., Hofmann 1997). Notice that oceanic plateau basalts (Ontong Java) record a mantle source somewhat less depleted in highly incompatible elements than MORB and that many oceanic island (OIB) and island arc basalts are derived from sources enriched in these elements.

Also characteristic of incompatible-element distribution diagrams are anomalies: elements that deviate from their neighbors. For instance, typical arc sources are strongly depleted in Nb and Ta and enriched in Sr, which is a feature also characteristic of continental crust. The Nb–Ta depletion originates in subduction zones where elements such as Th, K, Rb, and La are transferred in a fluid phase from the descending plate into the overlying mantle wedge, but Nb and Ta, being rather insoluble in the fluid phase, remain behind in the descending plate (McCulloch and Gamble 1991; Keppler 1996). This creates a "negative" Nb–Ta anomaly in the mantle wedge from which arc basalts are derived. Notice that average OIB has a positive Nb–Ta anomaly (Fig. 5.1).

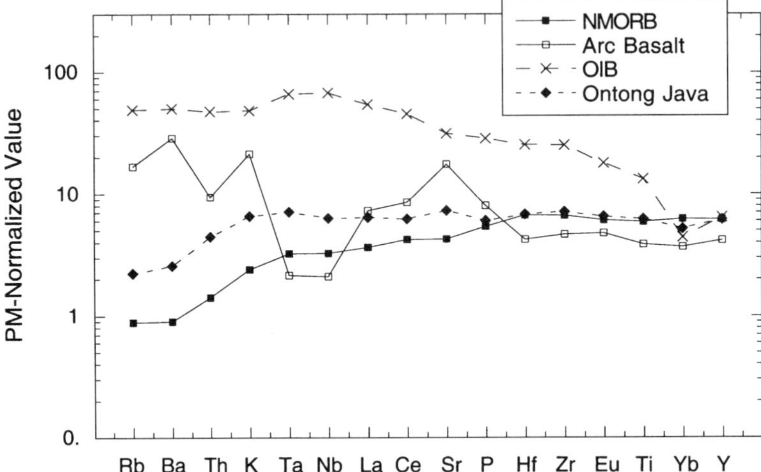

Figure 5.1. Incompatible element distributions in basalts from different tectonic settings. Data normalized to primitive mantle from Sun and McDonough (1989). NMORB = normal ocean-ridge basalt; OIB = oceanic island basalt. Data from Sun and McDonough (1989), McCulloch and Gamble (1991), and Mahoney et al. (1993).

As discussed later, this is generally interpreted to mean that the plume sources of OIB contain recycled oceanic crust, which is enriched in Nb–Ta owing to previous losses of neighboring elements in subduction zones.

Identifying Oceanic Mantle Components with Isotopic Tracers

An Overview

As shown by results from oceanic basalts, at least four, and perhaps as many as six, isotopic end members may exist in the oceanic mantle (Hart 1988; Hart et al. 1992). These are depleted mantle (DM), the source of normal ocean-ridge basalts (NMORB); HIMU, distinguished by a high $^{206}Pb/^{204}Pb$ ratio, which reflects a high U/Pb ratio ($\mu = {}^{238}U/^{204}Pb$) in the source; and two enriched sources (EM), which reflect long-term enrichment in light REE in the sources. The isotopic end member EM1 has relatively low $^{206}Pb/^{204}Pb$ and moderate $^{87}Sr/^{86}Sr$ ratios compared with EM2, which has intermediate $^{206}Pb/^{204}Pb$ and high $^{87}Sr/^{86}Sr$ ratios. Both EM components have high $^{207}Pb/^{204}Pb$ relative to $^{206}Pb/^{204}Pb$ compared with MORB. A possible fifth component, primitive mantle (PM) has been suggested by some investigators. However, when several isotopic systems are considered from the same suite of samples, conclusive evidence for a PM reservoir is not found. Another possible end member, FOZO or C, defined by isotopic mixing arrays in multidimensional space, is described more fully below.

The existence of at least four oceanic mantle end members is now well documented. What remains to be verified is the origin and location of each end member and their mixing relations. As discussed below, much progress has been made on these questions by also using rare gas isotopic data and trace element ratios. Also, the mixing hierarchy of components in basalts from single islands or island chains can provide useful information on the location of the components in the mantle.

Depleted Mantle

Depleted mantle (DM) is mantle that has undergone one or more periods of fractionation involving extraction of basaltic magmas. This type of mantle is known to underlie ocean ridges and probably extends beneath ocean basins, although it is not the source of oceanic island magmas. The "depleted" isotopic character (low $^{87}Sr/^{86}Sr$ and $^{206}Pb/^{204}Pb$ and high $^{143}Nd/^{144}Nd$) and low contents of Rb, U, and Th in NMORB require the existence in the Earth of a widespread depleted mantle reservoir. Rare gas isotopic compositions also suggest that this reservoir is highly depleted in rare gases compared with other mantle components. Although most of the geochemical variation within NMORB can be explained by magmatic processes such as fractional crystallization, variations in isotopic ratios demand that the depleted mantle reservoir be heterogeneous, at least on scales of 10^2–10^3 km. This heterogeneity may be caused by small amounts of mixing with enriched mantle reservoirs.

Hafnium isotopes provide a means of distinguishing the modern depleted reservoir (DM) from old depleted reservoirs in plume sources. In Iceland, for instance, two depleted reservoirs have been described by Kempton et al. (2000). Basalts from the Iceland plume define a sublinear array in ϵHf–ϵNd space, and the most depleted mantle components have the highest ϵHf and ϵNd values (Fig. 5.2). Although the shallow depleted mantle reservoir from which NMORB comes has relatively high ϵHf and ϵNd values, the depleted reservoirs recorded in MORB along the Mid-Atlantic ridge both north and south of the Iceland plume extend to even higher ϵHf values ($\geq$17). This feature requires long-term depletion in Lu in this reservoir (^{176}Lu is the parent isotope of ^{176}Hf) (Kempton et al. 2000). Unlike the NMORB reservoir, the high ϵHf reservoir must have been isolated from the convecting upper mantle for 2.5 or more Ga. The existence of this old depleted mantle reservoir, which has been identified in

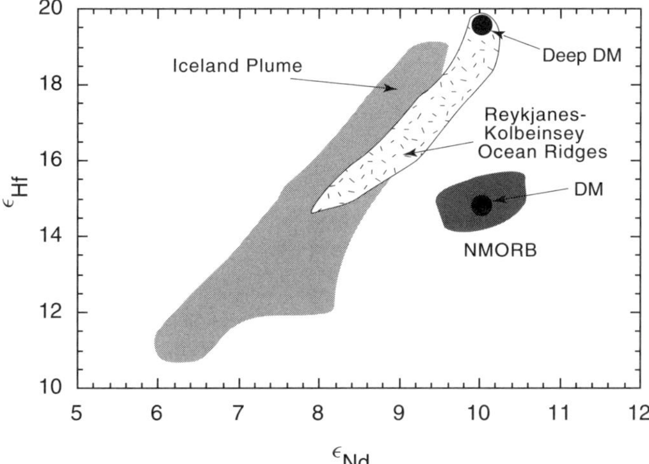

Figure 5.2. Epsilon Hf versus ϵ_{Nd} graph showing results from basalts derived from the Iceland plume and adjacent segments of the Mid-Atlantic Ridge (Reykjanes–Kolbeinsey Ridges). $\epsilon_{Hf} = (\{[^{176}Hf/^{177}Hf]/[^{176}Hf/^{177}Hf] \; -1) \times 10^4$; $\epsilon_{Nd} = (\{[^{143}Nd/\,^{144}Nd]_{sample}/[^{143}Nd/^{144}Nd]_{CHUR}\} -1) \times 10^4$. DM = shallow depleted mantle; Deep DM = deep depleted mantle. After Kempton et al. (2000).

several Atlantic plumes, means that a portion of the deep mantle was depleted in the Archean and has remained at least partially isolated ever since. It is possible that this depleted mantle reservoir is the complement to a large volume of continental crust that formed in the Late Archean, as described in Chapter 8.

However, not all investigators agree that a deep, old, depleted mantle reservoir is necessary to explain the compositions of basalts coming from mantle plumes. Hanan et al. (2000), for instance, have proposed that deep and shallow depleted mantle components can be distinguished using incompatible trace element distributions together with Hf and Pb isotopes. They suggest that Iceland basalt compositions can be explained by mixing of three mantle components: two deep components enriched in incompatible elements, one with relatively high and one with relatively low $^{206}Pb/^{204}Pb$ ratios, and a shallow depleted MORB source. The enriched radiogenic Pb source represents the Iceland plume and the depleted radiogenic Pb source may be an EM1 source with recycled subcontinental lithosphere. The model proposed by these investigators does not require a deep, depleted component in the Iceland plume as does the model of Kempton et al. (2000).

HIMU Mantle

The extreme enrichment in ^{206}Pb and ^{208}Pb in some oceanic island basalts requires the existence of a mantle source enriched in U+Th relative to Pb, and mantle isochrons suggest an age for this HIMU source on the order of 2.0–1.5 Ga. This reservoir is also enriched in radiogenic ^{187}Os (Hauri and Hart 1993). Because HIMU has $^{87}Sr/^{86}Sr$ ratios similar to NMORB, however, it has been suggested that it may represent subducted oceanic crust in which the [U+Th]/Pb ratio was increased by preferential loss of Pb in volatiles escaping upward from descending slabs during subduction. Supporting a recycled oceanic crust origin for HIMU are relative enrichments in Ta and Nb in many HIMU oceanic island basalts (Fig. 5.1). Devolatilized descending slabs should be relatively enriched in these elements because neighboring incompatible elements like Th, U, K, and Pb have been partially lost to the mantle wedge. Thus, the residual mafic part of the slab that sinks into the lower mantle and becomes incorporated in mantle plumes should be relatively enriched in Ta and Nb. Although recycled oceanic crust is widely agreed upon for HIMU, Kamber and Collerson (1999) have recently proposed that HIMU represents metasomatized upper mantle or oceanic lithosphere; on the basis of helium and Pb isotope distributions.

Enriched Mantle

Enriched mantle components are mantle reservoirs enriched in incompatible elements such as Rb, Sm, U, and Th compared with primitive mantle contents of these elements. At least two enriched components are required to explain the isotopic and trace element distributions in the sources of oceanic basalts (Zindler and Hart 1986): EM1 with moderate $^{87}Sr/^{86}Sr$ ratios and low $^{206}Pb/^{204}Pb$ ratios and EM2 with high $^{87}Sr/^{86}Sr$ ratios and moderate $^{206}Pb/^{204}Pb$ ratios. Both have low $^{143}Nd/^{144}Nd$ ratios. Among the numerous candidates proposed for EM1 are old oceanic mantle lithosphere ($\pm$ sediments) that has been recycled back into the mantle, metasomatized lower mantle, and recycled plume heads (Hart et al. 1992; Hauri et al. 1994; Kamber and Collerson 1999; Gasperini et al.

2000). Also, recent studies of Hf isotope distributions in Hawaiian basalts have been interpreted to indicate the presence of old pelagic sediments in their plume sources (Blichert-Toft et al. 1999). In addition, old continental lithospheric mantle, lower continental crust, or both have been proposed for the Indian Ocean and South Atlantic EM1 hotspot volcanics (e.g., Mahoney et al. 1992; Peate et al. 1999; Douglass et al. 1999). In these latter localities, low but variable $^{206}Pb/^{204}Pb$ ratios occur in the volcanics (Walvis ridge, SW Indian ridge, Marion hotspot), and a large range of $^{207}Pb/^{204}Pb$ and $^{208}Pb/^{204}Pb$ ratios are also found relative to the $^{206}Pb/^{204}Pb$ ratio (e.g., Mahoney et al. 1995; Douglass et al. 1999). These variations clearly indicate that EM1 can no longer be considered to represent one mantle component but must represent several components. All of these components must be depleted in the more highly incompatible element U relative to Pb on a time-integrated basis compared with the average depleted mantle from which NMORB comes, and they must have a time-integrated Rb/Sr similar to that of primitive mantle.

If there is a single EM2 component, it has isotopic ratios closer to average upper continental crust or to modern subducted continental sediments (i.e., $^{87}Sr/^{86}Sr > 0.71$ and $^{143}Nd/^{144}Nd \approx 0.5121$). Subducted continental sediments are favored by some investigators for this end member because EM2 commonly contributes to island arc volcanics in which continental sediments have been subducted, such as the Lesser Antilles and the Sunda arc (Hauri et al. 1994). However, based on helium and Pb isotope distributions in Hawaiian lavas, Kamber and Collerson (1999) proposed that EM2 is subcontinental lithosphere and that it has a variable composition.

Helium Isotopes

One of the most important observations in oceanic basalts is that their helium isotope ratios differ according to tectonic setting. There are two isotopes of helium: ^3He, which is a primordial isotope that was incorporated in the Earth as it accreted, and ^4He, an isotope produced by radioactive decay of U and Th isotopes. Plume-related basalts in oceanic areas have relatively high $^3He/^4He$ ratios – often more than 20 times that of air ($R/R_A \geq 20$), whereas MORB generally has R/R_A values of 7–9 (Hanan and Graham 1996). In some Pacific MORB basalts, R/R_A values increase with increasing $^{206}Pb/^{204}Pb$ ratios, whereas in the Atlantic the opposite trend is observed (Hanan and Graham 1996). Also, in the Atlantic and Indian Oceans, helium isotope ratios tend to be higher in basalts with EM1 characteristics. These high ratios, which apply to MORB and OIB, may be sampling two different EM1 reservoirs, both with low U/Pb and high Th/Pb ratios.

Two classes of models have been suggested to explain the origin of the high $^3He/^4He$ reservoir in the mantle. Some investigators interpret the high ratios to reflect recycled oceanic lithosphere in the deep mantle (Albarede 1998). Such lithosphere should have a high $^3He/^4He$ ratio because partial melting at ocean ridges extracts almost all of the U and Th from the mantle source. Because these elements are responsible for accumulation of ^4He over time, causing the $^3He/^4He$ ratio to decrease in the source, without U and Th the depleted oceanic lithosphere would acquire a high $^3He/^4He$ ratio. Alternative models call upon primitive, unfractionated sources deep in the mantle that still retain their original high $^3He/^4He$ ratios.

Basalts from the northern rift zone in Iceland show a well-defined gradient in helium isotope ratios increasing by a factor of 3 from the northernmost samples to central Iceland (Breddam et al. 2000). A circular region with high $^3He/^4He$ ratios is located over the southeastern part of Iceland, and it coincides with a regional gravity minimum and with the low seismic-velocity zone in the upper mantle previously discussed. Breddam et al. (2000) suggested the region of elevated $^3He/^4He$ ratios outlines the center of the Iceland plume and is coincident with the plume tail, which samples the deepest and relatively undegassed portion of the mantle source.

The Dupal Anomaly

With some important exceptions, most ocean islands showing EM-type mantle components occur in the Southern Hemisphere (Hart 1988). Also, NMORB in the South Atlantic and Indian Ocean exhibit definite contributions from EM sources, unlike NMORB from other ocean-ridge systems. This band of EM enrichment in the Southern Hemisphere is known as the Dupal anomaly (Fig. 5.3). Pb isotopic data indicate that the anomaly may have existed for billions of years, although not always as an intact geographic feature in the Southern Hemisphere (Hart 1984; Mahoney et al. 1998). With the exception of a few samples from the Arctic ridge, an intriguing feature of this anomaly is the apparent lack of any EM signatures in hotspot basalts from polar regions. Both shallow and deep origins have been proposed for the Dupal anomaly. Favoring a deep source is the correlation of the Dupal anomaly with body-wave anomalies in the mantle. Some of the Dupal maxima occur over the slowest seismic velocities in the

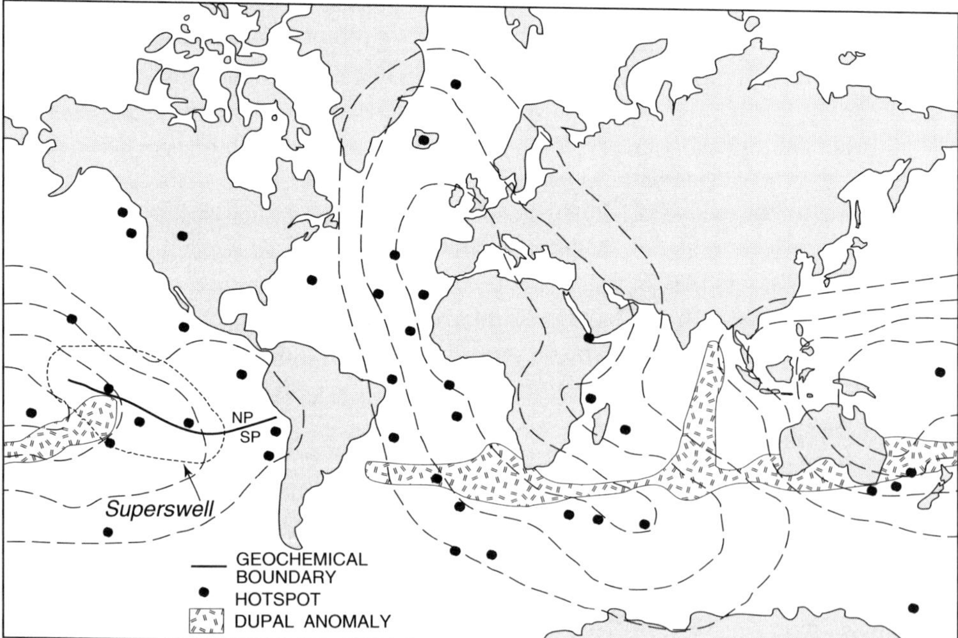

Figure 5.3. Map of modern hotspots showing the Dupal isotopic anomaly. Dashed lines are geoid anomalies. NP and SP, northern and southern isotopic provinces. From Vlastelic et al. (1999).

lower mantle, suggesting that the anomaly is an expression of material rising in plumes from the core–mantle boundary region (Fig. 5.3) (Hart 1988).

On the other hand, a shallow origin for Dupal is supported by EM components in South Atlantic and Indian Ocean NMORB and in island arc basalts in the Southwest Pacific (Mahoney et al. 1992). The best way out of this dilemma is for Dupal to have both deep and shallow sources. If EM2, which dominates the Dupal anomaly, is caused by modern continental sediments, the plume source could be from sediments that have been recycled through the deep mantle with sinking slabs. In contrast, the shallow source could be produced by subducted sediments that pass their geochemical signature to mantle wedges, and/or to the subcontinental lithosphere during plate devolatilization. Recent results from basalts along the Discovery section of the Mid-Atlantic Ridge (47°30′S) show a Dupal-type geochemical signature but with relatively low $^{206}Pb/^{204}Pb$ and $^{3}He/^{4}He$ ratios, which is a feature interpreted to represent recycled oceanic crust or delaminated subcontinental lithosphere (Sarda et al. 2000).

A recent reevaluation of Nd, Sr, and Pb isotope data from MORB in the Pacific basin shows the existence of two isotopic provinces with a boundary located at approximately 25°S along the East Pacific Rise (Vlastelic et al. 1999). The southern province, with slightly lower ϵ_{Nd} and $^{87}Sr/^{86}Sr$ values suggesting an important HIMU mantle source component, covers most of the South Pacific, including the Chile Rise south of latitude 25°S (Fig. 5.3; SP). The northern province (NP), which appears to sample enriched mantle components, includes part of the East Pacific Rise, Galápagos Ridge, and the Pacific basin north of this latitude. It is interesting that the two geochemical provinces approximately coincide with two large bathymetric provinces in the Pacific. The average depth to the axis of the East Pacific Rise in the northern province is about 2850 m, whereas in the southern province it is only 2450 m. Also, the geochemical boundary occurs near the northern end of the South Pacific superswell (centered at 10–30°S between longitudes 110 and 170°W), and extrapolation of the boundary approximately coincides with the axis of the South Pacific geoid anomaly.

Vlastelic et al. (1999) suggested two possible explanations for the geochemical boundary in the Pacific. The first is that mantle sources in the south associated with the superswell and geoid high, and hence with the mantle upwelling in the South Pacific, are contaminated with an HIMU component. In contrast, those to the north are contaminated with enriched (EM) mantle components. An alternative explanation is that small-scale heterogeneities in the mantle with EM characteristics are preferentially sampled by the small amounts of melting in the northern province with deeper ridge axes and cooler mantle. The higher extents of melting associated with the South Pacific upwelling include, in addition to the minor EM components, the HIMU component, which, being more refractory, requires higher temperatures to melt. One problem with the second explanation is that it is the low-degree, alkalic melts rather than high-degree melts that show the HIMU compositions in the South Pacific. Large variations are reported in the chemical and isotopic composition of basaltic glasses from oceanic crust in the eastern part of the Pacific superswell. These appear to result from convective mixing of an enriched mantle component upwelling near the center of the superswell with shallow depleted mantle (DM) (or varying degrees of melting of these two domains) (Janney et al. 2000).

Summary

By way of summary, perhaps the most significant chemical signatures recorded by the mantle are the complementary relationships between incompatible element enrichment in the continental crust, the probable siderophile element (Fe, Ni, Co, Cr, etc.) enrichment of the core, and the corresponding depletion of these elements in the depleted mantle reservoirs (Carlson 1994). It would seem that the key events controlling compositional variations in the Earth are early core formation followed by gradual or episodic extraction of continental crust to leave at least part of the mantle depleted in incompatible and siderophile elements. The shallow depleted mantle (DM) and the FOZO component (including a hidden eclogitic source?) may represent depleted mantle reservoirs remaining from these events. The other mantle components appear to reflect subducted inhomogeneities that were not sufficiently remixed into the convecting mantle to lose their geochemical and isotopic signatures. Many of these components, together with DM, appear to have been "fossilized" in the subcontinental lithosphere by accretion of spent plume material to the base of the lithosphere throughout geologic time (Menzies 1990). Geochemical evidence for layering in the mantle is equivocal and controversial. Although the data do not exclude chemical layering, the nonrandom occurrence of geochemical components in plume sources seems to require exchange between the upper and lower mantle.

Lithosphere and Crustal Contributions to Plumes

Introduction

Plume basalts erupted in oceanic areas pass through relatively thin lithosphere and crust with little or no interaction (Arndt and Christensen 1992). That continental flood basalts exhibit striking isotopic and geochemical differences from their oceanic counterparts, however, suggests that plume magmas beneath continents may be contaminated with subcontinental lithosphere, and/or continental crust, various investigators have suggested that some flood basalts are derived directly by melting of the mantle lithosphere from heat derived from plumes. The question of just how much interaction there is between plume-derived magmas and lithosphere or crust has proved very difficult to resolve. Because of the heterogeneity in both the subcontinental lithosphere and the continental crust and the complexity of processes involved in partial melting and assimilation, most isotopic and trace element studies have failed to identify various source components in flood basalts definitively. In the following sections we will review some of the major geochemical approaches used to constrain the number and types of source components in plume-related magmas.

Trace Elements

Overview

Primitive-mantle normalized spidergrams for oceanic island basalts (OIB) show broad enrichment in incompatible elements as a function of increasing incompatibility (Fig. 5.4). In addition, most have positive Nb–Ta anomalies. Although tholeiites from

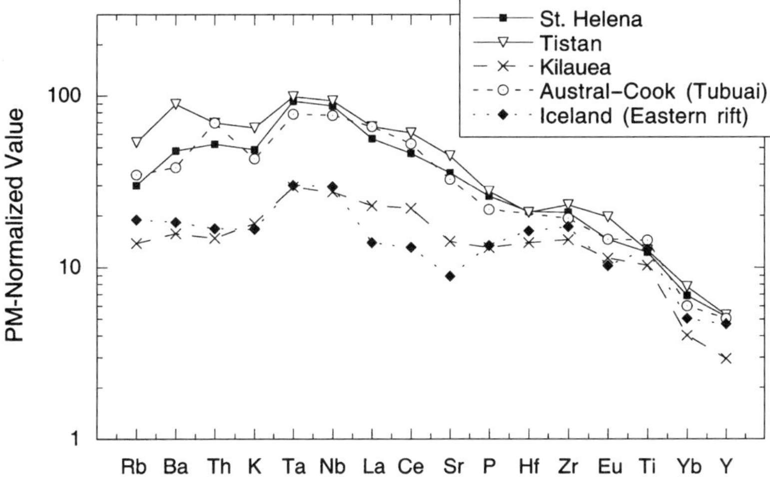

Figure 5.4. Incompatible element distributions in oceanic island basalts. Data from Budahn and Schmitt (1985), Weaver et al. (1987), Hemond et al. (1993), and Kogiso et al. (1997). Other information given in Figure 5.1.

Hawaii and Iceland have typical enriched patterns, their contents of the more incompatible elements are less than most other oceanic island basalts. It may not be a coincidence that these two locations have plumes with the highest buoyancy fluxes and probably the highest temperatures. This could lead to a greater degree of melting in the plumes, thus reducing the incompatible element contents of the melts without appreciably affecting the shape of the spidergrams, which reflect the composition of the sources.

Our database for oceanic plateaus is small, and basalts from two out of the three examples have incompatible element distributions similar to some MORB (Fig. 5.5). Ontong Java basalts show a slight depletion in the most incompatible elements. The differences between these patterns and those of the OIBs may reflect the differences

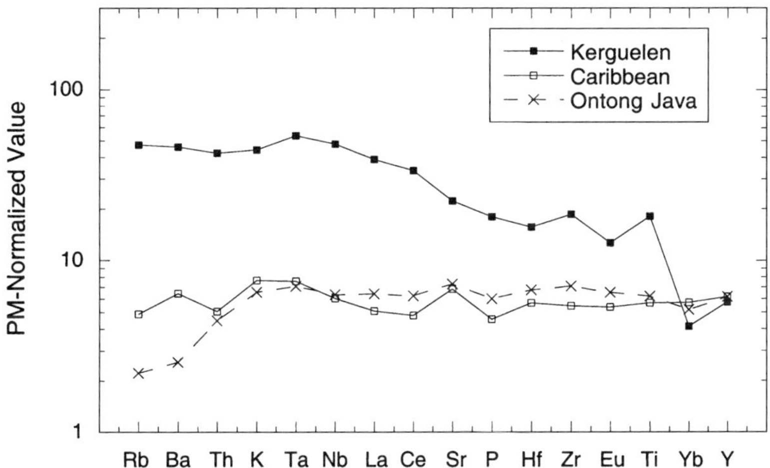

Figure 5.5. Incompatible element distributions in oceanic plateau basalts. Data from Neal et al. (1997), Yang et al. (1998), and Hauff et al. (2000). Other information given in Figure 5.1.

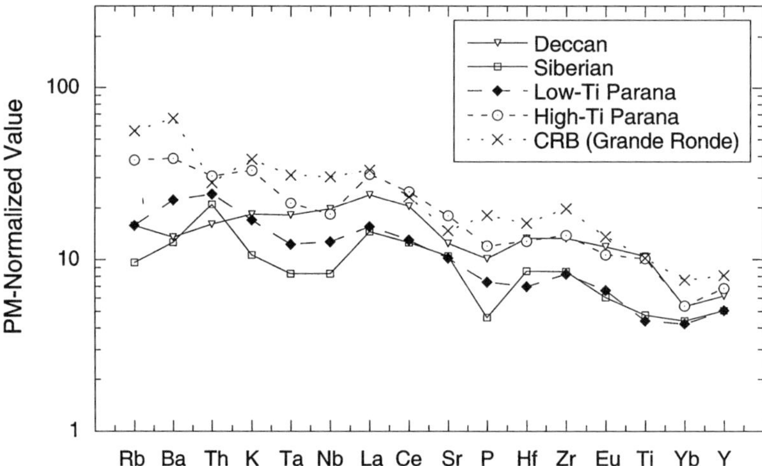

Figure 5.6. Incompatible element distributions in continental flood basalts. Data from Reidel (1983), Lightfoot et al. (1990), Peng et al. (1994), and Peate (1997). CRB, Columbia River basalt. Other information given in Figure 5.1.

in composition between plume tails (OIB) and plume heads (OPB). Plume heads, as discussed in the previous chapter, may be more susceptible to entrainment than tails, and if so, they could represent mixtures of deep mantle sources and shallow depleted mantle (DM). Such mixing could account for the MORB-like element patterns in Ontong Java and in the Caribbean plateau. Alternatively, the flat to depleted element patterns could reflect sources in the deep mantle from which the plumes are derived. Kerguelen basalts have distinct incompatible element distributions: they show enrichment in the most incompatible elements, small positive Nb–Ta anomalies, and low Yb and Y (Fig. 5.5). The first two characteristics are similar to OIBs, suggesting a similar mantle source. The low Yb and Y probably reflect garnet left in the source during partial melting, for garnet preferentially accepts these elements.

Although quite variable in detail, incompatible element distributions in flood basalts all show enrichment in the most incompatible elements. The Deccan basalts showing depletion in the most incompatible elements are an exception. In contrast to OIBs, most flood basalts also show small negative Nb–Ta anomalies (Fig. 5.6). As discussed below, the significance of these negative anomalies is a subject of disagreement: some investigators suggest that they result from crustal contamination, whereas others prefer a lithosphere source for flood basalts (or alternatively plume-derived basalts that are highly contaminated with lithosphere). As illustrated by the Paraná basalts, hi-Ti basalts are relatively enriched in incompatible elements compared with low-Ti basalts (Fig. 5.6).

Nb/U Ratios in the Mantle

The Nb/U (and Ta/U) ratios are similar in both OIB and MORB and average near 50, reflecting a similar ratio in depleted and enriched mantle reservoirs (Fig. 5.7). This relatively high ratio is different from that characteristic of primitive mantle (30), chondritic meteorites (32), and continental crust (10). This situation differs from other incompatible ratios such as Rb/K, Th/U, and La/Yb, which are all significantly lower

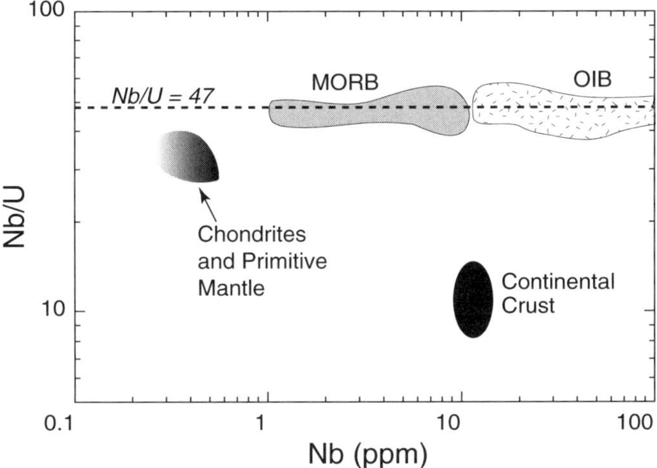

Figure 5.7. An Nb/U versus Nb graph for various basalts, chondritic meteorites, primitive mantle, and continental crust. After Hofmann (1986).

in MORB than in OIB. These lower ratios in MORB reflect a greater depletion in the depleted MORB source than in the OIB sources. Why, then, should the Nb/U ratio be different? If separation of continental crust from the mantle had left a variably depleted mantle, this region would continually mix to produce a mantle with homogeneous Nb/U ratios (Hofmann 1986; Hofmann et al. 1986). During the mixing, it is likely that the other incompatible element ratios also became homogenized.

Because OIBs do not show primitive mantle Nb/U ratios, they are not derived from primitive mantle sources. The uniform Nb/U ratios are also not consistent with the idea that OIB comes from a lower mantle source (via plumes) and that MORB comes from an isolated upper mantle source that was depleted by continental crust formation. Finally, for the same reasons, the incompatible element enrichment in OIB sources cannot result from recycled continental crust because this would significantly lower the Nb/U ratio in the OIB source. Hofmann et al. (1986) proposed a model whereby extraction of continental crust, most of which occurred at 2.7 and 1.9 Ga (Condie 1998), left a residue in the mantle with a high Nb/U ratio. Mixing of the mantle during or both during and after this massive crustal formation rehomogenized the mantle to a uniform Nb/U ratio. The depleted upper mantle source of MORB formed by partial melting of this uniform mantle at ocean ridges and production of new oceanic lithosphere, which is a process that fractionated K and Rb, Th and U, and La and Yb, but not Nb and U. The oceanic lithosphere is then recycled into the mantle at subduction zones without appreciably changing the Nb/U ratio. Hence, both the depleted upper mantle (the MORB source) and the deep plume source (recycled oceanic lithosphere) for OIB maintain the same Nb/U ratio.

Although this model has been widely accepted, there are two problems that must be solved if it is to survive. First, as oceanic crust dehydrates in subduction zones, U is a soluble element that is carried with other soluble elements into the mantle wedge. This leaves the residual oceanic crust with a high Nb/U ratio, and thus when recycled oceanic crust shows up in OIB plume sources, it should have a high Nb/U ratio, which it does not. A possible way around this problem is for U to be added back to the deep plume sources by recycling of continental sediments so as to maintain an approximately constant Nb/U ratio in the mantle (Plank and Langmuir 1998).

Another problem with a simple recycling model for deep mantle plume sources is that HIMU basalts have higher Nb/Zr ratios than other oceanic basalts. Because Nb and Zr are not fractionated from each other during subduction, the high Nb/Zr ratios in HIMU basalts are not expected. Kogiso et al. (1997) showed that the Nb/Zr fractionation cannot be explained by recycling of crust into the mantle or by partial melting processes in the upper mantle. Although perovskite can fractionate Nb during possible partial melting in the D″ layer above the core, it should also fractionate Nb from other incompatible elements like lead and thorium (Kato et al. 1988), and this is not observed in HIMU basalts. However, other high-pressure phases in D″ may fractionate only Nb (and Ta), a possibility that needs to be evaluated by ultra-high-pressure experimental work.

Th/Ta and La/Yb Ratios

Condie (1994) showed that a graph of Th/Ta versus La/Yb is useful in characterizing basalts from various tectonic settings. On the basis of data from young basalts, these ratios can also be used to characterize the mantle components in the previous section discussed above (Hart and Zindler 1989; Hart et al. 1992). Both Th/Ta and La/Yb ratios are near one in depleted mantle (DM), whereas the HIMU mantle component (mantle with high U/Pb ratios) has a high La/Yb ratio (ca. 18) and a low Th/Ta ratio (ca. 1.5) (Fig. 5.8). Enriched mantle components have similar La/Yb ratios of about 25 and Th/Ta ratios near 2 (EM1 = 2, EM 2 = 1.8). Although well defined on many isotopic plots, HIMU and enriched mantle components are not readily

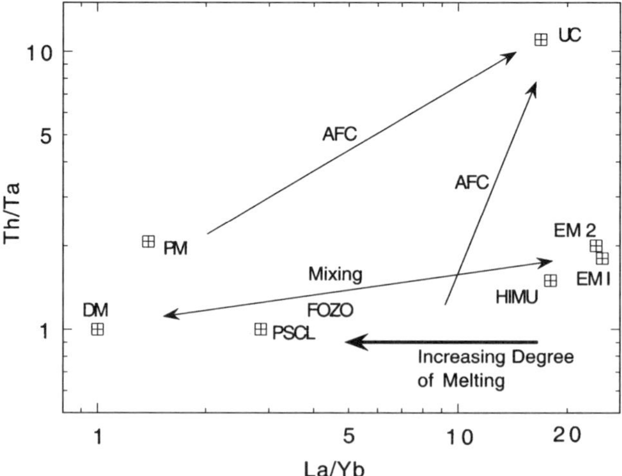

Figure 5.8. An Th/Ta versus La/Yb graph showing the distribution of mantle components (after Condie 1994). DM = depleted mantle; PM = primitive mantle; PSCL = post-Archean subcontinental lithosphere; FOZO = focal zone mantle; EM1 and EM2 = enriched mantle sources; HIMU = high-mu source; UC = upper continental crust; AFC = assimilation-fractional crystallization trajectory.

distinguished from each other on the Th/Ta–La/Yb plot. Sr-Nd-Pb isotopic data suggest the presence of a worldwide plume component common in oceanic island basalts referred to as FOZO that is located in the lower mantle, as discussed later in the chapter (Hart et al. 1992; Hauri et al. 1994). Although FOZO falls on or near a mixing line between DM and HIMU–EM in both isotopic space and on the Th/Ta–La/Yb diagram (Fig. 5.8), helium isotope data preclude its being a simple mixture of these end members (Hauri et al. 1994). Also shown on Figure 5.8 is the average composition of post-Archean subcontinental lithosphere as estimated from the average composition of spinel lherzolite xenoliths (PSCL) (McDonough 1990). That PSCL falls near the DM–EM mixing line is consistent with the possibility that the subcontinental lithosphere includes both DM and various enriched components (HIMU, EM1, EM2) as well as an unknown amount of FOZO, which also falls near the mixing line. Thus, the lithosphere may change in composition from place to place and should not necessarily be considered as a "pure" end member in modeling lithosphere contributions to magma sources.

One important limitation in using the Th/Ta–La/Yb diagram to constrain magma sources in the mantle is the fact that both of these ratios can be raised by assimilation-fractional crystallization and by decreasing degrees of melting in the mantle (McKenzie and O'Nions 1991). Model calculations suggest that assimilation-fractional crystallization (AFC) can raise Th/Ta and La/Yb ratios by up to a factor of 3 or 4 for assimilation of upper continental crust and up to a factor of 2 for lower continental crust (Condie 1997c). Two representative AFC trajectories are shown in Figure 5.8. The La/Yb ratio of a basaltic magma is also elevated if garnet is left in the melting residue. That some basalts with high La/Yb ratios have positive ∈Nd ratios suggests they come from relatively depleted mantle sources, and for these basalts, the La/Yb ratio is not indicative of the source La/Yb ratio but rather of a small degree of melting, leaving garnet in the restite (Lassiter and DePaolo 1997). With increasing degree of melting of garnet lherzolite, the La/Yb ratio decreases (Fig. 5.8), which, for tholeiitic magmas can change the ratio by up to a factor of two as the result of garnet left in the restite.

Oceanic island basalts show a wide range in La/Yb ratio (3–25), yet only a moderate range in Th/Ta (0.7–2) (Fig. 5.9). Consistent with the isotope data, Austral–Cook and Tristan basalts have a mixture of EM and HIMU components, and St. Helena may define a mixing line with HIMU at one end. Iceland (and perhaps Hawaii) has a depleted mantle component, although not necessarily the shallow component DM. The source of this component is not clear. It may not be DM entrained in the upper mantle but could be a heretofore unrecognized depleted component in the deep mantle (Kempton et al. 2000). Both Hawaii and Iceland may have a strong FOZO component in their plume sources (Fig. 5.9). The wide range in La/Yb, yet small range in Th/Ta in Iceland and Ascension basalts, probably reflects varying degrees of melting, leaving garnet in the restite. As expected, none of the OIBs shows contamination with continental crust. Unlike OIBs, oceanic plateau basalts (except for Kerguelen) overlap with NMORB, which suggests an important depleted component in their sources (Fig. 5.10). This may be a characteristic of plume heads that have entrained significant amounts of DM during their ascent, or alternatively, the depleted component may reside in the deep mantle at or near the plume sources.

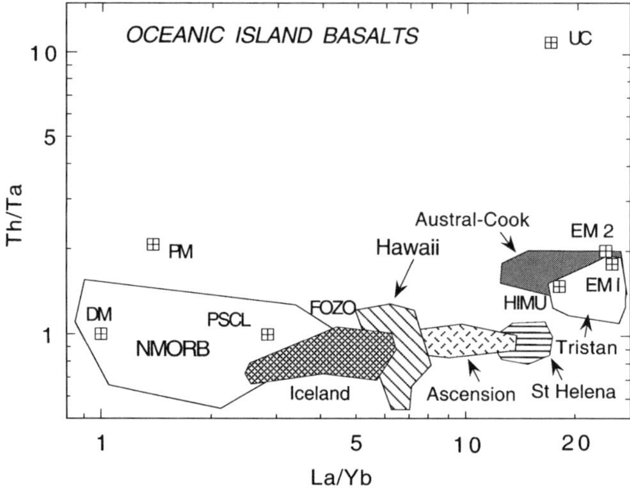

Figure 5.9. A Th/Ta versus La/Yb graph showing the distribution of oceanic island basalts. Data sources given in Figure 5.4. See Figure 5.8 for other information.

When considered together with major element and isotope distributions in continental flood basalts, the elevated Th/Ta and La/Yb ratios in these basalts must reflect, at least in part, smaller degrees of melting of their mantle sources than are characteristic of most plateau basalts (Fig. 5.11) (Lassiter and DePaolo 1997). These distributions are consistent with the lower temperatures in plume heads compared with plume tails feeding OIBs. As discussed later, some continental flood basalts are contaminated by upper continental crust, which is also consistent with their relatively high Th/Ta and La/Yb ratios. This may explain the unusually high ratios in basalts from the Kerguelen oceanic plateau (Fig. 5.10), thus supporting the idea that Kerguelen basalts were erupted on continental crust. Of the flood basalts shown, only the Paraná high-Ti group shows

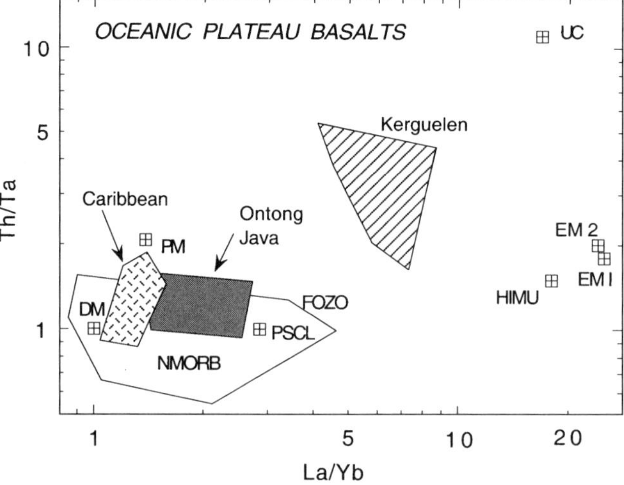

Figure 5.10. A Th/Ta versus La/Yb graph showing the distribution of oceanic plateau basalts. Data sources given in Figure 5.5. See Figure 5.8 for other information.

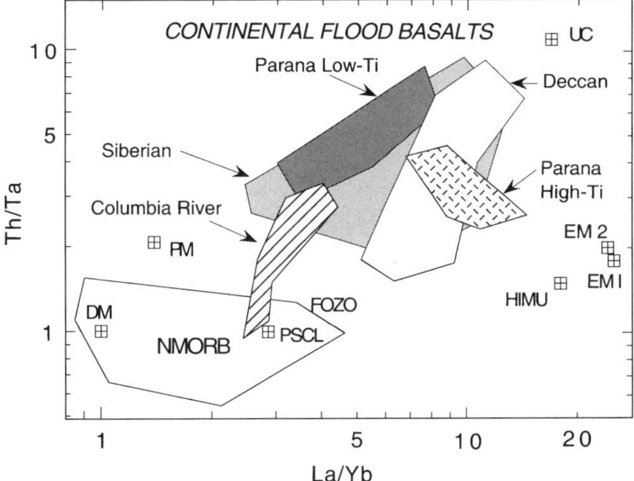

Figure 5.11. A Th/Ta versus La/Yb graph showing the distribution of continental flood basalts. Data sources given in Figure 5.6. See Figure 5.8 for other information.

any contribution from enriched or HIMU mantle sources, and only the Columbia River basalts show a depleted, possibly post-Archean subcontinental lithosphere (PSCL) component. It is also important to note in Figure 5.11 that most continental flood basalts cannot have a significant contribution from subcontinental lithosphere if the PSCL xenolith composition is representative of this lithosphere.

Nd and Sr Isotopes

Radiogenic isotopes provide a powerful means of tracking the source compositions of basalts. The growth rate of an isotopic ratio (such as $^{87}Sr/^{86}Sr$ and $^{143}Nd/^{144}Nd$) in a mantle source rock is proportional to the age of the rock and to the parent–daughter ratio. Nd isotopic ratios are generally expressed as ϵ_{Nd} values, which are a measure of the deviation of the $^{143}Nd/^{144}Nd$ in a given basalt from the primitive mantle (CHUR) value (see definition in caption to Fig. 5.12). When a mantle domain partially melts, if the daughter element is incompatible (as are Sr, Nd, and Pb), it carries the isotopic signature of the source into the melt and thus provides a very important tracer of the source composition and age. As illustrated in Figure 5.12 for the Nd and Sr isotopic systems, basalts from different tectonic settings carry very different isotopic signatures. In general, $^{143}Nd/^{144}Nd$ correlates negatively with $^{87}Sr/^{86}Sr$, which is predictable from the relative incompatibilities of the elements involved (Rb $<$ Sr and Sm $>$ Nd). Hence, mantle regions with low Rb/Sr ratios, such as the DM source of MORB, have low $^{87}Sr/^{86}Sr$ ratios and relatively high $^{143}Nd/^{144}Nd$ ratios (high ϵ_{Nd}). Continental crust, which is enriched in Rb relative to Sr and in Nd relative to Sm, shows the opposite isotopic signatures. The position of OIB–OPB sources between the depleted MORB source (DM) and continental crust (UC) is consistent with this source's being a mixture of these end members – perhaps resulting from recycling continental crust into the mantle (Hauri et al. 1994; Hofmann 1997). As discussed below, however, the situation is much more complicated.

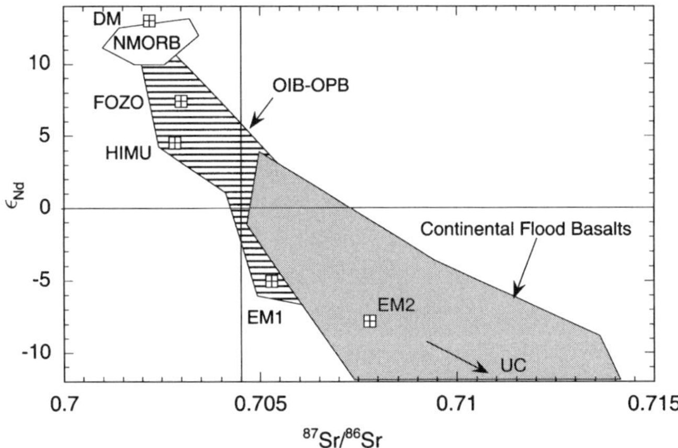

Figure 5.12. Epsilon Nd versus $^{87}Sr/^{86}Sr$ ratio for basalts from various tectonic settings. After Hofmann (1997). Also shown are mantle components: DM = depleted mantle; FOZO = focal zone mantle; EM1 and EM2 = enriched mantle sources; HIMU = high-mu source; UC = upper continental crust. OIB = oceanic island basalts; OPB = oceanic plateau basalts; NMORB = normal ocean-ridge basalt. $\epsilon_{Nd} = (\{[^{143}Nd/^{144}Nd]_{sample}/[^{143}Nd/^{144}Nd]_{CHUR}\} - 1) \times 10^4$. Upper Continental Crust: $\epsilon_{Nd} = -22$; $^{87}Sr/^{86}Sr = 0.725$.

Basalts from the Iceland and Hawaiian plumes have positive ϵ_{Nd} values, some of which overlap MORB distributions (Fig. 5.13). As with the Th/Ta–La/Yb ratios, this relationship again emphasizes a depleted component in the plume sources of these rocks. Whether this depleted component is shallow DM or a component in the deep mantle is still uncertain (Kempton et al. 2000). St. Helena is an example of basalts from an HIMU-dominated source, and Tristan basalts come from an EM1-type source (Fig. 5.13). Basalts from the Austral–Cook chain define an array extending from positive to negative ϵ_{Nd} values suggestive of a source with mixed depleted mantle and EM2 components. The Nd–Sr isotope relationships also allow Iceland, Ontong Java, and the

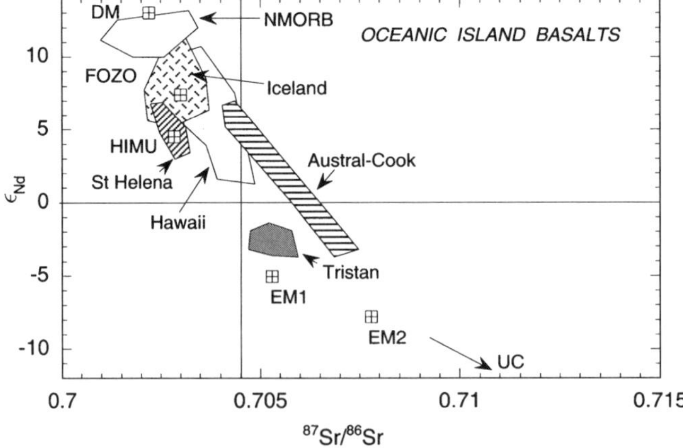

Figure 5.13. Epsilon Nd versus $^{87}Sr/^{86}Sr$ ratio for basalts from oceanic islands. References and data sources given in Figure 5.4. Other information given in Figure 5.12.

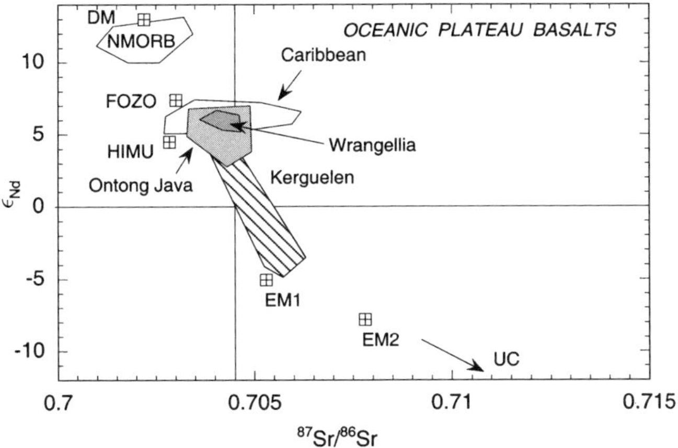

Figure 5.14. Epsilon Nd versus $^{87}Sr/^{86}Sr$ ratio for basalts from oceanic plateaus. References and data sources given in Figure 5.5. Other information given in Figure 5.12.

Caribbean plateau to have significant contributions of FOZO and perhaps HIMU in their sources (Figs. 5.13 and 5.14). The large range in $^{87}Sr/^{86}Sr$ ratios in Caribbean basalts may reflect interaction with seawater (Hauff et al. 2000). Kerguelen data, on the other hand, define a possible mixing trajectory between FOZO, or DM or both and an enriched source such as EM2 or continental crust.

Continental flood basalts have negative ϵ_{Nd} values and relatively high $^{87}Sr/^{86}Sr$ ratios, which are again suggestive of contamination with continental crust or subcontinental lithosphere (Hooper 1997) (Fig. 5.15). As exemplified by the Deccan and Siberian traps, many flood basalts show evidence for contamination by more than one enriched reservoir. In the Deccan case, it would appear that all magmas were contaminated by a common crustal component and that upper basalt flows in the province were contaminated again by another mafic source located perhaps in the mid-crust (Peng et al. 1994).

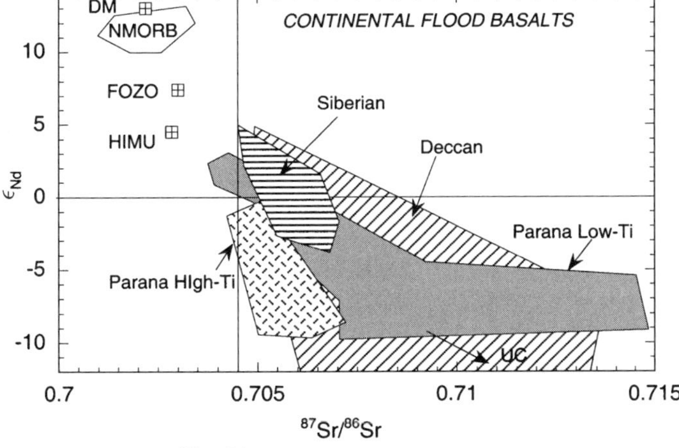

Figure 5.15. Epsilon Nd versus $^{87}Sr/^{86}Sr$ ratio for continental flood basalts. References and data sources given in Figure 5.6. Other information given in Figure 5.12.

High- and Low-Ti Basalts

A feature of many flood basalt provinces that is still not well understood is the existence of high- and low-Ti basaltic magmas. In at least two provinces, the Paraná and the Karoo, these magma types have distinct geographic distributions (Figs. 3.16 and 3.20). Although geochemical studies indicate the two basalt groups cannot be related by fractional crystallization, there is considerable debate as to whether they represent different degrees of melting of the same source or contamination of plume-derived magmas by the subcontinental lithosphere (Peate 1997; Sharma 1997; Hawkesworth et al. 2000). Fodor (1987) proposed that low-Ti magmas are produced near a plume axis by high degrees of melting and that high-Ti magmas are produced by low degrees of melting around the margin of the plume. Arndt et al. (1993) proposed a similar model in which the high-Ti group is produced in a plume beneath thick lithosphere and the low-Ti group beneath thin lithosphere. These melting models, however, cannot explain the isotopic differences between high- and low-Ti basalts (Fig. 5.15) (Peate 1997). Others have addressed this problem with models that involve different sources. Because the low-Ti group is not found in plume-related basalts from oceanic areas, some investigators have been led to suggest they come from the subcontinental lithosphere or that they are produced by crustal contamination (Hergt et al. 1991; Turner and Hawkesworth 1995).

The high- and low-Ti basalts from the Paraná province define two populations on most geochemical and isotopic diagrams (Figs. 5.11 and 5.15). In Th/Ta–La/Yb space, both groups seem to have a continental component, but neither can have much contribution from the subcontinental lithosphere if the mantle xenolith average (PSCL) shown in Figure 5.11 is representative of this lithosphere. The high-Ti group shows an inverse correlation of Th/Ta with La/Yb, perhaps reflecting input from enriched or HIMU mantle sources. The higher La/Yb ratio in the high-Ti group suggests garnet in the restite and thus favors a smaller degree of melting, as suggested by Sharma (1997) for the Siberian high-Ti basalts. On an ϵ_{Nd}- $^{87}Sr/^{86}Sr$ diagram, the low-Ti group again suggests a crustal contribution with high $^{87}Sr/^{86}Sr$ ratios and negative ϵ_{Nd} values (Fig. 5.15). The high-Ti group, although also allowing some crustal input, may have a significant EM1 component in its source. On the incompatible element spidergram, both groups have negative Nb–Ta anomalies suggestive of contamination by continental crust (Fig. 5.6). Hawkesworth et al. (2000) have proposed a model whereby both groups of Paraná–Etendeka basalts can be produced from the same mantle lithosphere source undergoing different amounts of extension and thus different degrees of decompression melting.

Although the origin of high- and low-Ti flood basalts is still not resolved, most trace element and isotopic data demand two sources for the two groups (Sharma 1997). High-Ti magmas appear to have been generated by low degrees of melting of garnet lherzolite, leaving garnet in the restite, whereas low-Ti magmas reflect a high degree of melting of a different mantle source in which garnet is not left in the restite. Although the high-Ti source is undoubtedly a plume head, the low-Ti source may be either a shallower level in a plume head with a different composition or a component in the subcontinental lithosphere. The distinct geographic distribution of the two groups supports the latter possibility. In both cases, crustal contamination is necessary to explain high Th/Ta ratios, and, in the low-Ti basalts, also the high $^{87}Sr/^{86}Sr$ ratios.

Oxygen Isotopes

Although few data are available from plume-related basalts, oxygen isotope distributions offer a potentially important tool to distinguish mantle from crustal components in the sources. This is because ^{18}O is greatly enriched in the continental crust compared with the mantle. As reflected by fresh NMORB glasses, DM has a $\delta^{18}O$ value of about 5.7‰, which is similar to the value for uncontaminated oceanic plume sources of about 6‰ (Mattley, Lowery, and Macpherson 1994). Some continental flood basalts, such as the Columbia River and Paraná basalts, have $\delta^{18}O$ values significantly higher than mantle values, which suggests they have been contaminated by continental crust (Carlson 1984; Harris et al. 1990). Phenocrysts of plagioclase from some of the lower basalts in the Deccan traps have $\delta^{18}O$ values up to 7.4‰, which is a finding again suggestive of contamination by continental crust (Peng et al. 1994).

Osmium Isotopes

The Re–Os isotopic system has proved to be one of the most useful in sorting out the various source components in basalts and especially in distinguishing between plume, lithosphere, and crustal components (Schiano et al. 1997). During mantle melting, Os is highly compatible, whereas Re is moderately incompatible. This is in contrast to other isotopic systems in which both parent and daughter elements are incompatible during melting. The differences in Re and Os compatibility result in continental crust having a high Re/Os ratio relative to the upper mantle, and hence, as the crust ages, it develops very high $^{187}Os/^{188}Os$ ratios (generally expressed as γ_{Os} as defined in Fig. 4.9). Subduction of continental crust (generally as sediments) would thus impart distinctively high γ_{Os} values to part of the mantle. That oceanic island basalts have variable, but generally high γ_{Os} values suggests that their sources contain variable amounts of recycled crustal materials (Fig. 5.16) (Pelgram and Allegre 1992). In contrast to continental crust, studies of xenoliths of spinel lherzolite that come from the subcontinental lithosphere have shown that lithosphere has low γ_{Os} values (≈ -10) owing to the long-term depletion in Re (Carlson and Irving 1994). Hence, contamination of crustal and subcontinental lithosphere components will have opposite effects on Os isotopes. On an -$\epsilon_{Nd} - \gamma_{Os}$ diagram, mixing lines for these two components deviate in opposite directions from a plume end member (Fig. 5.16).

Only a relatively small number of flood basalt provinces have been analyzed for osmium isotopes. Samples with the most positive ϵ_{Nd} values are assumed to reflect the plume component, as represented by some of the picrites from the West Greenland flood basalts (Schaefer et al. 2000). As seen if Figure 5.16, data suggest that basalts from the Siberian and Karoo provinces are contaminated with up to 20% of subcontinental lithosphere (Ellam et al. 1992; Horan et al. 1995). A few Siberian flood basalts also show evidence of crustal contamination. In terms of volume, much of the Siberian province

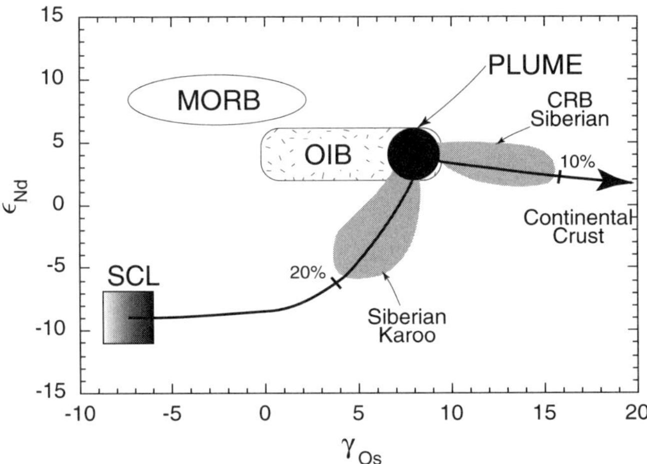

Figure 5.16. Epsilon Nd versus γ_{Os} diagram showing flood basalts in relation to various terrestrial reservoirs. After Ellam et al. (1992) and Horan et al. (1995). SLC = subcontinental lithosphere; MORB = ocean-ridge basalts; OIB = oceanic island basalts; CRB = Columbia River basalts. Heavy black lines are mixing trajectories with percentage contamination by SCL and continental crust noted. Arrow points to continental crust.

is dominated by plume-derived magma, and not until relatively late in the history of the eruptions did the lithosphere become an important source component. In contrast to Siberian and Karoo basalts, the highly radiogenic Os isotope signature characteristic of some of the Columbia River basalts ($> +10$) indicates 2–10% contamination by continental crust with no contribution from the lithosphere (Fig. 5.16) (Chesley and Ruiz 1998).

Summary

Three lines of evidence have been used to favor a lithospheric source for flood basalts. First, low-Ti basalts, which are an important component in many flood basalts, are unknown in oceanic island basalts. This observation has been interpreted by some investigators to indicate that the source of the low-Ti basalts is the subcontinental mantle lithosphere. Second, the geographic distribution of high- and low-Ti basalts within flood basalt provinces in Gondwana continents is consistent with the idea that high-Ti basalts come from plume heads and low-Ti basalts from lithosphere around the margins of plume heads (Gibson et al. 1995). And lastly, Os and Nd isotopic studies as discussed above, suggest mixing of two mantle end members: A γ_{Os}–ϵ_{Nd} end member interpreted to be a mantle plume, and a low γ_{Os}–ϵ_{Nd} end member interpreted as subcontinental lithosphere (Ellam et al. 1992).

The major problem with a dominant lithospheric source for plume-related basalts is production of enormous volumes of magma in short periods of time. In particular, the long time scale required for conduction of heat from a plume into overlying lithosphere precludes this lithosphere as a major source for preextensional volcanism (Arndt and Christensen 1992; Arndt et al. 1993). Although the hydrous melting model described previously may provide a means of producing large volumes of magma, available evidence suggests the subcontinental lithosphere is low in water (Cordery et al. 1997). If this were not the case, pyroclastics rather than flows should dominate in flood basalt eruptions. This conclusion is also supported by the low water content of the chilled margins of picrite dykes that are feeders to Karoo basalts. Yet another rather robust argument against a dominantly lithospheric magma source comes from xenoliths of subcontinental lithosphere. These xenoliths are composed principally of anhydrous phases. And finally, lithospheric mantle sources cannot explain the large volumes of mafic igneous rock erupted and intruded in oceanic plateaus. The oceanic lithosphere upon which these plateaus are built is anhydrous and relatively free from metasomatic components.

Although both mantle plume and lithosphere sources are still viable candidates for flood basalts, most geochemical and isotopic evidence supports a dominant plume source, allowing for variable, but generally small, contributions from subcontinental lithosphere and continental crust.

Mixing in the Mantle

When viewed in Nd–Sr–Pb isotopic space, arrays defined by various oceanic islands tend to cluster along two- and three-component mixing lines with DM, HIMU, EM1, and EM2 at the corners of a tetrahedron (Fig. 5.17). There is, however, a notable lack of mixing arrays joining EM1, EM2, and HIMU, which seems to rule out random mixing in the mantle (Hart et al. 1992; Hauri et al. 1994). The mixing arrays are systematically oriented, originating from points along the EM1–HIMU and EM2–HIMU joins and converging in a region within the lower part of the tetrahedron, but not at DM. This convergence suggests the existence of another mantle component that appears in oceanic mantle plumes. This component, which is called FOZO (or C) after focal zone from the converging arrays, is a depleted mantle component (Hart et al. 1992) (Fig. 5.17).

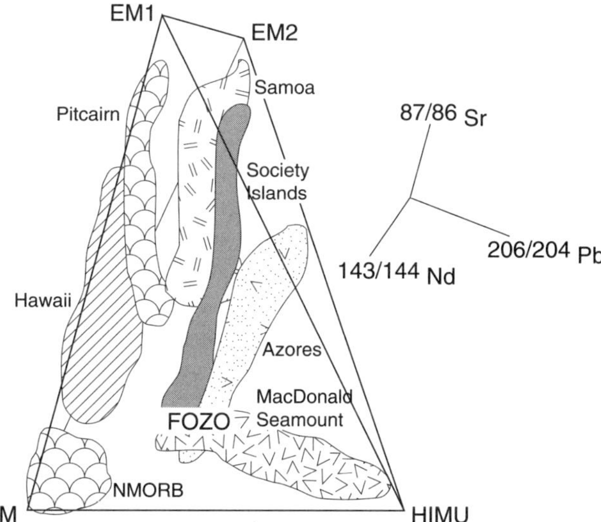

Figure 5.17. The mantle tetrahedron showing the distribution of oceanic basalts in Sr–Nd–Pb isotopic space. See Figure 5.12 for abbreviations. Modified after Hauri et al. (1994).

Helium isotope ratios also progressively increase towards FOZO, and because ^{3}He is not formed by radiogenic decay but was incorporated in the Earth during planetary accretion, the high ^{3}He/^{4}He ratios are commonly interpreted to mean that the FOZO component is relatively primitive and probably resides deep in the mantle. Alternatively, as mentioned above, the high ^{3}He/^{4}He ratios in FOZO may represent a recycled oceanic lithosphere in this source.

The absence of any mixing arrays emanating from the EM1–EM2 join or from the EM1–EM2–HIMU face of the tetrahedron indicates that there was some process that caused juxtaposition of EM components and HIMU that predates the mixing with FOZO. The probability that most or all hotspot sources do not mix with the shallow, depleted mantle DM is consistent with most plume entrainment occurring in the lower mantle.

The isotopic mixing arrays do not tell us which components occur in the plume sources and which have been entrained in the mantle. A high ^{3}He/^{4}He ratio in FOZO in contrast to a low ^{3}He/^{4}He ratio in DM may indicate that FOZO is in the lower mantle and does not mix with DM in the upper mantle. It is not possible from isotopic data alone to determine if the EM–HIMU components are the plume sources in the D″ layer and FOZO is entrained in the lower mantle as the plumes rise, or the other way around (FOZO is the plume and EM–HIMU are entrained). When coupled with geophysical data discussed in Chapters 1 and 4, however, the former model seems most acceptable.

Although the interpretation that linear arrays in isotope space represent mixing arrays is widely accepted, it has not gone unchallenged. Morgan (1999) proposed that these arrays are magma extraction trajectories produced by progressive melting of a heterogeneous mantle source characteristic of a particular hotspot. He has suggested that melt extraction trajectories form when some plume sources differ in the amount of previous melt extraction from other plumes. Magma modeling shows that sequential melts along a melt extraction trajectory trace a one-dimensional path through isotope space independent of the number of distinct isotopic components in the initial source mixture. Later partial mixing of the melts along a melt extraction trajectory will tend to preserve the linear array.

Although isotopic, rare gas, and trace element distributions in oceanic basalts demand a heterogeneous mantle, the nature, scale, preservation, and history of mantle heterogeneities remain problematic. Experimental studies on diffusion rates of cations in mantle minerals provide a basis to constrain the minimum scale of mantle heterogeneities. Results suggest that, even in the presence of a melt in which diffusion rates are high, heterogeneities larger than 1 km can persist for several billion years (Zindler and Hart 1986; Kellogg 1992). Geochemical heterogeneities in ultramafic portions of ophiolites occur on scales of centimeters to kilometers, which indicates that small-scale heterogeneities survive in the upper mantle. Geochemical and isotopic variations in basalts from a single volcano require heterogeneity on the scale of several kilometers, and variations in MORB along ocean ridges or of arc basalts along single volcanic arcs demand heterogeneities on scales of 10^2–10^3 km. At the large end of the scale is the Dupal anomaly (Fig. 5.3), which requires global-scale anomalies that have survived for billions of years. Numerical stirring models show that, if the viscosity of the lower mantle is 1–2 orders of magnitude greater than the upper mantle, mantle heterogeneities

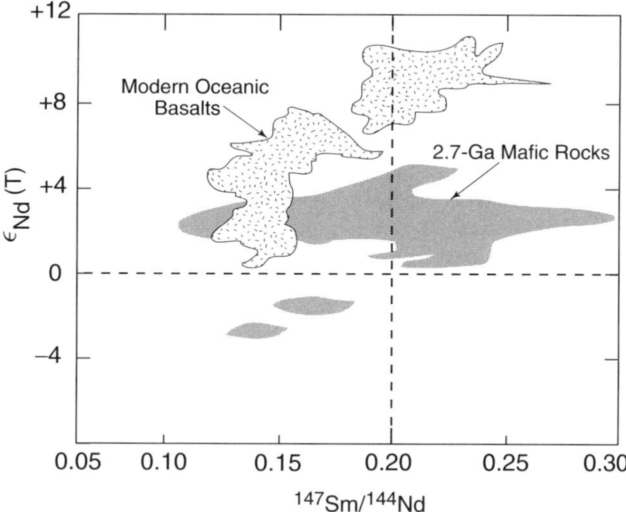

Figure 5.18. Distribution of Late Archean and modern basalts on an ϵ_{Nd} (T) versus $^{147}Sm/^{144}Nd$ graph. After Blichert-Toft and Albarede (1994). *T* (age) = 0 Ma for modern basalts and 2.7 Ga for Archean rocks.

should survive for up to 2 Gyr as suggested by isotopic data from oceanic island basalts (Davies and Richards 1992). These models, furthermore, allow heterogeneities of a variety of scales to survive for long periods.

Because the mantle was hotter in the Archean, and hence probably convected more rapidly and chaotically, it is of interest to see if mantle heterogeneities were preserved in the Archean mantle. Nd isotopic ratios and Sm/Nd ratios provide a means of comparing the Archean mantle with the modern mantle by analyzing for these quantities in greenstone basalts. The mantle array on an ϵ_{Nd}–$^{147}Sm/^{144}Nd$ plot varies significantly from the modern arrays (Fig. 5.18). For a similar overall range in Sm/Nd ratio, the Archean array varies by only one epsilon unit compared with at least 8 epsilon units in modern arrays (Blichert-Toft and Albarede 1994). These results demonstrate that long-term depletion and enriched mantle domains had developed in the mantle by the Late Archean. The absence of a larger range in ϵ_{Nd} values in the Archean mantle must mean that heterogeneities with variable Sm/Nd ratios did not survive long enough to be recorded in the $^{143}Nd/^{144}Nd$ isotopic ratio. This, in turn, suggests faster mixing in the Archean mantle and is thus consistent with hard turbulence (see Chapter 4) and a higher rate of plume generation.

New Ideas on Mantle Convection

One of the long-standing controversies in Earth sciences is the nature of mantle convection. Perhaps the greatest challenge in understanding mantle convection is to compile, evaluate, and reconcile the wide range of observations and constraints from many disciplines, and especially from geochemistry and geophysics. The isotopic and geochemical data discussed in this chapter, as well as heat production considerations, suggest the existence of distinct mantle reservoirs that have retained their identity for at least 2 Gyr. A major reservoir boundary is commonly placed at the 660-km discontinuity (Hofmann

1997). The noble gas record in plume-derived basalts, which requires storage of primordial gases in the deep mantle, suggests that the deep mantle has not been extensively degassed, which is a feature difficult to reconcile with whole-mantle convection. Heat flow arguments also do not favor whole-mantle convection. The sum of the heat produced by continental crust and depleted mantle extending all the way to the core–mantle boundary accounts for only a small fraction of the present-day heat flow at the Earth's surface (Albarede 1998). This seems to require a high heat-producing reservoir in the lower mantle, which again is difficult to account for if the entire mantle is convecting.

On the other hand, seismic tomography indicates that most or all descending plates penetrate the 660-km discontinuity and sink to, or close to, the core–mantle boundary (van der Hilst et al. 1997). Numerical computer models furthermore suggest that it is difficult to maintain an impermeable boundary at the 660-km discontinuity or any place else in the mantle. Models indicate that differences between the upper and lower mantle should disappear in a few hundred million years as long as plates continue to penetrate the boundary (Albarede and van der Hilst 1999). Also, high-pressure mineral physics experiments together with seismic observations indicate that the Clapeyron slope of the 660-km phase transition is too small to cause long-term stratification of the mantle. In addition, no convincing evidence exists for a thermal or compositional boundary at the 660-km discontinuity, of which one or both would develop if the upper and lower mantle convected separately.

Can any model accommodate all of these apparently conflicting observations? The recent model of Kellogg et al. (1999) modified by Albarede and van der Hilst (1999) seems to hold promise. In this model, the mantle to a depth of approximately 2000 km is relatively uniform in major element composition and largely represents a depleted and outgassed reservoir (Fig. 5.19). A transition zone at a depth of 400–1000 km divides the

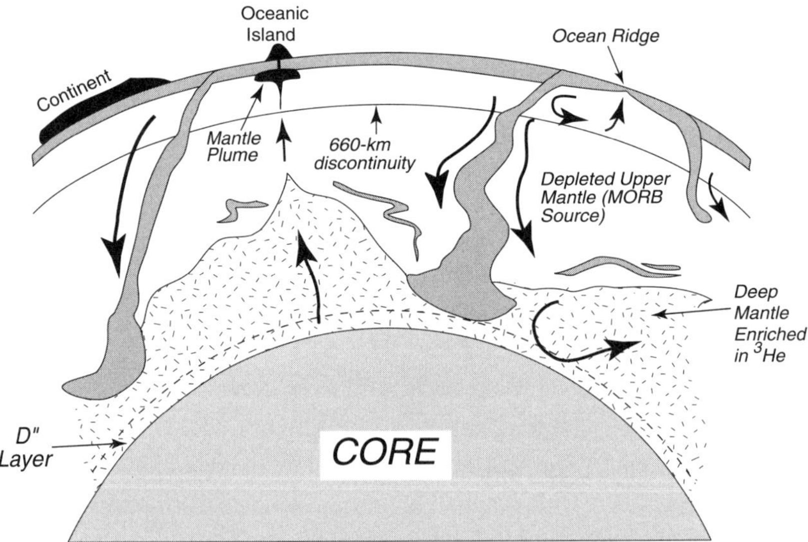

Figure 5.19. Convection model for the Earth involving a thick zone deep in the mantle that is convectively isolated from the middle and upper mantle. Mantle plumes may be produced in the D″ layer at the core–mantle interface and at the upper boundary of the deep mantle layer at local high spots. Modified after Kellogg et al. (1999).

mantle into two regimes with different time scales of mixing: (1) the upper mantle (DM), which is well mixed, is the source of MORB, and (2) the lower mantle, which mixes at a much slower rate. The bottom layer of the mantle, ranging from about 1600 km deep to near the core–mantle interface, has a very high density and contains undegassed material enriched in radiogenic heat-producing elements. Mantle plumes that give rise to oceanic island and oceanic plateau basalts come from the upper boundary of this deep domain rather than from the D″ layer, as assumed in most plume models. It is along this boundary where descending plates accumulate (Fig. 5.19). The plates are highly depleted in such incompatible elements as K, Rb, Ba, Th, and U owing to the subduction process. Numerical models of Kellogg et al. (1999) show that if the lower mantle layer is only 4% more dense than the overlying mantle, it will be dynamically stable and will develop substantial topographic relief, as shown in Figure 5.19. Although such relief strongly reduces the ability to identify this interface by seismic tomographic methods, recent numerical modeling of these data supports the existence of an isolated lower mantle (van der Hilst and Karason 1999; Tackley 2000). The lower mantle layer is thin or absent beneath descending plates and thick far from these plates. Although some mantle plumes may arise from D″, most are generated at the thermal boundary zone at the top of the lower mantle. This region, which contains fragments of older crust and lithosphere mixed with lower mantle, imparts the distinctive EM and HIMU geochemical signatures to some rising plumes and their derivative basalts. If there is a depleted component in the deep mantle as suggested by Kempton et al. (2000), this component may also reside in this dense lower-mantle layer.

Whether this model will survive the test of rapidly accumulating new data remains to be seen. However, it appears that any model for mantle convection must be consistent with the existence of a compositionally distinct and gravitationally stabilized lower mantle.

6

Mantle Plumes and Continental Growth

Introduction

Oceanic plateaus and large aseismic ridges are difficult to subduct (Vogt et al. 1976; Nur and Ben-Avraham 1982; Burke et al. 1978; Burke 1988); therefore, they may be accreted to continents (Abbott and Mooney 1995; Saunders et al. 1996; Kerr et al. 1997a). Some investigators have suggested they represent a major component of continental growth (Abbott 1996; Condie 1997c). Buoyancy models by Cloos (1993) indicate that oceanic plateaus up to about 17-km thick are easily subductable, although they may be partially obducted during collision.

When oceanic plateaus and aseismic ridges (hereafter collectively referred to as oceanic plateaus) reach about 30 km in thickness, they become too buoyant to subduct (Cloos 1993; Abbott 1996). Those formed during the last 200 Myr comprise a significant volume of crust in oceanic basins (3.7×10^6 km^3) (Schubert and Sandwell 1989), and if these were all accreted to the continents, the total volume of continental crust would increase by about 5%. Although several remnants of oceanic plateaus accreted to continental crust have been described, there is a curious sparsity of large accreted oceanic plateaus. This is especially puzzling because many relatively large ocean plateaus exist in the oceans today, and these should collide and at least partially accrete to the continents in the future. Of the several reasons for a paucity of oceanic plateaus in the geologic record, Saunders et al. (1996) favor recycling of oceanic plateaus into the mantle at subduction zones. This is due, in part, to the inversion of mafic plateau crust into dense eclogite, which readily sinks into the mantle. However, oceanic plateaus are presumably underlain by depleted, buoyant lithosphere, which should tend to keep them from subducting even with some eclogite in their roots. As an alternative explanation for the paucity of accreted oceanic plateaus, is it possible they have been overlooked in the geologic record? Accreted plateaus may be rapidly transformed into continental crust, making it difficult to recognize definitive mantle-plume geochemical signatures. Perhaps we should look to greenstones for remnants of oceanic plateaus. Greenstones, which are supracrustal successions containing an abundance of mafic volcanics, range in age from Early Archean to Cenozoic and appear to have formed in different tectonic settings, although oceanic settings dominate (Condie 1994). It may be possible to track the role of oceanic plateaus in the growth of continents using greenstones that are remnants of oceanic plateaus.

In this chapter we have two objectives. First, we will review examples of oceanic plateaus that, at least in part, have been accreted to the continents in the last 100 Myr. Second, we will discuss the possible role that mantle plumes may have played in the growth of continents, by magmatic underplating and by accretion of oceanic plateaus.

Accreted Oceanic Plateaus

Fragments of oceanic plateaus, often as obducted slices, have been recognized in terranes accreted to the American Cordillera and to eastern Asia during the Mesozoic (Isozaki et al. 1990; Condie and Chomiak 1996). In this section we will review some of the tectonic and geochemical features of these accreted oceanic plateaus. These results provide convincing evidence that large fragments of oceanic plateaus can be accreted to continental margins.

Caribbean Oceanic Plateau

Tectonic Overview

One of the most widely studied accreted oceanic plateaus is the Caribbean plateau, slices of which are exposed in Colombia and on various islands in the Caribbean (Fig. 6.1). The water-covered portion of the plateau, which formed between 90 and 75 Ma, has an area of about 6×10^5 km^2, but the original plateau may have been more than twice this size. As discussed in Chapter 3, seismic data indicate that the plateau varies in thickness from 8 to about 20 km. Most of the igneous rocks in the Caribbean plateau formed during the initial plume-head stage of a mantle plume in the Pacific, believed by some to be the Galápagos plume west of Ecuador. Since the Jurassic, the plateau has moved northeastward between North and South America as the Farallon plate subducted. Because the Caribbean plateau was too buoyant to subduct, it clogged the eastern trench, which led to a flip in subduction direction from east to west, and the Lesser Antilles arc formed along the eastern margin of the plateau (Fig. 3.5). The northern part of the Caribbean plate is bounded by a series of east-west trending transverse faults and the southern margin is a broad complex of convergence and right-lateral faulting (Fig. 6.1). As the Caribbean plateau collided with South America, Cuba, and Hispaniola, slices were obducted onto the continent and these are partially exposed today (Lapierre et al. 1997; Kerr et al. 1999).

As discussed in Chapter 2, at shallow exposure levels the Caribbean plateau is composed chiefly of pillowed and massive basalt flows with associated sills. Picritic lavas are locally important and komatiites (12–24% MgO) occur on Gorgona Island west of Colombia. Published $^{40}Ar/^{39}Ar$ ages from Caribbean plateau basalts in Haiti, Curacao, Gorgona, Beata Ridge, and Costa Rica show three peaks in eruption age: 90–88 Ma, 76–75 Ma, and a small peak at 55 Ma (Revillon et al. 2000). The distribution of ages suggests that the 90–88-Ma peak is volumetrically most important as it is found

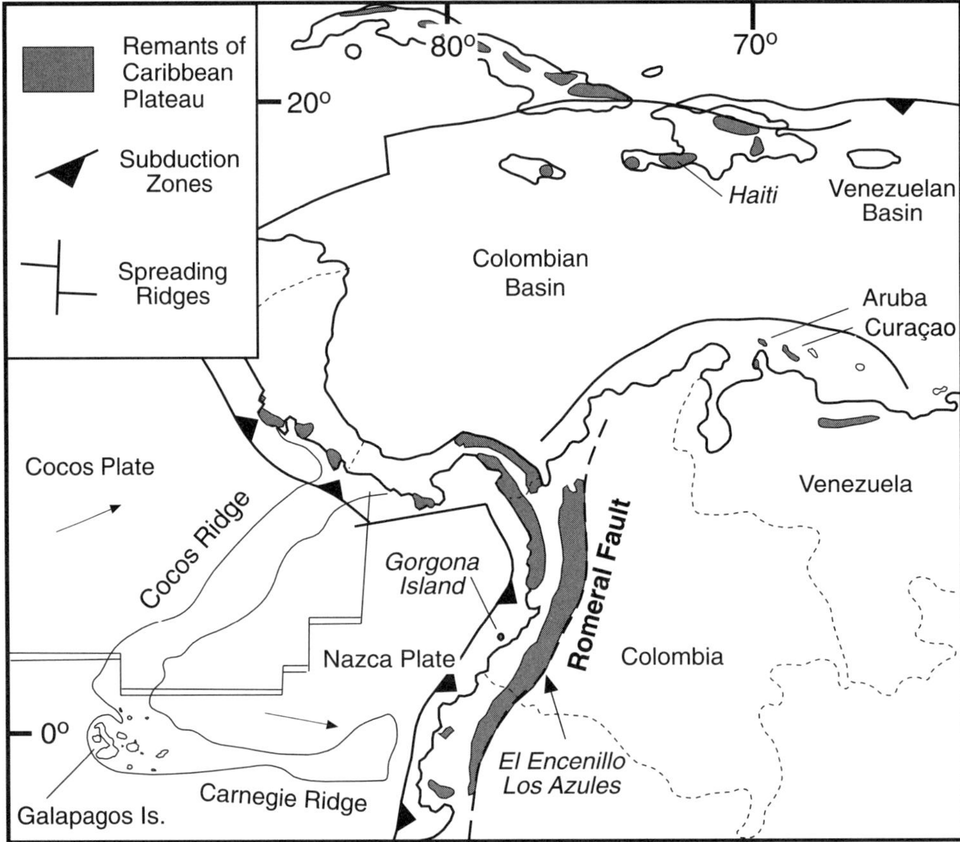

Figure 6.1. Map of the Caribbean region showing major outcrops of the Caribbean oceanic plateau. After Kerr et al. (1998).

throughout the Caribbean region as well as in Colombia and Gorgona Island. The ~75-Ma peak is also widespread, but the 55-Ma peak is recognized only on the Beata Ridge south of Hispanoila.

Coastal Ecuador also includes fragments of one or more Cretaceous oceanic plateaus (Reynaud et al. 1999). Basalts in these fragments are overlain by sediments of up to 98 Ma, an eruptive age greater than that of the Caribbean plateau. Furthermore, geochemical differences such as lower Th/Ta ratios indicate a mantle plume source different from the Caribbean source. It appears that the plume that generated the Ecuador plateau basalts must have been located in the Southeast Pacific – far south of the Galapagos plume that generated the Caribbean oceanic plateau.

Mantle Sources

Basalts from the Caribbean plateau exhibit the typical low Th/Ta ratios of other oceanic plateau basalts (Fig. 6.2). In addition, they have rather low La/Yb ratios and high ϵ_{Nd} values suggestive of a depleted mantle component in the plume source (Kerr et al. 1997b). It is also important to note that igneous rocks of all three age groups in the Caribbean plateau share the same geochemical and isotopic characteristics, indicating a similar

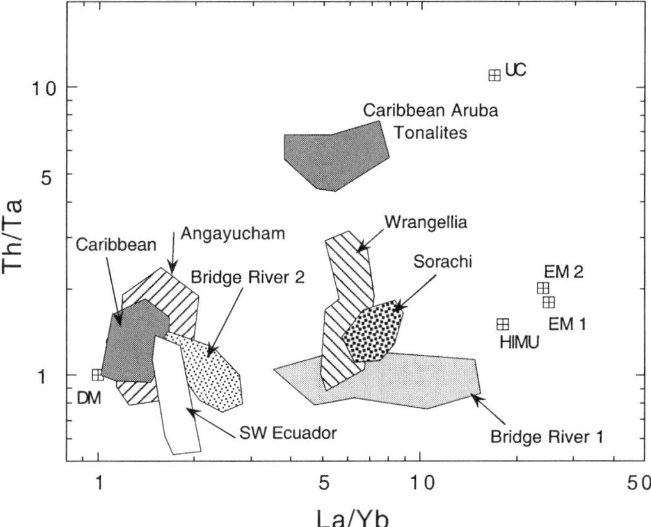

Figure 6.2. A Th/Ta versus La/Yb graph showing the distribution of basalts from accreted oceanic plateaus. Data sources: Pallister et al. (1989), Dostal and Church (1994), Barker et al. (1994), Kimura et al. (1994), Lassiter et al. (1995), Kerr et al. (1996a,b), White et al. (1999), and Reynaud et al. (1999). See Figure 5.8 for other information.

source. One of the most peculiar features of the chemistry of Caribbean lavas is the large range of $^{87}Sr/^{86}Sr$ ratios (Fig. 5.14). Most, but not all, of the relatively high Sr isotope ratios result from assimilation of altered oceanic crust, which acquired its high Sr isotope ratios from reaction with seawater (Revillon et al. 2000). Linear correlations between radiogenic isotopes (Nd, Sr, and Pb) and highly incompatible elements indicate mixing of two or more components in the mantle source (Hauff et al. 2000). Lavas enriched in incompatible elements and in radiogenic daughters may represent recycled oceanic crust, whereas the depleted lavas come from depleted mantle – perhaps entrained in the deep mantle by the plume that gave rise to the Caribbean plateau. High $^{3}He/^{4}He$ ratios in Caribbean picritic lavas from Costa Rica reflect a deep mantle component with either primitive or depleted mantle geochemical characteristics.

An important and intriguing question is how magmas with similar geochemical and isotopic characteristics in the same geographic locations in the Caribbean plateau formed in two or three pulses over 35 Myr. Possible explanations include the following (Revillon et al. 2000).

1. The two or three magmatic pulses reflect three separate plumes from the same mantle source, each of which formed independently and arrived at the same spot beneath the Caribbean plate. The chances of this happening seem rather remote.
2. The magmatic episodes reflect a "pulsating" plume in which a single plume hits the bottom of the lithosphere repeatedly and episodically, or for some reason, magma production in a plume head is episodic.
3. Melting for the second and third volcanic episodes reflects extension of the lithosphere unrelated to the mantle plume.

Both the second and third possibilities are viable alternatives for the Caribbean plateau.

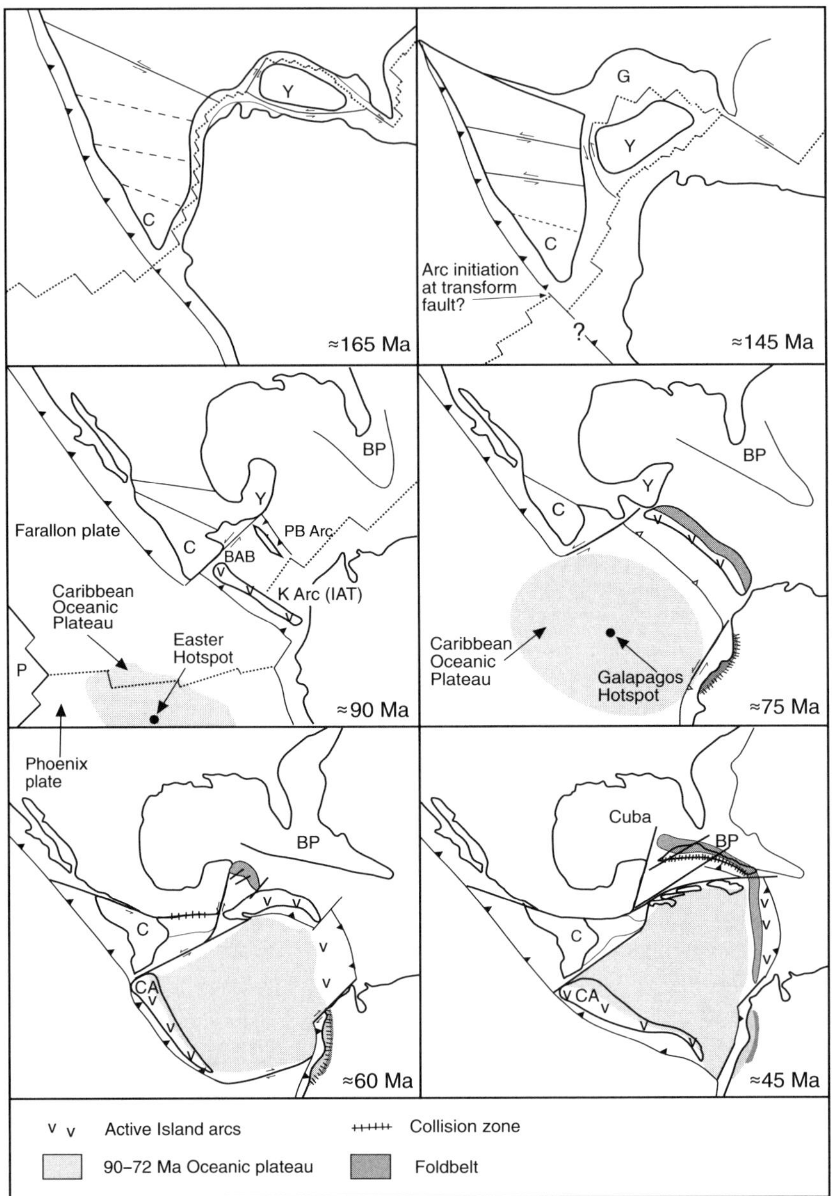

Figure 6.3. Plate reconstructions for the Caribbean region during the last 165 Myr. Modified after Kerr et al. (1999). Y = Yucatan; C = Chortis; G = Gulf of Mexico; BP = Bahamas platform; BAB = back arc basin; PB = oceanic arc; CA = Central America; K = Cretaceous arc.

Tectonic History of the Caribbean Plateau

The tectonic history of the Caribbean region began about 165 Ma when South America was rifted from North America (Burke et al. 1978; Burke 1988) (Fig. 6.3). At this time, Yucatán was rifted from the Gulf Coast of North America, giving rise to the Mississippi embayment. By 110 Ma, an oceanic arc formed along the edge of the Farallon plate as it began to subduct beneath the proto-Caribbean plate in the vicinity of what is today Central America. With time, this arc, some of which is preserved in Cuba today, moved into the gap between the two continents (Kerr et al. 1999). The northern ophiolite melange in Cuba with high-pressure minerals may represent the accretionary prism of the arc. Throughout the Cretaceous, the Antillean arc migrated northeastward as the Farallon plate grew. At about 90 Ma, a plume head arrived at the lithosphere possibly beneath the now subducted Phoenix plate (possibly the Easter plume) (Fig. 2.1) and the plume-derived magmas rapidly formed much of the volume of the Caribbean plateau (Revillon et al. 2000) (Fig. 6.3). The Galapagos plume head impacted the Caribbean plate about 75 Ma, producing a second period of volcanism and plateau growth. The Caribbean plateau was carried through the gap between the Americas, and when it reached the leading edge between 80 and 60 Ma it clogged the subduction zone, resulting in a flip in subduction direction. After this time, the Atlantic plate was consumed beneath the Caribbean plate, and the Lesser Antilles arc formed along the eastern margin of the Caribbean plate. At about the same time, the Middle America trench propagated southward, and a new arc, the Central American arc, developed on the trailing edge of the Caribbean plateau (60–45 Ma) (Fig. 6.3). As the Caribbean plate resisted subduction continuing into the Tertiary, it collided with the northwestern edge of South America, and slices of the plateau were obducted onto the continent in Colombia.

Accreted Oceanic Plateaus in the American Cordillera

Wrangellia Terrane

Several accreted terranes in the American Cordillera have been recognized as remnants of oceanic plateaus. Among the best known are Wrangellia and Angayucham, remnants of two different oceanic plateaus that formed during the Triassic and then were accreted to the northwestern coast of North America during the Cretaceous. Wrangellia comprises several large allochthonous blocks, the largest of which are exposed on Vancouver Island in British Columbia and in the Wrangell Mountains in southeastern Alaska (Fig. 6.4). Wrangellia also contitutes a thick sequence of oceanic flood basalts and associated hyaloclastic breccias (Karmutsen Formation) resting unconformably on late Paleozoic arc volcanics and sediments (Fig. 6.5) (Jones et al. 1977; Richards et al. 1991). Pelagic chert and limestone immediately underlying Wrangellia basalts indicate a deep-water environment with no major volcanism or tectonic activity. Just before eruption of the plateau basalts, the basement was rapidly uplifted, as indicated by the local occurrence of subaerial basalts at the base of the succession (Richards et al. 1991). That both subaerial and submarine basalts occur in Wrangellia volcanic sections indicates thick crust and isostatic uplift from plume- derived magmas such that the plateau emerged above sea level at times. The southern part of the plateau comprising the Karmutsen Formation on Vancouver Island is dominantly submarine, whereas the Nikolai greenstone in Alaska appears to have been primarily subaerial (Barker et al. 1994). On the basis of well-dated pelagic fossil assemblages, the Triassic volcanism in Wrangellia is constrained to < 5 Myr at about 230 Ma (Lassiter et al. 1995). Assuming a minimum volume for Wrangellia plateau basalts of 1×10^6 km^3 and an eruption

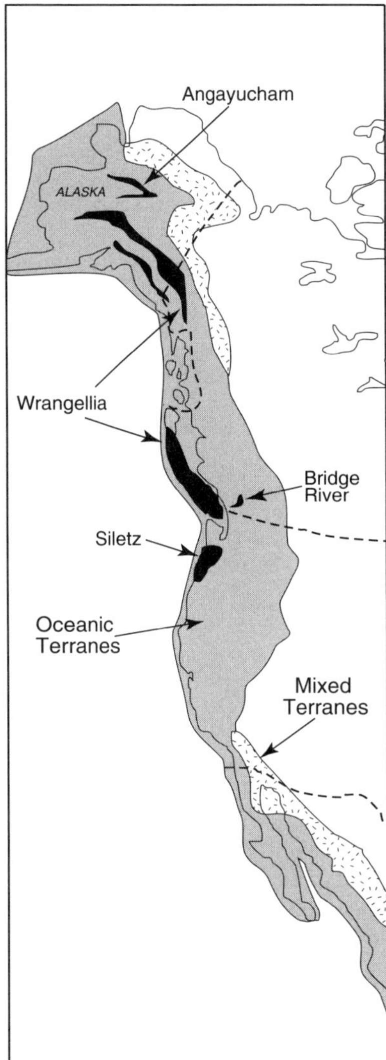

Figure 6.4. Map showing distribution of accreted juvenile crust in the North American Cordillera. Oceanic terranes represent remnants of oceanic arcs, ophiolites, and oceanic plateaus. Mixed terranes include tectonic domains of both oceanic and continental affinities. Terranes mentioned in the text are shown in black.

interval of 2.5×10^6 yr (Panuska 1990), eruption rates were at least 0.4 km^3/yr and may have been considerably greater. After cessation of volcanism, Wrangellia appears to have subsided in response to thermal contraction of the lithosphere and loading with later sediments. Postvolcanic sediments change laterally from shallow, tidal carbonates into deep-water limestones and shales, as represented by sediments of the McCarthy Formation in Alaska and the Parson Bay Formation on Vancouver Island (Fig. 6.5).

Basalts erupted during the early stages of Wrangellian volcanism are more highly fractionated than later basalts, and these also show chemical and isotopic evidence for assimilation of older arc crust and perhaps mantle lithosphere (Lassiter et al. 1995). The degree of assimilation appears to decrease with age of eruption so the younger, more voluminous phases record more accurately the composition of the mantle plume. ϵ_{Nd} values are consistently high (+6 to +7) and Sr isotope ratios group around 0.7034 (Fig. 5.14), suggesting an important depleted component in the Wrangellian plume

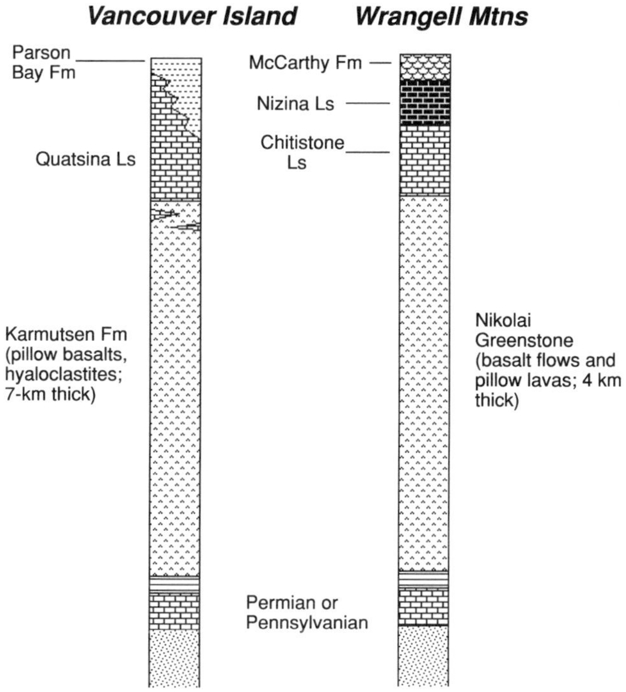

Vancouver Island **Wrangell Mtns**

Parson Bay Fm
Quatsina Ls
Karmutsen Fm (pillow basalts, hyaloclastites; 7-km thick)
McCarthy Fm
Nizina Ls
Chitistone Ls
Nikolai Greenstone (basalt flows and pillow lavas; 4 km thick)
Permian or Pennsylvanian

Figure 6.5. Generalized stratigraphic columns of Triassic rocks of Wrangellia in southern British Columbia (Vancouver Island) and Alaska (Wrangell Mountains). After Richards et al. (1991).

source. Relatively high Th/Ta ratios of some of the basalts, however, reflect contamination of the basalts with older arc crust (Fig. 6.2). The intermediate La/Yb ratios of Wrangellian basalts suggest that the plume head contained a mixture of depleted and enriched (HIMU, EM) source components. Although the Wrangellian basalts have several chemical and isotopic features in common with Ontong Java and Caribbean plateau basalts, the relatively low $(CaO/Al_2O_3)_8$ ratios at moderate $(FeOT)_8$ contents suggest melting at greater depths in the mantle source than characteristic of most other oceanic plateau basalts (Fig. 6.6) (Lassiter et al. 1995). It is probable that the presence of a relatively thick arc lithosphere beneath Wrangellia at the time of volcanism prevented rifting and limited the rise and degree of decompression melting of the plume head.

Angayucham Terrane

The Angayucham terrane in and south of the Brooks Range in northwestern Alaska represents remnants of a highly segmented Triassic oceanic plateau tectonically mixed with arc remnants of late Paleozoic age and with accreted oceanic islands and seamounts of Early Jurassic age (Fig. 6.4) (Barker et al. 1988; Pallister et al. 1989). The Triassic plateau succession, which is deformed and variably metamorphosed, comprises pillow basalt, diabase sills and dykes, mafic hyaloclastic breccia, and radiolarian chert. Like the Wrangellian basalts, the Angayucham basalts have relatively low $(CaO/Al_2O_3)_8$ ratios at moderate $(FeOT)_8$ values suggestive of a relatively deep source (Fig. 6.6). Incompatible element distributions in Angayucham basalts, as shown by the

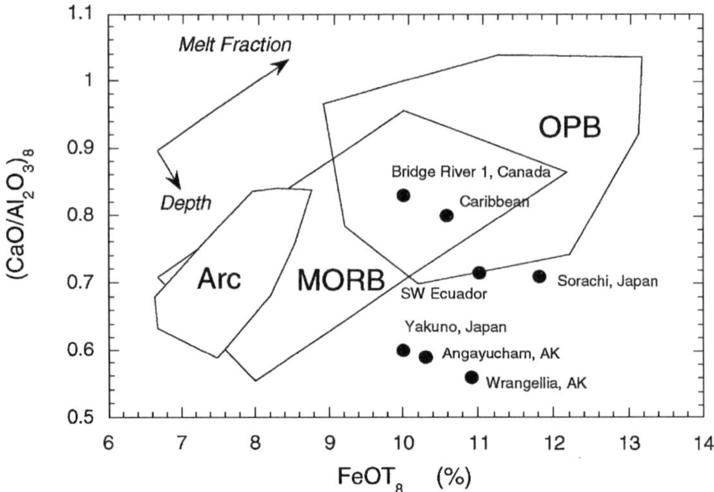

Figure 6.6. $(CaO/Al_2O_3)_8$ versus $(FeOT)_8$ graph showing the average compositions of basalts from accreted oceanic plateaus. See Figure 4.17 for other information.

Th/Ta–La/Yb relationships (Fig. 6.2), indicate the presence of a major depleted component with a low La/Yb ratio in the plume head similar to the Caribbean plateau source. This confirms that Wrangellia and Angayucham basalts come from different Triassic plumes.

A tectonic model presented by Pallister et al. (1989) suggests that during the Late Triassic an oceanic plateau formed west of the late Paleozoic coastline of the Brooks Range in, what is today, northern Alaska. During the Early Jurassic, after cessation of volcanism in the Angayucham oceanic plateau, seamount–island volcanism began in response to a new mantle plume. The volcanic islands developed on or adjacent to the Triassic oceanic plateau. As indicated by the oldest tonalites in the Koyukuk arc terrane, subduction began during the Middle to Late Jurassic and continued into Early Cretaceous. The Koyukuk oceanic arc developed outboard of the seamount–oceanic plateau. As this arc approached the continental margin, the older oceanic plateau and seamount chain was partially underplated and obducted onto the Paleozoic continental margin.

Bridge River Terrane

The Bridge River Complex in southern British Columbia also appears to be, in part, a remnant of an obducted oceanic plateau (Dostal and Church, 1994; Church et al. 1995) (Fig. 6.4). This complex comprises a succession of deformed pillow basalt, mafic breccia, and associated sills and dykes, with variable amounts of pelagic chert and argillite, and local thin bands of pelagic limestone. Although the base of the unit is not exposed, ultramafic rocks in tectonic contact could represent basement rocks. One of the puzzling features of the complex is the long time span of 170 Myr represented by the succession. The lower part of the complex comprises pelagic chert with Mississippian radiolarian fossils and is intruded by mafic to tonalitic bodies of Pennsylvanian to Permian age. The upper part of the section, which is associated with basalts that have oceanic plateau geochemical affinities, contains chert with radiolarians of Middle Triassic to Middle Jurassic age. It is likely that an unconformity or a fault separates the two assemblages.

Two populations of basalt are recognized in the Pioneer Formation of the Bridge River Complex. Both are in the Triassic–Jurassic part of the complex. The first (Bridge River 1, Fig. 6.2), appears to have a plume source with mixed depleted and enriched mantle components. The second (Bridge River 2) reflects a depleted source, perhaps isolated in the deep mantle. The BR1 population has a very high $(CaO/Al_2O_3)_8$ ratio at a moderate $(FeOT)_8$ value suggestive of high degrees of melting in the plume head, similar to the Caribbean plateau (Fig. 6.6).

Siletz Terrane

The Siletz terrane comprises a large fraction of the continental margin of Oregon and Washington in the northwestern United States (Snavely, 1987) (Fig. 6.4). Most of the exposed parts of this terrane include basalts and associated mafic sills that appear to be remnants of an Eocene oceanic plateau that collided and became sutured to the continent some 50 Ma. Overlying the terrane are sedimentary–volcanic basins up to 7-km thick that record episodic convergence and forearc volcanism after terrane accretion. Side scan sonar and seismic reflection data provide a clear picture of deformation at the subduction zone deformation front (MacKay et al. 1992; Trehu et al. 1995). Deformation changes from seaward- to landward-verging thrusts at about 44° 52′ N. The Siletz terrane is the basement beneath the continental shelf and acts as a subduction zone "backstop", and a deep seismic reflection is interpreted as the base of this terrane. These data suggest that the Siletz terrane is an approximately 10-km-thick slice of an oceanic plateau that underplated the accretionary prism.

Accreted Oceanic Plateaus in Japan

Japan is an amalgamation of Paleozoic and Mesozoic accreted oceanic terranes (Kimura et al. 1994). Although lithologic assemblages of many of these terranes are suggestive of an oceanic plateau tectonic setting (Taira et al. 1989; Isozaki et al. 1990; Maruyama 1997; Tatsumi et al. 2000), few have been described and chemically analyzed. The Permian Yakuno Complex in southern Honshu is a thick succession of submarine basalts and associated gabbros and ultramafic rocks with minor pelagic sediments accreted to Japan in Late Permian (Isozaki 1997). Mafic granulites in this complex suggest original thicknesses of 15–30 km consistent with it, representing a remnant of an oceanic plateau. Similar, but poorly documented, accreted oceanic terranes of late Paleozoic/Triassic age occur in Japan along a belt with a strike length of more than 2000 km and widths of 200–300 km. Many Carboniferous greenstones in the Chichibu and Chugoku belts in southern Japan have lithologic assemblages and basalt geochemistry that suggest they are accreted oceanic islands or plateaus (Tatsumi et al. 2000). Cretaceous and Early Tertiary accretionary complexes extend from northern Japan through Hokkaido and into Sakhalin in Russia (Fig. 6.7). The complexes include two different units: the Sorachi Group, a unit composed chiefly of mafic volcanic

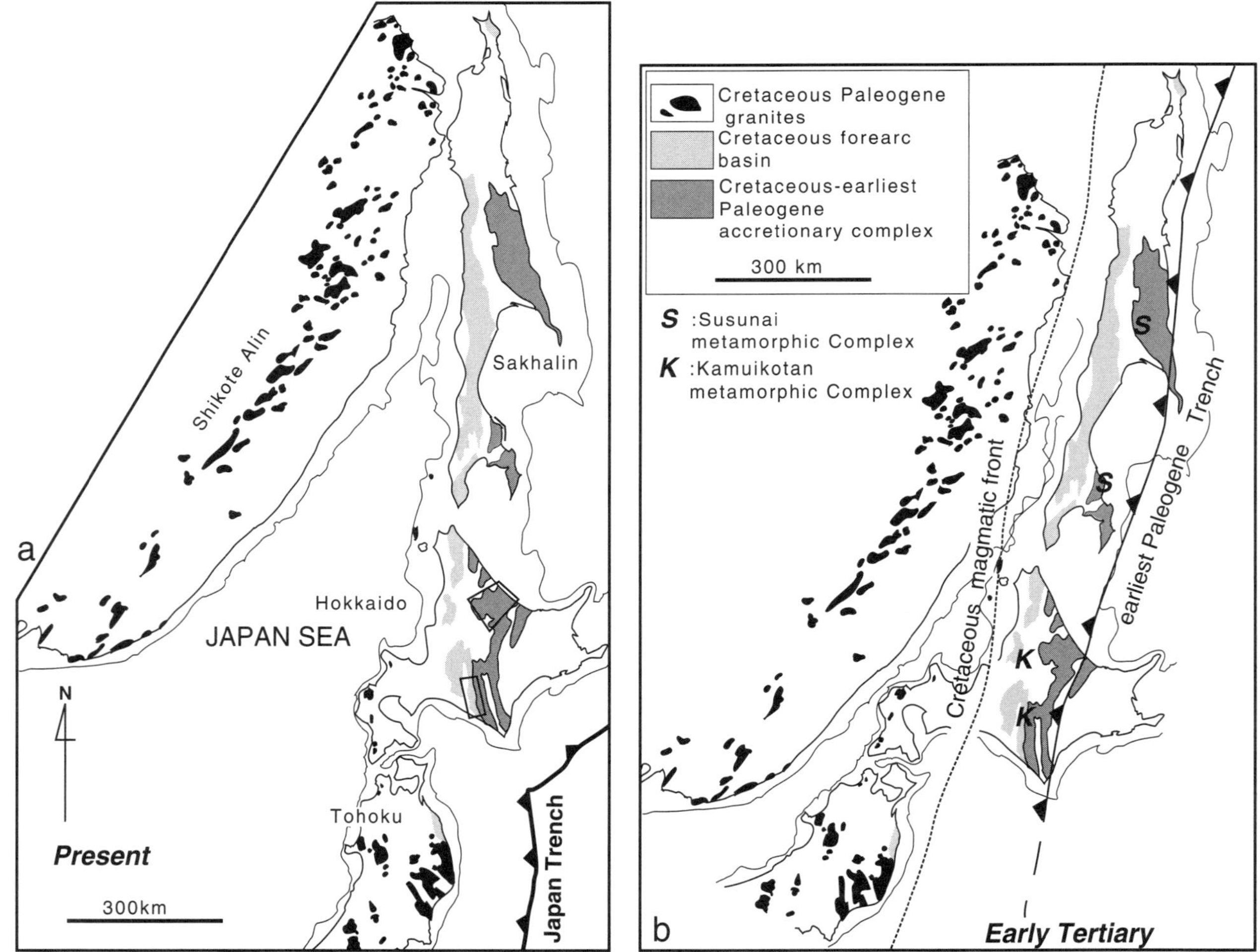

Figure 6.7. Map of northern Japan and Sakhalin showing distribution of Cretaceous and Early Tertiary oceanic terranes accreted to the continent. (a) Present and (b) Early Tertiary reconstruction. Courtesy of Gaku Kimura.

and intrusive rocks with oceanic plateau affinities; and a serpentine melánge containing high-pressure metamorphic rocks (Kimura et al. 1994). The Sorachi Group, of Late Jurassic age, includes dominantly pillow basalts and associated sills and dykes with pelagic sediments – chiefly radiolarian chert. This group is overlain by late Mesozoic coral reef limestones. Structural studies indicate these accretionary complexes are a series of stacked thrust sheets reflecting westward subduction and obduction of oceanic terranes. The Sorachi Group contains 60–70% of basalts and associated mafic intrusives, and the amount of hyaloclastite is the order of 50% of the volcanic section. Large vesicles in some of the pillow basalts suggest eruption in shallow water. Basalts of the Sorachi Group range from tholeiites to picrites with some alkaline varieties. They have geochemical affinities to oceanic plateau basalts and have been interpreted as remnants of an accreted oceanic plateau (Kimura et al. 1994). These rocks also have relatively high $(CaO/Al_2O_3)_8$ and $(FeOT)_8$ values suggestive of high degrees of melting in a plume head at moderate depths (Fig. 6.6). On the Th–Ta-La–Yb graph, the Sorachi basalts are similar to the most depleted basalts from Wrangellia and likewise reflect a plume source containing both depleted and enriched mantle (Fig. 6.2).

From paleomagnetic results it is possible to reconstruct the approximate location of the Sorachi oceanic plateau at 140 Ma. The paleomagnetic trajectory analysis for accreted terranes in northern Japan and Sakhalin suggest they formed in what is now the mid-Pacific basin, near the equator, where the positive geoid anomaly exists today. The coral reef carbonates capping the shallow part of the Sorachi Group also reflect very low latitudes in agreement with the paleomagnetic data. As illustrated on Figure 6.8, the Sorachi plateau may have been one of two or more oceanic plateaus that formed near the triple junction of the Pacific, Farallon, and Kula plates in the Early Cretaceous.

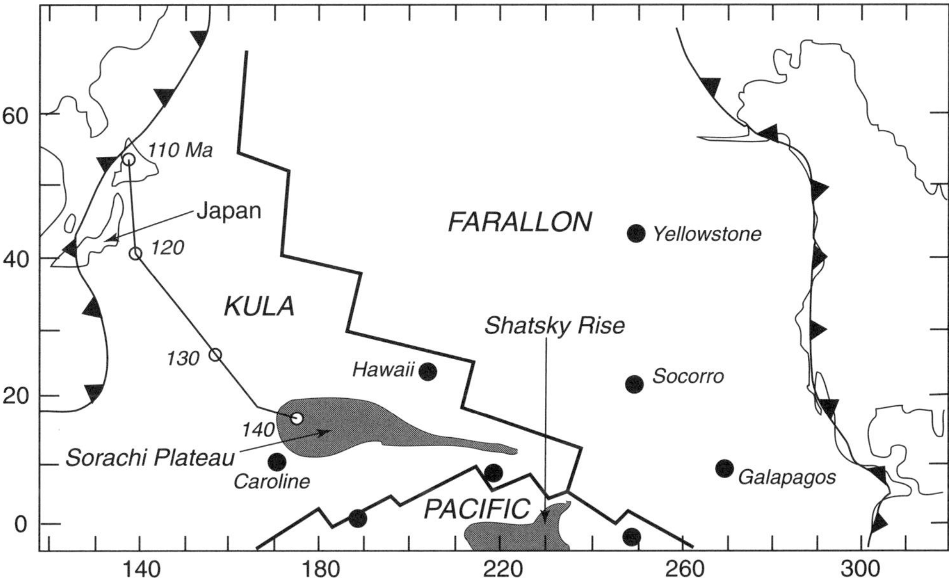

Figure 6.8. Plate reconstruction of the Pacific region about 140 Ma. Position of the Sorachi oceanic plateau calculated from the polar wander path; ages given in Ma. (Modified after Kimura et al. (1994).)

How Do Continents Grow?

Although most investigators agree that production of continental crust is related to subduction, just how continents are produced in arc systems is not well understood. Oceanic terranes such as island arcs and oceanic plateaus may be important building blocks for continents as they collide and accrete to continental margins (Kimura et al. 1993; Abbott and Mooney 1995; Condie and Chomiak 1996). However, the largely mafic character of these terranes (Kay and Kay 1985; DeBari and Sleep 1991), in contrast to the felsic composition of the upper continental crust indicates that these terranes must have undergone dramatic changes in composition to become important building blocks of the continents.

Various mechanisms have been suggested for the growth of the continents, the most important of which are (1) magma additions by crustal underplating and (2) collision and accretion of oceanic terranes to continental margins (Rudnick 1995). Magma from the mantle may be added to the crust by underplating, involving the intrusion of sills and plutons, and by overplating of volcanic rocks (Fig. 6.9). Magma additions occur in a variety of tectonic environments, the most important of which are arcs, rifted continental margins, and beneath flood basalt provinces. Note that the last two environments are related to mantle plumes. Large volumes of juvenile magma from the mantle are added to oceanic and continental margin arcs (Condie 1997a, Condie 1998). Major continental growth by this mechanism can occur during seaward migration of subduction zones, when arc magmatism keeps up with slab migration. Field relationships in exposed lower crustal sections, such as the Ivrea Zone in Italy and the Musgrave block of western Australia, suggest that lower crustal mafic granulites are intrusive gabbros and diabases. Their existence confirms the idea that underplating of mafic magma may be an important

a. MAGMATIC OVER- AND UNDERPLATING

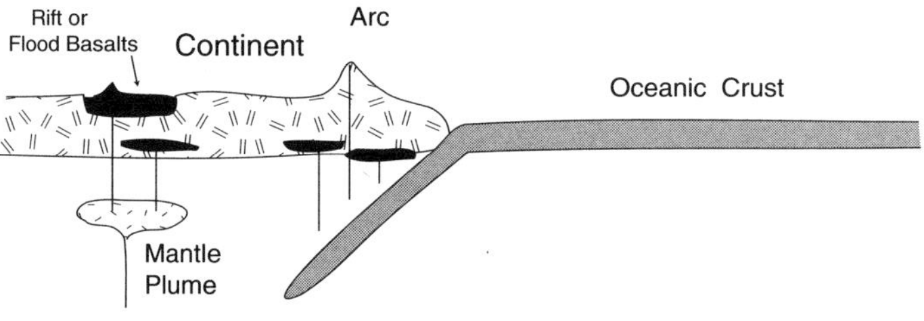

b. TERRANE COLLISIONS

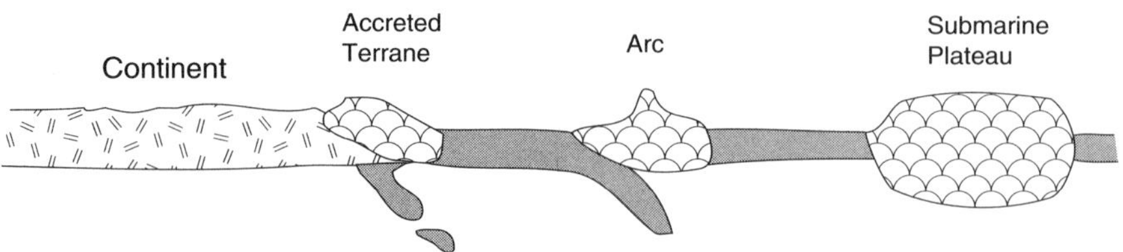

Figure 6.9. Growth of continents by magmatic over- and underplating and by terrane collisions.

mechanism by which the continental crust has thickened (Rudnick and Fountain 1995). Also, a high-velocity layer at the base of some Precambrian shields has been interpreted as a mafic underplate (Durrheim and Mooney 1991). Thus, both field and seismic evidence support a significant but unknown contribution to continental growth by crustal underplating of basaltic magma. As discussed below, mafic underplating is probably associated with either or both the breakup of supercontinents or the production of LIPS.

Continental growth also occurs when oceanic terranes collide with, and become sutured to, continents (Fig. 6.9). Collision of oceanic plateaus with continental margins may be an important mechanism of continental growth (Abbott and Mooney 1995; Condie 1997c; Barr et al. 1999). Most of the Cordilleran and Appalachian orogens in North America are collages of oceanic terranes added by collision either at convergent plate boundaries or along transform faults. Lithologic association, chemical composition, terrane life span, and tectonic history of Cordilleran terranes in northwestern North America are consistent with collisional growth of the continent in this area (Condie and Chomiak 1996; Patchett and Gehrels 1998). The Cordilleran crust appears to have formed mostly by accretion of oceanic terranes with pre-collisional histories of variable complexities and durations (Fig. 6.4). Although some of these terranes began to evolve into immature continental crust before accretion to North America, most probably began this evolution at, or not long after, they accreted to the continent. This was accomplished largely by incompatible element enrichment resulting from subduction-related processes associated with collisionally thickened crust (Arndt and Goldstein 1989). Similar mechanisms have been proposed for Archean continents (Percival and Williams 1989).

Plume-Related Underplating during Supercontinent Breakup

One of the first lines of evidence that the lower continental crust has grown, at least in part, by underplating of basaltic magma comes from seismology (Drummond and Collins 1986; Nelson 1991). Strong reflectors, which are widespread in the lower continental crust, have generally been interpreted as flows, sills, and layered mafic intrusions. Supporting this conclusion is that some shallow reflectors have been traced to the surface and identified (Percival et al. 1992). Single basalt flows can be up to 100 m thick and cover areas of several thousand square kilometers. Crustal xenolith studies and studies of uplifted segments of the lower crust clearly show mafic intrusions in the lower crust, some of which are considerably younger than their host rocks (Rudnick 1995; Rudnick and Fountain 1995; Condie 1997a). A bimodal distribution of acoustical impedance in the lower continental crust favors the existence of interlayered mafic and felsic units. Recent studies of the Parana-Etendeka and Ethiopian rifted margins show that felsic volcanics (ash flow tuffs) and volcaniclastic sediments may also be important components along some rifted continental margins (Menzies et al. 2000). Together with contemporary volcanism, rift flanks can be elevated up to 4 km above sea level, and thus may be important sediment sources. Prior to breakup, continental cratons in Pangea were close to sea level and hosted a variety of sedimentary basins.

Another potentially important source of juvenile crust is crustal underplating associated with the eruption of continental flood basalts. Our largest data source for mafic

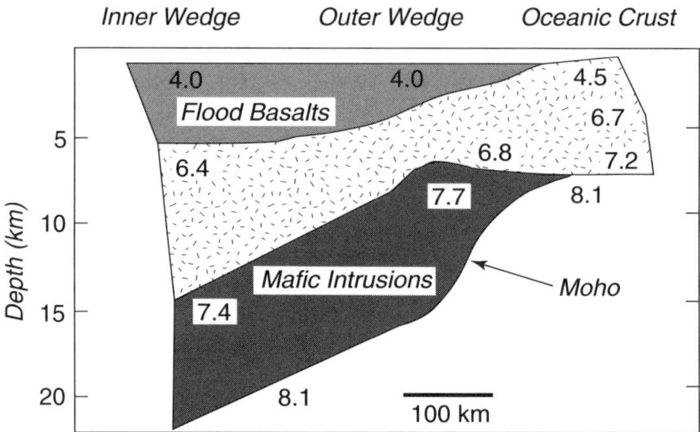

Figure 6.10. Seismic structure of the rifted coastline of East Greenland showing the three major seismic layers. Water and sediment layers not shown. Numbers are P-wave velocities in km/s. Modified after Eldholm and Grue (1994).

underplating of continental crust comes from the North Atlantic Igneous Province, where both seismic studies and deep sea drilling help constrain the volumes of underplated igneous rock (Eldholm and Grue 1994; Fitton et al. 1998). Results of these studies show that enormous volumes of mafic magma (both as basalt flows and intrusives) were emplaced along the transition between continental and oceanic crust during rifting of Laurentia from Baltica (Fig. 3.10). Along more than 2600 km of the propagating coastlines in the North Atlantic (coasts of Greenland and North Europe) during the Early Tertiary, both submarine and subaerial volcanism that occurred were mainly concentrated in a 1–2 Myr period, about 57 Ma (Eldholm and Grue 1994). Similar widespread volcanism, but more prolonged, occurred along the eastern Coast of North America during the opening of this part of the Atlantic in the Mesozoic (White et al. 1987).

Seismic studies show that the upper crust along the rifted continental margins of the North Atlantic comprises chiefly flood basalts and related intrusions interbedded with sediments of variable thickness. Combined seismic and drilling results indicate three important crustal layers in the rifted continental margins around the North Atlantic (Fig. 6.10): (1) an upper layer of flood basalts (partly subaerial) and sediments, (2) a middle layer of mafic dykes changing into gabbro with depth, and (3) a lower layer composed of mixed gabbro and associated ultramafic cumulates. This stratigraphy strikingly resembles the oceanic plateau model constructed from exposed slices of the Caribbean plateau discussed in Chapter 3 (Figs. 3.6 and 3.7). Ages of magnetic anomalies on the seafloor in the North Atlantic clearly show that many of the mafic wedges in the crust were accreted after continental breakup. In other areas, however, these wedges may precede or be synchronous with continental breakup.

Minimum calculated rates of magma emplacement along the Baltic–Greenland plate boundary are 0.6–2.4 km^3/yr, which are similar to extrusion rates for the Deccan and Ontong Java LIPs (Table 3.1). The estimated rate of crustal underplating, assuming a 3-Myr period, is 2.2 km^3/yr, which is considerably less than the rate estimated for Ontong Java of about 9 km^3/yr (Eldholm and Grue 1994). This rate is only somewhat greater than the production rate of mafic extrusion and intrusion for the entire South Atlantic Volcanic Province of only 0.1 km^3/yr (Hawkesworth et al. 1999).

Magma production models indicate that the North Atlantic Volcanic Province and similar volcanic provinces around rifted continental margins reflect melting chiefly within a mantle plume (White et al. 1987; Eldholm et al. 1995). The geochemical characteristics of basalts along rifted continental margins, discussed previously in Chapter 5, are interpreted by some investigators to indicate that mafic magmas are derived chiefly from a deep depleted component in mantle plumes and that depleted asthenosphere (DM) has played a relatively minor role in magma formation.

In summary, the following features characterize rifted continental margins (Menzies et al. 2000).

1. High-velocity reflectors along rifted continental margins represent a mixture of volcanic rocks (basaltic lavas and felsic ash flow tuffs), underplated mafic intrusive rocks, and volcaniclastic sediments.
2. Both the erupted and intruded mafic igneous rocks were emplaced in a short period of time (1–2 Myr).
3. Along some rifted margins, such as the Ethiopia and Parana- Etendeka margins, felsic volcanic components may be important.
4. Uplifts along rifted flanks of opening ocean basins may be important sources for detrital sediments in adjacent extensional basins.
5. Magmatism along rifted continental margins may predate rifting by several million years, may be synchronous with rifting, or may post-date rifting.
6. Prior to breakup, the continental landmasses were close to sea level and hosted a variety of sedimentary basins.
7. Most of the mafic magmas erupted and intruded along rifted continental margins appear to be derived from mantle plume sources, in part from a deep depleted plume component within these sources.

Accretion of Plume Heads to the Lithosphere

The subject of underplating of continental crust with plume magmas cannot be separated from the question of what happens to plume heads as they produce magmas. Are they recycled back into the asthenosphere or do they become part of the mantle lithosphere? Bearing on this question are results from a geochemical study of basalts from the Arabian–Nubian shield. Basalts from this region appear to have been derived from the same enriched source for the last 900 Myr (Stein and Hofmann 1992; Stein and Goldstein 1996). This is illustrated on an ϵ_{Nd}–$^{87}Sr/^{86}Sr$ diagram, where oceanic basalts erupted between 900 and 870 Ma exhibit somewhat enriched isotopic signatures with ϵ_{Nd} values of +3 to +6 and $^{87}Sr/^{86}Sr$ ratios of 0.7025 to 0.703 (Fig. 6.11). As discussed in Chapter 5, these isotopic compositions are typical of plume-generated basalts and especially those found in oceanic plateaus. If basalts are formed at different times from a common source, the younger basalts must have higher isotopic ratios owing to radioactive decay. Phanerozoic rift-related basalts in Israel erupted in the last 200 Myr have somewhat elevated $^{87}Sr/^{86}Sr$ ratios consistent with Rb/Sr ratios in the 900 Ma

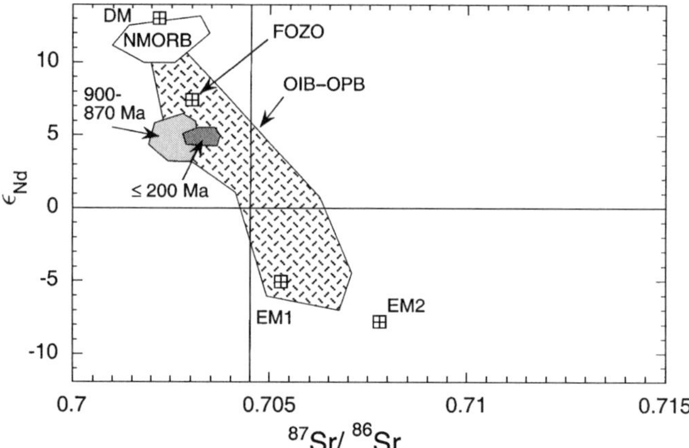

Figure 6.11. Epsilon Nd versus $^{87}Sr/^{86}Sr$ ratio showing distribution of 900–870-Ma plume-related basalts from the Arabian–Nubian shield compared with Israel rift basalts erupted in the last 200 Myr. After Stein and Goldstein (1996). DM = depleted mantle; NMORB = normal ocean = ridge basalt; FOZO = focal zone source; OIB = oceanic island basalts; OPB = oceanic plateau basalts; EM1 and EM2 = enriched mantle sources.

source from which the older basalts were derived (Stein and Goldstein 1996) (Fig. 6.11). Also supporting a common source are the ϵ_{Nd} ratios which remain unchanged as calculated for the Sm/Nd ratios of the source. The young basalt source can be distinguished isotopically from the depleted asthenosphere source (DM) with high ϵ_{Nd} values and low $^{87}Sr/^{86}Sr$ ratios. As the Red Sea opened some 30 Ma, this depleted mantle source was tapped. The Israel rift basalts erupted in the last 200 Ma, however, appear to have come from the same source as the Proterozoic basalts.

Stein and Hofmann (1992) suggest that the source of basalts in the Arabian–Nubian shield for the last 900 Myr has been a fossil plume head that was accreted to the lithosphere during formation of the Arabian–Nubian shield some 900–700 Ma. They suggest that the plume spread out at the base of the lithosphere, cooled, and became accreted to the base of the lithosphere. At later times, thinning of the lithosphere during periods of extension triggered partial melting and volcanism. During the initial stages of opening of the Red Sea, basaltic magmas were also derived from the accreted plume head. As the basin continued to open, depleted asthenosphere displaced the lithosphere and became the source of basaltic magmas as the degree of decompression melting increased. At the northern end of the Red Sea, where a new ocean basin is just beginning to open, modern basalts are still tapping the enriched lithospheric source.

Another example of plume-head accretion to the lithosphere is given for the Archean Slave craton in Chapter 7.

Oceanic Plateaus and Continental Growth

Oceanic Plateaus as Lower Continental Crust

Some or most oceanic plateaus have arc systems erupted along their margins, such as the Solomon arc along the southern margin of the Ontong–Java Plateau and the

Lesser Antilles arc along the eastern edge of the Caribbean Plateau. During collision and obduction, these arcs that are obducted at shallow levels onto continental crust. Although in some instances thick slices of oceanic plateaus can also be obducted (20 km for the Caribbean Plateau) or underplated (10 km for the Siletz terrane in Oregon), arcs should greatly dominate in obducted fragments. Perhaps this is the reason for the relative abundance of arc-related greenstones in the geologic record, as discussed in Chapter 7.

So what is the fate of oceanic plateaus that collide with continents and lose their surficial arcs by obduction? Perhaps they are accreted to continental margins and with time evolve into lower continental crust. Although crustal thickening during collision could result in the production of eclogites in the root zones of accreted plateaus, leading to minor recycling of plateaus into the mantle as suggested by Saunders et al. (1996), a significant volume of these plateaus may be accreted to the continents. This idea has important implications for continental development because the lower continental crust would comprise chiefly accreted and underplated oceanic plateaus, whereas the upper continental crust would form by subduction-related processes – perhaps beginning before accretion of oceanic plateaus to continental margins. Most of the upper continental crust, however, must develop after collision by subduction-related magmatism as discussed below. In any case, two important observations indicate that island arcs are not important components in the lower continental crust: (1) they are generally less than 25 km thick and as such should be subducted (Cloos 1993), and (2) lower crustal xenoliths contain too much Ni, Cr, and Co for an arc source (Rudnick 1992a; Condie 1994, 1997c; Abbott 1996).

One way to test the oceanic plateau accretion model is to compare the chemical and isotopic composition of lower crustal mafic xenoliths with that of oceanic plateau and arc basalts (Condie 1999). In principle, we can constrain the importance of mantle plume components in the lower crust from incompatible element distributions and isotopic compositions of lower crustal mafic xenoliths (Rudnick 1992a, 1995). Although, as discussed in Chapter 7, the Th/Ta ratio and Ni content of basalts can be used as tracers of plume and arc components, the potential for Th to be depleted during granulite–facies metamorphism makes it unsuitable for studying lower crustal rocks. For this reason, the La/Nb (or La/Ta) ratio may be a more reliable index, as suggested by Rudnick (1995). Like the Th/Ta ratio, the La/Nb ratio monitors the subduction geochemical component (i.e., depletion in Ta and Nb compared with neighboring incompatible elements on a primitive-mantle normalized graph), yet La, unlike Th, is not readily lost during granulite-grade metamorphism (Hansen et al. 1995). Also, as discussed in Chapter 5, contamination of plume-derived basaltic magmas by continental crust can raise Th/Ta and La/Nb ratios and, thus, caution must be exercised in using these ratios to distinguish plume and arc sources for basalts.

The La/Nb ratio in subduction-related volcanics is generally greater than one, with an average value of about 3, in contrast to the low La/Nb ratios of plume and ocean ridge volcanics (Rudnick 1995; Condie 1999) (Fig. 6.12(a)). There is a distinct minimum in La/Nb between the two groups at about 1.4, which serves as a boundary between plume and arc-related basalts. It is noteworthy that basalts from Wrangellia (Nikolai greenstone and Karmutsen Fm, see earlier in the chapter) have distinctly low La/Nb ratios,

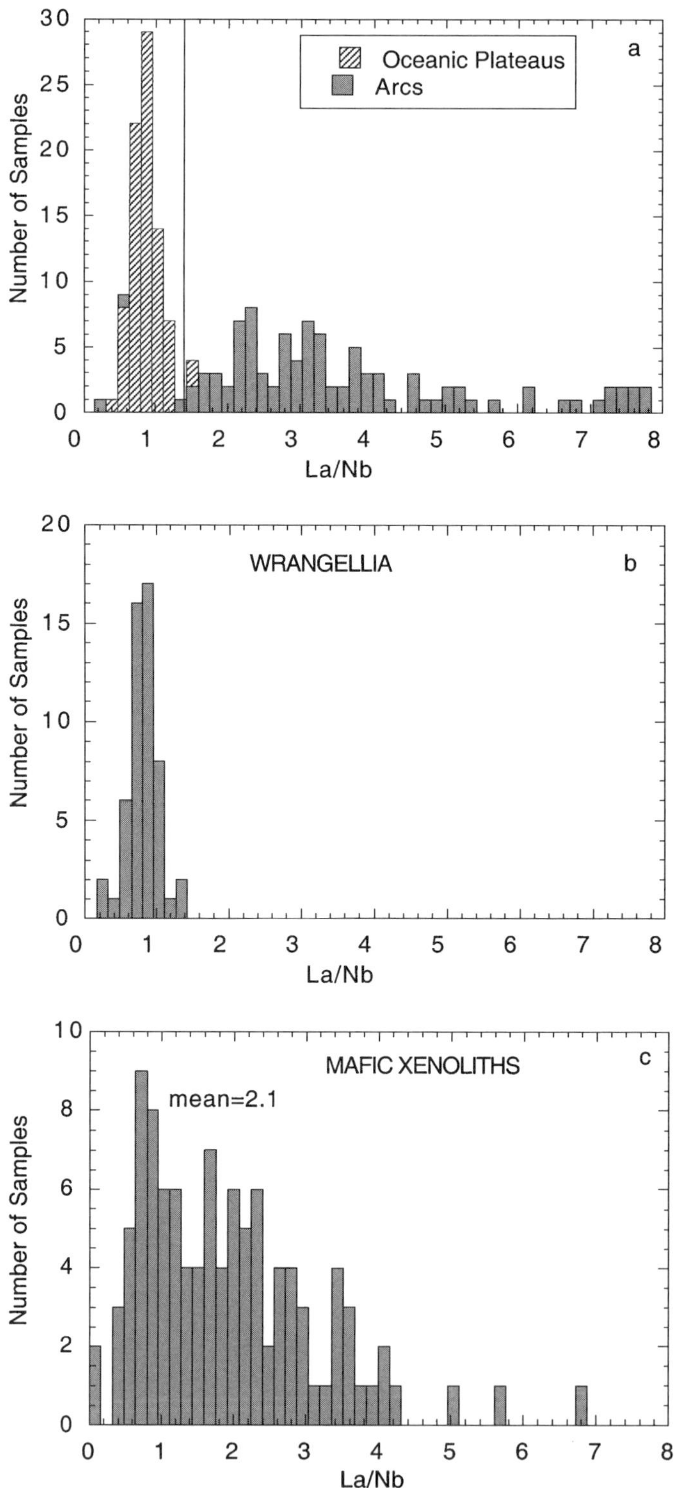

Figure 6.12. Some La/Nb ratios in various basalts. (a) oceanic plateaus and arcs; (b) Wrangellia; (c) mafic lower crustal xenoliths. References given in Condie (1999). Vertical line in (a) is at La/Nb = 1.4, the proposed boundary between arc and plateau + ocean-ridge basalts.

supporting an oceanic plateau origin for Wrangellia as suggested by other investigators (Richards et al. 1991; Lassiter et al. 1995) (Fig. 6.12(b)).

The distribution of La/Nb ratios in mafic, lower-crustal xenoliths is shown in Figure 6.12(c). At most localities, a wide range in La/Nb is evident in the xenolith population, often spanning the boundary between arc and plume-related magmas at an La/Nb value of 1.4. If this ratio is not changed by crustal contamination, the results suggest that both arc and plume components are present in the lower crust. In some suites, such as the Colorado Plateau suites in the Navajo and San Francisco volcanic fields, Chudleigh in eastern Australia, and the Eifel field in Germany, it would appear that the lower crust comprises chiefly subduction-related components, where most xenoliths have La/Nb ratios of more than 1.4 (Condie 1999). On the other hand, some suites such as those in Germany and the Eastern Mojave desert in California appear to sample chiefly Hessian plume components in the lower crust.

Similar results are apparent on an La/Nb–Ni_{40} graph when the entire mafic xenolith suite from a given location is considered as cogenetic (Fig. 6.13). The Ni_{40} value is the Ni content in ppm at a Mg number of 40, determined by plotting Mg number against Ni content for a suite of cogenetic xenoliths (Condie 1999). Mean values for most suites scatter between arc and plume sources. Again, Eastern Mojave and Hessian sites suggest a dominant plume source, and Chudleigh, San Francisco, Navajo, and Geronimo volcanic fields show arc-dominated sources, although with unusually high Ni_{40} values at some sites.

Some arc magmas are intruded into oceanic plateaus before plateau collision with a continent. For instance, both the Caribbean and Ontong Java plateaus have marginal arc systems on one or more sides, and arc magmas are probably intruded at deep levels in these plateaus around their margins (Tejada et al. 1996; Kerr et al. 1997b). Other arc magmas are clearly intruded into plateaus after their accretion to a continent,

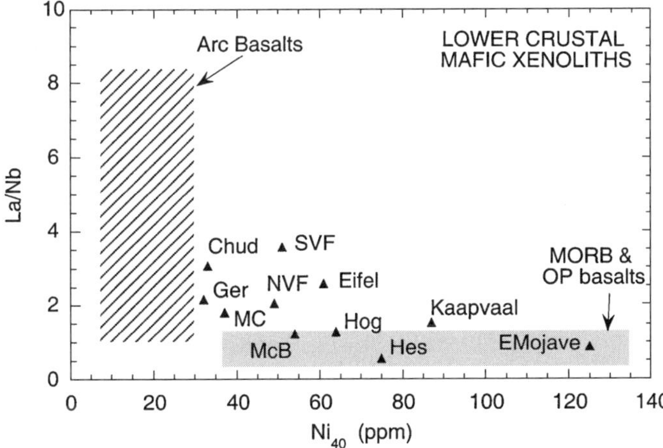

Figure 6.13. An La/Nb versus Ni_{40} diagram showing the distribution of lower crustal mafic xenoliths. References given in Condie (1999). Chud = Chudleigh, Australia; McB = McBride, Australia; Ger = Geronimo, Arizona; MC = Massif Central, France; SVF = San Francisco volcanic field, Arizona; NVF = Navajo volcanic field, Arizona/New Mexico; Hog-Hoggar, Algeria; Hes = Hessian, Germany; Eifel = Germany; Kaapvaal = South Africa; E Mojave = California. Ni_{40} is the Ni content at a Mg number of 40.

as for instance, in Wrangellia (Condie 1997c, 1999). Perhaps one can approximate the ratio of arc to plume components in the lower continental crust using the number of samples from a given xenolith site with La/Nb ratios above and below 1.4, respectively (see Fig. 6.12(a)). If the available xenolith database is representative of the post-Archean lower continental crust and an La/Nb value of 1.4 is used to separate arc from plume components, approximately one-third of the lower crust comprises plume components, and the remaining two-thirds are arc components (Fig. 6.12c). This, of course, is based on the assumption that the La/Nb ratio has not been changed during high-grade metamorphism and that the original basalts were not significantly contaminated with continental crust.

Another important aspect of this problem is that plume components can also be added to the lower crust by later underplating. Hence, the value of one-third for plume component in the lower crust is a maximum value for the amount of accreted oceanic plateau component, for some of the xenoliths may come from later underplated plume-derived basalts. The only way to distinguish xenoliths coming from accreted plateaus from those coming from plume-derived mafic underplate is by using isotopic ages: accreted plateau basalts should be no younger than the age of deformation in an orogen, whereas underplated material is generally younger than this age. Because it is difficult to date mafic xenoliths, we have few reliable isotopic dates. Nd and Pb isotopic data from some mafic xenolith localities suggest that the lower and upper crust are similar in age and, thus, both probably represent juvenile crust (Wendlandt et al. 1993; McGuire and Stern 1993; Huang et al. 1995). Geochemical data from these suites, however, suggest arc rather than plume components dominate in the lower crust. Sr, Nd, and Pb isotope studies from other mafic xenolith localities, however, clearly suggest later mafic underplating of the crust from plume sources (Downes et al. 1990, 1991; Rudnick et al. 1986; Rudnick 1990; Rudnick and Goldstein 1990; Kempton et al. 1990; Huang et al. 1995). Although some of the arc components may be crustally contaminated plume components, it is unlikely that all of the arc components in the lower crust have this origin. This is because the degree of contamination necessary to change the Th/Ta, La/Nb and various isotopic ratios from plume to arc signatures would also change the major element composition of the mafic rocks to intermediate compositions (Voshage et al. 1990), a change which is not observed in most lower crustal mafic xenoliths.

Considered as a whole, the mafic xenolith data suggest that a minimum of one-third of the post-Archean continental crust comprises mafic rocks from plume sources, either as accreted oceanic plateaus or as mafic underplate (Condie 1999). Because crustal and host magma contamination may raise such ratios as Th/Ta and La/Nb in xenoliths, the value of one-third is a minimum value for the post-Archean lower crust.

Making Continental Crust from Oceanic Plateaus

Although it is likely that an unknown volume of the continents consists of plume-generated magma, either as accreted oceanic plateaus or as mafic underplate, the question of whether felsic upper crustal igneous rocks can be generated by partial melting of the mafic roots of oceanic plateaus (or plume underplate) is problematic. At several locations in the Caribbean basin, basalts of the Caribbean plateau are intruded with near- contemporary felsic plutons (Lewis and Jimenez 1991; White et al. 1999). These occurrences provide an opportunity to study possible genetic links between mafic and felsic components and, in particular, to test mafic oceanic plateau igneous rocks as possible sources for felsic intrusions. In a more general way, one can test the idea that partial melting of mafic oceanic plateau crust can produce felsic upper continental crust.

On the island of Aruba in the southern Caribbean (Fig. 6.1), a suite of tonalitic rocks similar to the Archean TTG (tonalite–trondhjemite–granodiorite) suite intrudes mafic rocks of the Caribbean plateau. Plateau basalts at this location erupted 91–88 Ma followed by felsic pluton intrusion at 85–82 Ma (White et al. 1999). Sr and Nd isotopic compositions of Aruba tonalities overlap values of the mafic plateau rocks, which they intrude. However, as illustrated by Th/Ta-La/Yb relationships, incompatible element distributions in the tonalite are more like arc rocks than plume-derived rocks. In particular, tonalities have strong negative Nb–Ta anomalies not found in the Caribbean plateau igneous rocks. Similar differences are reported in the Duarte Complex in Hispaniola, which is also a remnant of the Caribbean plateau that is intruded with granitoids of similar age (Lewis and Jimenez 1991). Unless some Nb-bearing phase remains in the restite, these differences strongly suggest that the felsic plutons intruding the Caribbean plateau are not derived from partial melting of the mafic roots of this plateau. The striking similarity in incompatible element distributions in the granitoids to those of arc-derived magmas suggests that in both the Aruba and Duarte Complexes, the felsic magmas are subduction related and produced during post-accretionary arc magmatism.

Subduction-derived magmas fractionate and may be intruded into accreted oceanic plateaus as well as erupted on top of them – a process that with time may lead to a felsic composition of the upper continental crust. The lower crust, however, must remain dominantly mafic, although with some intruded felsic arc components. Modeling of upper crust production by this mechanism requires recycling of significant amounts of restite or cumulate into the mantle. However, the average thickness of continental crust cannot accommodate the large quantity of restite/cumulate required by fractional crystallization or partial melting models (Arndt and Goldstein 1989; DeBari and Sleep 1991; Yanagi and Yamashita 1994; Condie and Chomiak 1996).

Discussion of Oceanic Plateau Accretion

One characteristic feature of oceanic plateaus is a high seismic velocity layer (V_P = 7.2–7.8 km/s) in the lower crust (Nur and Ben-Avraham 1982; Farnetani et al. 1996). Model lithologic sections of Ontong Java and the Caribbean oceanic plateaus indicate that the lower crust consists of a large proportion of mafic and ultramafic cumulates, as described in Chapter 3. These cumulates, together with frequent sills, are probably responsible for the high velocities in the middle and lower crust of oceanic plateaus. Many, if not most, Proterozoic cratons also have a high velocity layer in the lower crust (Durrheim and Mooney 1991), and lower crustal mafic xenoliths typically have high measured seismic velocities (Rudnick and Jackson 1995). Strong seismic reflectors, also common in the lower continental crust, could also be caused by layered cumulates and sills. Although these observations are consistent with the idea that oceanic plateaus are an important component in the lower continental crust, they are not conclusive because later underplating of the continents by mantle plume magmas may also produce high-velocity cumulates and sills. For instance, such underplating is widely recognized along passive continental margins, as previously discussed.

As discussed above, if oceanic plateaus are important in the lower crust, mafic xenoliths derived from the lower crust should carry the geochemical signatures of oceanic plateau basalts. Unfortunately, results are complicated by arc-derived basalts and MORB contained in oceanic plateaus, upper crustal contamination of basaltic magmas, remobilization of elements during high-grade metamorphism, and possibly by later plume- and asthenosphere-derived magmas injected into the lower crust (Downes et al. 1990; Rudnick 1992b; Downes 1993; Rudnick and Fountain 1995). Also, at given xenolith localities, the range of compositions can be large, reflecting a mixture of plume and arc-derived components in the lower crust. Clearly, it is going to be a challenging problem using Nd, Sr, and Pb isotopes to sort out accreted plateau sources from later underplated plume magmas in lower crustal xenolith suites.

Still another aspect of the xenolith argument is that deep-seated, detrital metasedimentary xenoliths, although minor in number, are recognized in most lower crustal suites (Rudnick 1992; Downes 1993; Mattie et al. 1997). That these sediments are not deposited on oceanic plateaus cannot be used to negate the existence of oceanic plateaus in the lower crust because rocks of diverse sources can be tectonically mixed during plateau-continent and plateau-arc collisions.

Another attractive feature of the oceanic plateau accretion model is that it can account for the common lack of a positive Eu anomaly in lower crustal rocks (Rudnick 1992b; Mattie et al. 1997) because a positive Eu anomaly is predicted if lower crustal rocks are cumulates or restites in which plagioclase remains behind during formation of the upper crust. In the oceanic plateau accretion model, the upper and lower crust have two quite different sources and much of the restite/cumulate produced during formation of the upper crust is recycled into the mantle, rather than remaining in the lower crust. A mixture of cumulate–restite and liquid in the lower crust results, on average, in little if any Eu anomaly, although in specific lower crustal xenolith suites both positive and negative Eu anomalies have been reported (Rudnick 1992b).

The most convincing test for oceanic plateau accretion as a process of continental growth is to identify young accreted oceanic plateaus that are in the process of evolving into continental crust. It is well established that the Mesozoic Cordilleran crust in northwestern North America is relatively immature and, on the whole, more mafic in composition than Precambrian cratonic crust (Condie and Chomiak 1996; Patchett and Gehrels 1998). Wrangellia, the dismembered oceanic plateau accreted to northwestern North America in the Late Cretaceous, may provide a test for this continental evolution model (Condie 1997c). On the basics of results from several seismic reflection sections from the LITHOPROBE project in Canada (Clowes et al. 1992; Varsek et al. 1993), Wrangellia can be traced laterally at depth into central British Columbia (Fig. 6.14). The eastern margin of Wrangellia is a major transcurrent fault system (the Fraser–Pasayten fault system) separating the accreted oceanic plateau from the Precambrian craton. Although greenstone basalts with both arc and plume geochemical characters occur in Wrangellia, the average composition of these basalts clearly is plume-like in

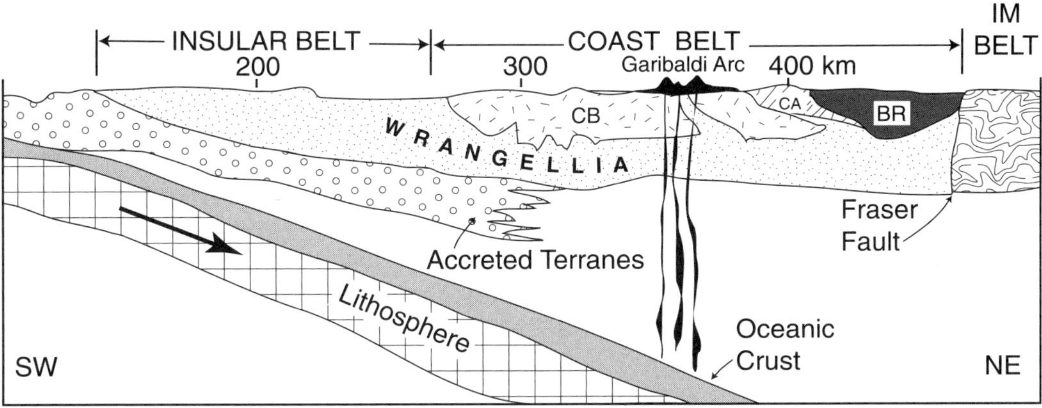

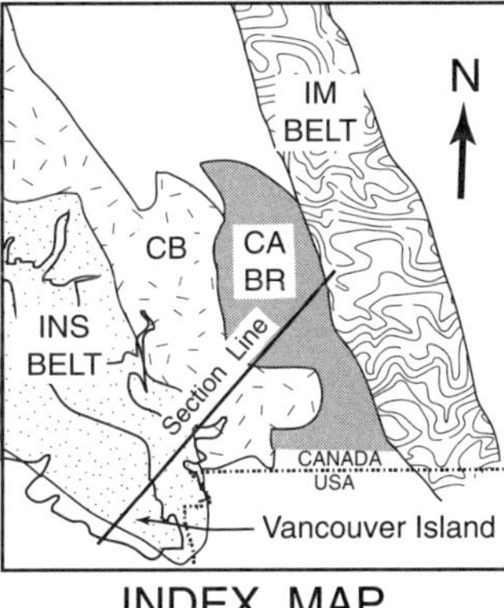

Figure 6.14. Speculative geologic cross section of southeastern British Columbia, Canada, showing possible extent of Wrangellia. Modified after Clowes et al. 1992 and Condie 1997c. INS = Insular belt (Alexander + Wrangellia); CB = Coast Range batholith; CA = Cadwallader terrane; BR = Bridge River terrane; IM = Intermontane belt.

character. Since accretion to the North American craton, the upper crust in Wrangellia has been intruded by the Coast Range batholith, a complex of felsic plutons carrying a subduction-zone geochemical signature (Nb and Ta depletion) and having juvenile isotopic characters (Samson and Patchett 1991; Brandon and Smith 1994). Also, the Cascade–Garabaldi arc system and its predecessors have been erupted on the surface of Wrangellia. With the emplacement of the Coast Range batholith and the eruption of arc volcanics, the upper crust of Wrangellia has become more felsic and appears to be evolving into more typical upper continental crust. Perhaps Wrangellia provides us with a young and still evolving example of how continental crust is formed from two sources: the lower crust from an accreted oceanic plateau and the upper crust from subduction-related processes.

What the Future Holds

Our ongoing studies of continents suggest that continents, subduction, and water go hand in hand. However, the details of how arcs, which are largely mafic, evolve into felsic continental crust still remain problematic. In any case, hydrous melting of mafic sources is important to form felsic magmas to make continental crust. What about the role of mantle plumes in making continents? Although geochemical arguments clearly favor arcs in making continental crust, we cannot yet eliminate plume contributions. First and foremost, more effort is needed to see through melting, crustal contamination, and metasomatic effects in lower crustal rocks to sort out plume and arc components. If the bottom line eventually shows that plumes are of little importance in the growth of continents, we are then faced with an even more difficult and intriguing question: Why? If thick oceanic plateaus are not accreted to continents, what is their fate? If the lower crust does not contain an important underplated mantle plume component, how did it escape underplating when plumes have bombarded the continental as well as the oceanic lithosphere through time? Answers to these questions should be very important in understanding how continents evolve and how they are related to the convecting mantle.

Another fascinating question is, what does the oceanic plateau model imply about the origin of the first continents? The earliest continents, in fact, may have started out life as oceanic plateaus (Abbott and Mooney 1995). Some may have formed by collision of ocean-ridge and oceanic plateau crustal blocks, followed by subduction zones developing around their margins. Partial melting of the deep mafic roots of the thickened blocks or of mafic rocks in descending slabs around their margins (Martin 1994) may have led to the production of felsic magma that was emplaced at shallow crustal levels, forming the first felsic crust. Because buoyant subduction is thought to have been more important in the Archean than afterwards (Abbott 1996), and because Archean oceanic crust may have been similar in thickness to oceanic plateaus (up to 20 km) (Sleep and Windley 1982), both oceanic crust and oceanic plateaus may have been accreted to the margins of the first Archean continents. Such collisions may have been very important during the Late Archean when the first supercontinent formed (Rogers 1996; Condie 1994, 1997c, 1999). If these ideas survive, they clearly imply that mantle plumes were important in nucleating continents, then subduction took over in producing the felsic component in continents. Why do we not see this process occurring today, or do we see it and not yet recognize it?

This brings us to another question that will be returned to in Chapter 8. Has mantle plume activity been episodic and, if so, what should we look for in the geologic record that records this episodicity?

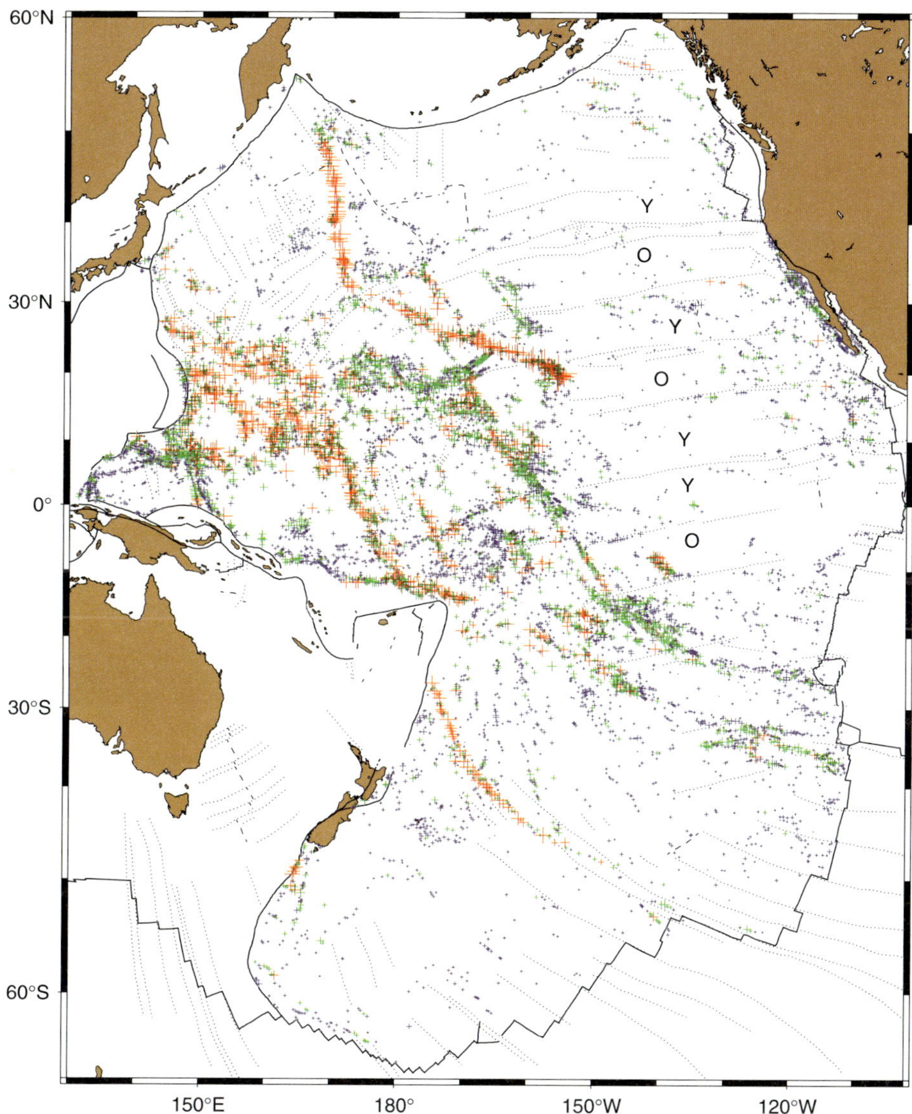

Plate 1. Distribution of seamounts in the Pacific basin based on satellite gravity altimetry data. The size of each cross is proportional to the maximum vertical gravity gradient. Small seamounts, blue; intermediate seamounts, green; and large seamounts, orange. Bold lines are plate boundaries, dotted lines are fracture zones, and dashed lines are troughs. Reproduced with permission from Wessel and Lyons (1997). Copyright © 1997 by the American Geophysical Union.

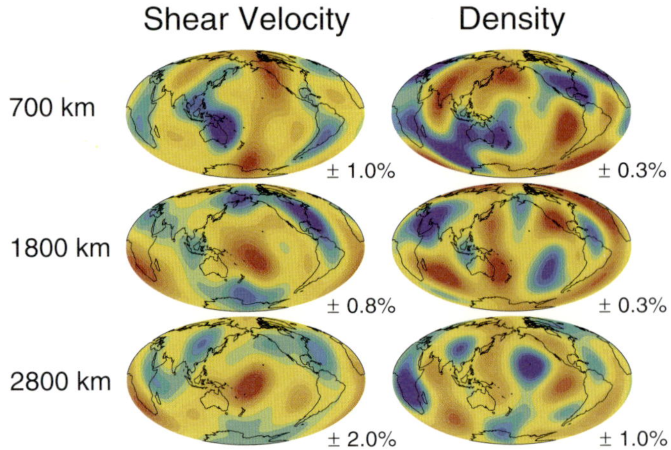

Plate 2. Shear wave velocity and corresponding density model for three depths in the mantle. Blue = higher than average velocity and density; red = lower than average velocity and density. Map scale given in percentages. Courtesy of Miaki Ishii.

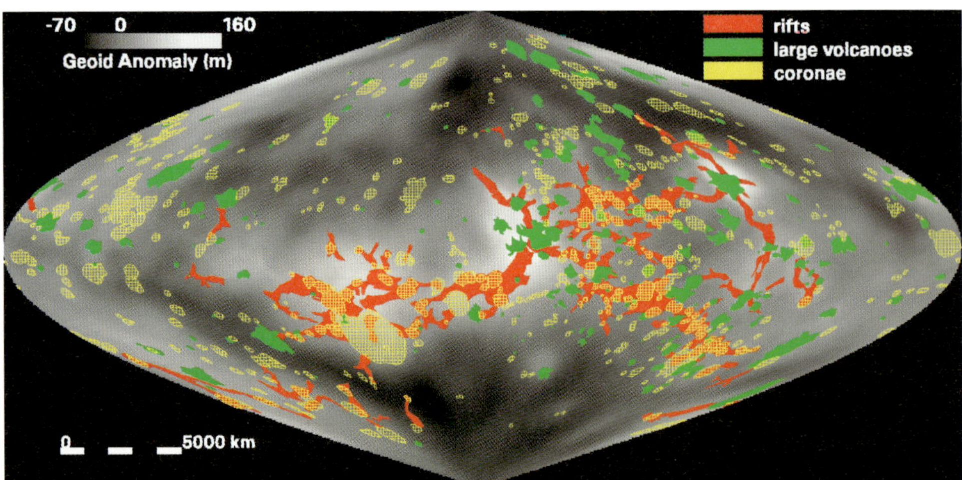

Plate 3. Sinusoidal projection of the Venusian surface centered at 180° longitude. Many large shield volcanos rest on geoid highs, and rifts tend to connect geoid highs. Most coronae lie along the rift systems but away from geoid highs and lows. Reproduced with permission from Herrick (1999). Copyright © 1999 by the American Geophysical Union.

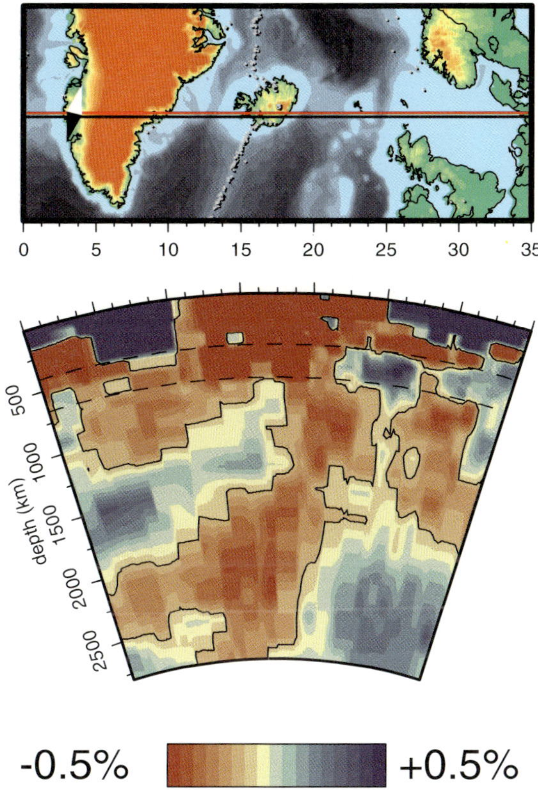

Plate 4. S-wave anomaly cross section beneath Iceland showing a low-velocity anomaly that extends to the base of the mantle. Reproduced from Bijwaard and Spakman (1999), with permission. Copyright © 1999 by Elsevier Science. Red = slow, yellow = normal, and blue = fast velocities.

7

Mantle Plumes in the Archean

Introduction

There is considerable interest in the role that mantle plumes may have played in Archean ($\geq$2.5 Ga) magma production, crustal underplating, production of oceanic plateaus, and in cooling of the mantle (Abbott 1996; Tomlinson and Condie 2001). Mantle temperatures must have been higher in the Archean, and hence the sinking of buoyant oceanic lithosphere into the mantle can account for only a small fraction of Archean heat loss (Bickle 1986; Davies 1992). Although plumes today account for no more than about 10% of the Earth's heat loss, the question of whether plumes were more important in cooling the mantle during the Archean is an important unknown in terrestrial thermal history (Davies 1993). If plumes were more important in the early stages of Earth history, as suggested by Fyfe (1978), perhaps the Archean Earth was more like Venus is today.

In this chapter we will review methods that have been used to identify Archean mantle plumes by using igneous rocks derived from plume sources and discuss the results in terms of mantle evolution.

Tracking Plumes into the Archean with Greenstones

Overview

How do we evaluate the role of mantle plumes in the early history of Earth? Three general approaches to this question are well established. First, it is well known that rock associations in oceanic plateaus and flood basalts differ from those in oceanic crust, oceanic islands, and arc systems (Condie 1997a; Kerr et al. 2000). In principle, we should be able to track plumes through the geologic record by using greenstones, flood basalts, and giant dyke swarms. A second approach is to use the chemical composition of mafic igneous rocks in greenstone successions (Condie 1997a; Kerr et al. 2000). As previously discussed, incompatible element distributions are particularly sensitive to different sources within the mantle. And finally, the distribution and composition of komatiites in the geologic record may give us important information on the depth, temperature, and importance of plume magma sources with time. We will now review each of these approaches.

Greenstone Lithologic Associations

If we are to use greenstones to track ancient mantle plumes, we must be able to distinguish between the various tectonic settings in which greenstones form. We can generally identify greenstones formed in continental settings (such as rifts, hotspots, cratons) by their distinct field, lithologic, and geochemical characters (Condie 1994; Thurston 1994; Kerr et al. 2000). However, to distinguish among greenstones formed in various oceanic settings is more difficult. From our database on modern oceanic plateaus and recently accreted oceanic plateaus (such as Wrangellia and the Caribbean plateau), oceanic plateaus are composed chiefly of pillow basalts and related hyaloclastic deposits, and mafic sills and various cumulate rocks (including ultramafics) become progressively more important with depth (Saunders et al. 1996; Kerr et al. 1997a). In addition, young oceanic plateaus, such as Ontong Java, are overlain by a thin veneer of pelagic carbonates and chert (Berger et al. 1992). Ophiolites contain similar rock types, and indeed it is difficult to distinguish plateau-related from ophiolite-related greenstones if the distinct ophiolite stratigraphy (including sheeted dykes and harzburgites) is not preserved.

Thurston and Chivers (1990) showed that Archean greenstones comprise several different lithologic assemblages. Most appear to represent arc assemblages (Condie 1994). However, two assemblages may represent remnants of LIPs and thus reflect mantle plume sources. Both are largely or exclusively submarine. Archean *mafic plain assemblages* contain a large proportion of pillow basalt with variable amounts of komatiite and small amounts of chemical sediments such as chert and banded iron-formation (BIF). Felsic-to-intermediate volcanic rocks are uncommon in these greenstones. *Platform assemblages*, which commonly overlie felsic basement, also comprise chiefly submarine basalts and komatiites. In some cases, a basal unconformity is preserved with tonalite conglomerates and sandstones (such as Steep Rock and North Caribou Lake greenstone belts, western Superior Province). In addition, platform successions contain carbonates, BIF, and minor amounts of felsic volcanic rocks. In some cases, platform volcanic sequences contain detrital zircons inherited from the basement and thus appear to be variably contaminated with upper continental crust (Tomlinson et al. 1999). The mafic plain succession is thought to represent a remnant of an oceanic plateau, whereas the platform succession may have been erupted through, or deposited on, continental crust.

Most Archean greenstones are terranes or superterranes that contain several to many greenstone blocks amalgamated to make one greenstone "belt" (Thurston 1994; Condie 1994; de Wit and Ashwal 1995; Polat et al. 1998). The giant 2.7-Ga Abitibi greenstone belt in the Superior Province of Canada, for instance, contains 10 or more tectonically juxtaposed blocks, each with its own distinct rock sequences and deformational history (Desrochers et al. 1993; Xie et al. 1993). These are generally grouped into one of five "domains" (Fig. 7.1). Several of the Abitibi domains comprise the mafic plain association and are interpreted as remnants of oceanic plateaus and arcs that collided 2.7 Ga (Polat et al. 1998, 1999). Four greenstone successions from the Kinojevis Group

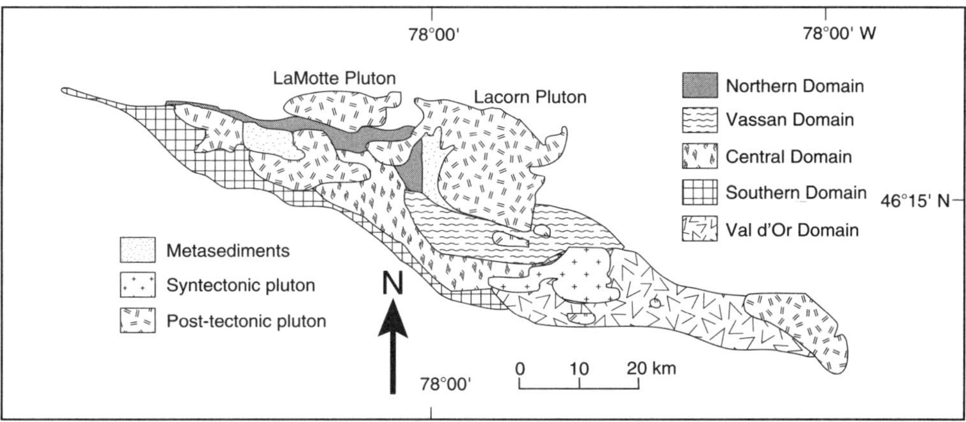

Figure 7.1. Generalized map showing distribution of accreted tectonic domains in the Late Archean Abitibi greenstone belt, southern Ontario. Modified after Desrochers et al. (1993).

have been described as plume-generated magmas (Xie 1996). Some arc volcanism, such as that in the Val d'Or domain, occurred after collision of most of the Abitibi domains (Desrochers et al. 1993). Numerous greenstones in the western Superior Province have mafic plain associations and appear to be the products of plume magmatism (Tomlinson et al. 1999; Tomlinson and Condie 2001; Kerr et al. 2000). For example, the 2.97-Ga Lumby Lake greenstone succession in western Ontario is dominated by mafic volcanic rocks, mostly pillow basalts, but also contains komatiite flows, chemical and clastic sediments, and minor felsic volcanics (Fig. 7.2). Basalt flows commonly have massive lower portions and pillowed tops; individual flows are occasionally separated by thin units of BIF or chert – probably of hydrothermal origin. Komatiite flows typically have spinifex-textured tops and cumulate bases and are associated with pyroclastic komatiites.

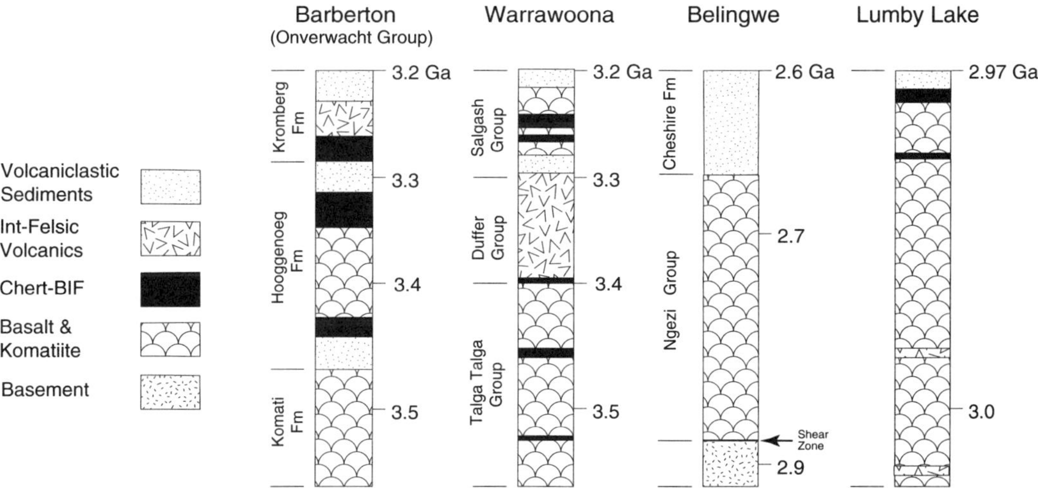

Figure 7.2. Generalized stratigraphic sections of Archean mafic–plain associations. Data from Kusky and Kidd (1992), Barley (1993), Lowe (1999), and Tomlinson et al. (1999).

Another example of a plume-related greenstone with oceanic plateau affinities is the Belingwe greenstone in Zimbabwe. This 2.7-Ga greenstone succession includes up to 6.5 km of komatiitic and basaltic flows of the Ngezi Group (Fig. 7.2) (Nisbet et al. 1993; Kusky and Kidd 1992). Pillow lavas are common throughout the Ngezi Group, but only the Zeederbergs Formation contains abundant hyaloclastic breccias and tuffaceous beds. Unconformably overlying this mafic plain sequence is the Cheshire Formation, a sequence of siliclastic sediments, chert, and carbonates that may represent hydrothermal deposits associated with plume volcanism.

Of the few known Early Archean greenstones (≥3.5 Ga), the two best preserved examples (Barberton and Pilbara) both comprise, in part, mafic plain associations. The 3.5-Ga Barberton greenstone in South Africa, which again is an amalgamation of terranes, is composed of up to 10 km of a succession of pillowed basalts, komatiites, and associated chert and carbonate (Lowe et al. 1985; Lowe 1999) (Fig. 7.2). Both sediments and submarine volcanics were formed in a broad basin – perhaps on top of an oceanic plateau. The development of weathered surfaces on some flows indicates at least localized subaerial exposure during eruption of the magmas. Komatiite flow sequences usually end with komatiitic tuffaceous units, which, in turn, are overlain by shallow-water chemical sediments (chert and carbonates). Low-temperature hydrothermal alteration resulted in widespread silicification of komatiitic flow tops. In contrast, basalt eruptions were largely subaqueous flows with little evidence of pyroclastic eruptions. Another characteristic of the Barberton succession is that the volcanism is strikingly bimodal with small volumes of dacitic to rhyolitic tuff associated with the basalts and komatiites.

In the Pilbara craton in Western Australia, the Warrawoona greenstone succession (3.46 Ga) consists of three sequences (Krapez 1993; Barley 1993). The lower Talga Talga sequence, composed chiefly of submarine basalt, gabbro, and diabase intercalated with minor cherty sediments, is a mafic plain succession of probable plume origin (Fig. 7.2). Overlying this sequence is the Duffer sequence composed chiefly of calc–alkaline volcanic rocks and associated sediments probably of arc origin. And finally, this succession is capped with the Salgash sequence, comprising again a mafic plain succession of basalts, komatiites, and cherty sediments. Thus, two of the three greenstone successions in the Warrawoona Group appear to have plume affinities.

From a total of 51 Archean greenstones for which lithologic proportions are available, 65% have arc affinities and 35% oceanic plateau or MORB affinities (Condie 1994, Condie 1997c). Out of 96 post-Archean greenstones, only about 10% have oceanic plateau or MORB affinities. Thus, in greenstones of all ages, lithologic proportions show that arc-types greatly exceed oceanic plateau and MORB types in abundance. However, Archean greenstones clearly contain a larger proportion of examples with plume affinities than post-Archean greenstones.

Greenstone Geochemistry

One way to track the relative importance of mantle plumes in the Archean is from the geochemical signatures in greenstone basalts and lower crustal xenoliths (Condie 1994, 1997c, 1999; Campbell et al. 1989; Campbell 1998; Kerr et al. 2000). As discussed

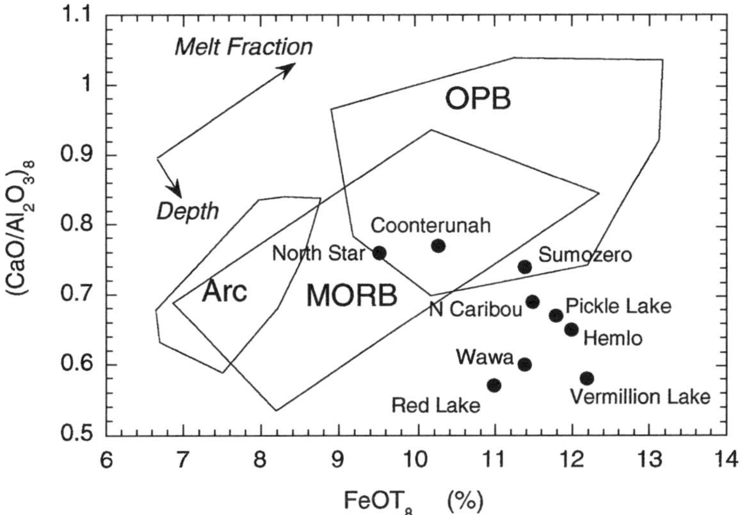

Figure 7.3. $(CaO/Al_2O_3)_8$ versus $(FeOT)_8$ graph showing the average compositions of basalts from mafic–plain Archean greenstone associations. Data from Kerrich et al. (1999), Tomlinson and Condie (2001), and Green et al. (2000). See Figure 4.17 for other information.

in Chapter 3, oceanic plateau basalts (OPB) generally show flat, incompatible element distributions with slight depletion in the most incompatible elements in some sequences (Fig. 5.1). In contrast, arc-related basalts show negative Nb–Ta anomalies and, in some instances, negative Th, P and Ti anomalies. The most striking distinction between arc-related and oceanic nonarc-related basalts is the negative Ta–Nb anomaly (McCulloch and Gamble 1991), which may be used to constrain the tectonic setting of oceanic basalts in Archean greenstone belts (Condie 1994). In addition, young oceanic plateau basalts typically have relatively flat, chondrite-normalized REE patterns in contrast to MORB, which are depleted in light REE. The CaO/Al_2O_3 and iron contents of basaltic magmas are sensitive indicators of the depth and degree of melting (Chapter 4). Average values of these indicators calculated at $MgO = 8\%$ are shown in Figure 7.3 for examples of Archean greenstones with mafic-plain assemblages. Although these greenstones overlap the oceanic plateau field and show a range in melt fraction and depth of melting, most appear to come from plume sources deeper than those of young oceanic plateaus.

As shown in previous chapters, the Th/Ta–La/Yb diagram is useful in constraining the source composition of basaltic magmas. Contamination of plume-derived magmas with upper continental crust raises Th/Ta and La/Yb ratios, and samples generally fall upon mixing arrays between average Archean upper continental crust and a mafic end member in the OPB or NMORB fields (Fig. 5.8). In some instances, it is possible to identify crustal contamination trends in Archean greenstone belts (Sun et al. 1989; Condie 1994). In these cases, OPB can acquire a "pseudosubduction" geochemical signature by upper crustal contamination. Some Archean greenstone belts have been described as arc or back–arc-type based on limited or ambiguous geochemical data, and these may also represent crustally contaminated, plume-derived magmas. The only reliable way to sort out crustally contaminated, plume-derived basalts from subduction-related basalts is by lithologic association. Arc-related basalts, which can be interbedded with

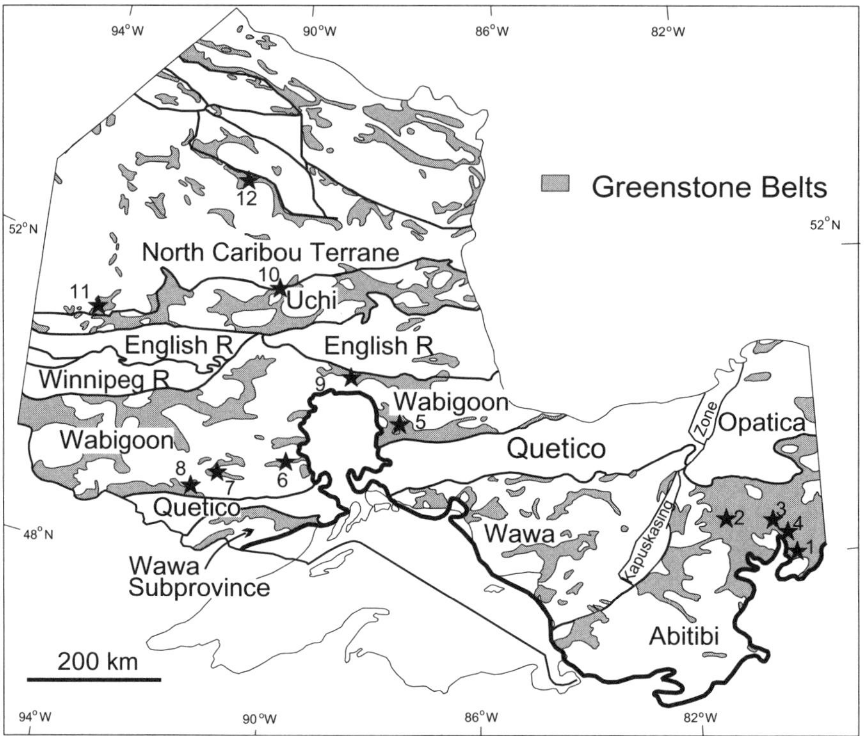

Figure 7.4. Map showing distribution of Archean greenstone belts in western Canada. Numbered localities are greenstones with plume affinities based on both lithologic and geochemical data. After Tomlinson and Condie (2001). 1, Boston township; 2, Tisdale township; 3, Munro township; 4, Kinojevis Group; 5, southern Onaman–Tashota terrane; 6, Heaven Lake belt; 7, Lumby Lake belt; 8, Steep Rock belt; 9, DiAlton Lake–Toronto Lake belt; 10, Pickle Lake belt; 11, Red Lake belt; 12, North Caribou Lake belt.

plume-related basalts (Kerrich et al. 1999; Hollings et al. 1999) are typically associated with intermediate to felsic volcanics and graywackes. Field and geochemical data provide strong evidence that plume and arc processes operated synchronously or intermittently at the same location during development of many Archean greenstones.

Archean basalts from the mafic plain assemblage typically have Th/Ta–La/Yb distributions similar to young oceanic plateau basalts (Tomlinson et al. 1999; Tomlinson and Condie 2001). In the western Superior Province in Canada (Fig. 7.4), about 60% of the basalts fall in or near the field of modern oceanic plateau basalts, and 11% fall in the field of ocean island basalts on the Th/Ta–La/Yb diagram (Figs. 5.7–5.11 and 7.5). Several of the 3.0–2.9-Ga greenstone assemblages in the northwestern Superior Province are identified as plume-generated sequences that erupted through thin continental basement, and these are generally of similar age (Hollings and Kerrich 1999; Tomlinson and Condie 2001). Some basalts and komatiites in the Red Lake greenstone belt are inferred to have interacted with older felsic crust based on geochemical evidence of crustal contamination (Tomlinson et al. 1998). These define the high Th/Ta and La/Yb part of the field for the western Superior Province. As mentioned previously, many of the Abitibi greenstone successions appear to represent accreted oceanic plateau

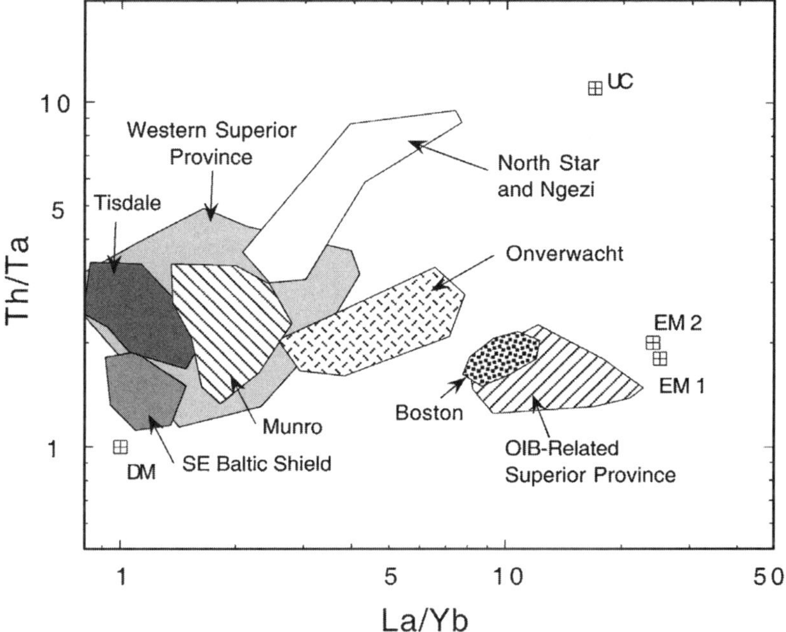

Figure 7.5. A Th/Ta versus La/Yb graph showing the distribution of Archean basalts from greenstones with mafic-plain assemblages. Data sources: Puchtel et al. (1999) and Tomlinson and Condie (2001). See Figure 5.8 for other information and comparison with modern basalts.

and oceanic island fragments (Desrochers et al. 1993; Xie et al. 1993). The La/Yb and Th/Ta ratios for basalts from these Abitibi sequences show a greater range of values compared with those of modern plume-generated basalts, as evidenced by data from Tisdale, Boston, and Munro townships (Fig. 7.5) (Tomlinson and Condie 2001).

On the basis of stratigraphy and geochemistry, the Kostomuksha and Sumozero–Kenozero greenstone belts in the southeastern Baltic shield (2.9–3.0 Ga) are interpreted to represent portions of obducted oceanic plateaus (Puchtel et al. 1998, 1999). These successions are dominated by mafic volcanics with lesser amounts of komatiite, gabbro, and komatiitic pyroclastic rocks. The Kostomuksha sequence is capped by chemical sediments and felsic volcanic rocks. In terms of Th/Ta ratios, komatiitic and basaltic lavas from these belts are similar to modern oceanic plateau basalts (Fig. 7.5). The La/Yb ratios are slightly lower than modern OPB, reflecting more depleted mantle sources.

Basalts from the Onverwacht and Warrawoona Groups also show oceanic plateau affinities on the Th/Ta–La/Yb plot (Fig. 7.5). The $\approx$ 3.46-Ga North Star Basalt is the lowest stratigraphic unit in the Talga Talga sequence of the Warrawoona Group (Fig. 7.2) (Barley 1993). Lavas in this unit range from massive to pillowed and are commonly associated with hyaloclastic breccias, which indicates subaqueous eruption. North Star basalts plot in and above the OPB field on the Th/Ta–La/Yb diagram (Fig. 7.5). The trend of high Th/Ta ratios leading toward upper continental crust suggests crustal contamination (Condie 1994). Basalts from the 3.5-Ga Onverwacht Group in the Barberton belt of South Africa also show increasing Th/Ta with La/Yb but with much lower Th/Ta ratios than the North Star basalts. This may reflect a plume source with enriched components (EM1, 2). Basalts from the Ngezi Group of the Belingwe belt in Zimbabwe plot within the North Star field on Figure 7.5, which is again suggestive of plume-generated magmas contaminated by upper continental crust.

Geochemical signatures of Archean plume-generated greenstones may be more diverse than those of most modern oceanic plateau volcanic sequences. This is partly due to crustally contaminated plume magmas with elevated Th/Ta ratios, but some of the oceanic Archean greenstone sequences show higher Th/Ta ratios and lower La/Yb ratios than young oceanic plateau basalts. The relatively high Th/Ta ratios in some Abitibi greenstone sequences may reflect recycling of continental crust within mantle plumes, whereas lower La/Yb ratios may be ascribed to light REE depletion of the source owing to previous melt extraction. High La/Yb ratios in komatiites and basalts that result from heavy REE depletion combined with negative Zr and Hf anomalies (see the next section) suggest the presence of majorite garnet in the restite and melting at depths of 300–600 km (Xie et al. 1993).

Komatiites

Overview

Komatiites are high-temperature, ultramafic volcanic rocks containing more than 18% MgO (Arndt and Nesbitt 1982). Although the tectonic setting in which komatiites form is still uncertain, most investigators agree that their high liquidus temperatures require an anomalously hot mantle plume source (Campbell et al. 1989; Xie and Kerrich 1994; Herzberg 1995). Komatiites are rare after the Archean and are especially rare in Phanerozoic greenstones. The only well-documented examples in the Phanerozoic are the komatiites on Gorgona Island west of Ecuador, which appear to be related to the Caribbean oceanic plateau (Storey et al. 1991). They were probably derived from the hot plume tail of the Galápagos hotspot about 90 Ma (Kerr et al. 1996b). Komatiites are common in Archean mafic-plain greenstones, where they comprise up to 30% of the succession (Fig. 7.6). In most of these greenstones, however, they comprise 5% or less of the succession, and in Archean arc-type greenstones they are absent (de Wit and Ashwal 1995). After the Archean, komatiites decrease rapidly in abundance.

Heads It's Basalts, Tails It's Komatiites

Komatiites with high MgO (25%) and low water contents ($\leq$4%) have estimated liquidus temperatures of 1410–1420 °C (Grove et al. 1997). For those Archean komatiites with the highest MgO contents (27–30%), implied dry eruption temperatures are on the order of 1600 °C (Arndt). If the magmas contain appreciable water, as suggested by some investigators, these temperatures could be lower by 200–250 °C (Grove et al. 1997). However, if Archean komatiitic magmas were relatively dry, the temperature difference between the sources of associated basalt and komatiite is 400–450 °C. This equates with potential source temperatures of komatiites of the order of 1800 °C or more compared with basaltic magma source temperatures of 1400 °C (Campbell et al. 1989). This implies either or both (1) the Archean mantle was hotter than the modern mantle or (2) komatiites came from anomalously hot regions within the mantle.

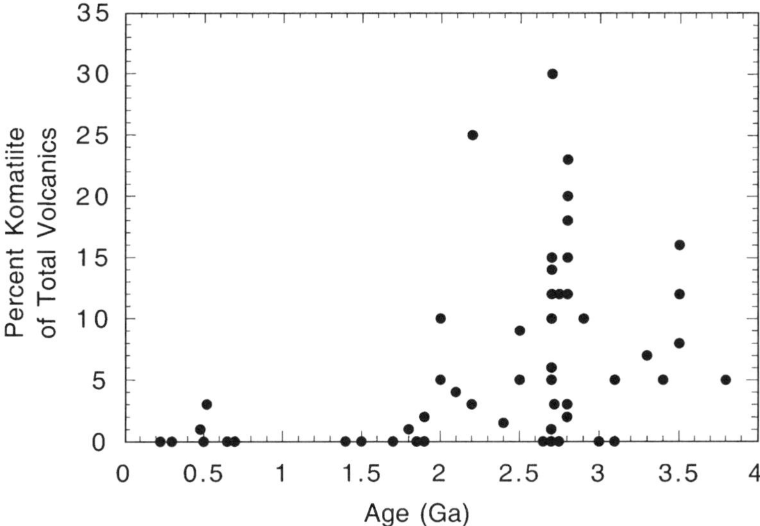

Figure 7.6. Abundance of komatiite in greenstones as function of age. After de Wit and Ashwal (1995).

To accommodate the difference in dry liquidus temperature between tholeiitic basalt and komatiite with a single mantle plume source, Ian Campbell and colleagues at the Australian National University published a paper in *Nature* with the title "Heads It's Basalts, Tails It's Komatiites" (Campbell et al. 1989). A mantle plume source for komatiites provides a local high-temperature region in the plume tail, but at the same time the plume head, which has been cooled with entrained mantle, provides a more normal temperature source for associated basaltic magmas. This model supports the idea that komatiites are associated with mantle plumes and that Archean mafic-plain greenstone assemblages, which contain komatiites, are the products of plume-generated magmatism.

Geochemistry

Two groups of komatiites are recognized in the Archean. Al-undepleted komatiites have near chondritic Al_2O_3/TiO_2 and Ti/Zr ratios as well as relatively flat, heavy REE patterns; these komatiites show flat or depleted incompatible element distributions (Fig. 7.7). In contrast, Al-depleted komatiites have low Al_2O_3/TiO_2 and Ti/Zr ratios with fractionated REE patterns and usually show relative enrichment in incompatible elements (Arndt 1994). Some greenstones contain one or the other type of komatiite, whereas other greenstones like the Abitibi and Lumby Lake greenstones in the Superior Province of Canada include both Al-depleted and Al-undepleted types, often in the same succession (Xie et al. 1993; Tomlinson et al. 1999). Al-depleted komatiites come from sources in which garnet (probably the high-pressure garnet, majorite) remains in the restite after melting, whereas Al-undepleted komatiites come from sources in which garnet does not remain in the restite (Jochum et al. 1991). With high-precision trace element analyses it is possible to detect relatively small anomalies on primitive-mantle normalized plots – anomalies that have proved to be extremely important in constraining the depths at which komatiitic magmas are produced. Negative Nb–Ta anomalies are common in some komatiites and appear to reflect contamination with

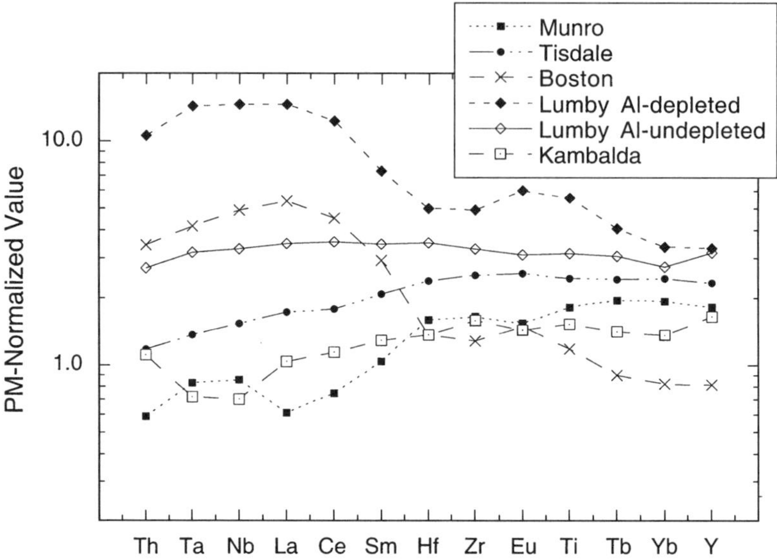

Figure 7.7. Incompatible element distributions in Archean komatiites. Data from Xie et al. (1993), Arndt (1994), Fan and Kerrich (1997), and Tomlinson et al. (1999). Other information given in Figure 5.1.

continental crust, as illustrated by komatiites from Kambalda in Western Australia (Fig. 7.7). Many Al-depleted komatiites show negative Zr–Hf anomalies, which probably reflect majorite garnet fractionation (Fig. 7.7, Boston and Lumby Lake). In contrast, Al-undepleted komatiites either show no Nb, Ta, Zr, or Hf anomalies or show positive Nb–Ta and Zr–Hf anomalies (Lumby Lake and Munro) (Fan and Kerrich 1997). If these elements were not remobilized during metamorphism or alteration, the latter feature may reflect perovskite addition.

The Nb/Nb*, Zr/Zr*, and Hf/Hf* ratios are very useful when plotted against $(La/Yb)n$ ratios in distinguishing three major evolutionary paths recorded in komatiites: majorite fractionation, perovskite addition, and crustal contamination (Fig. 7.8). The starred values in each ratio are interpolated values on primitive-mantle normalized plots. When plotted against the $(La/Yb)n$ ratio, Nb, Zr, and Hf evolutionary paths deviate from the cross-over point (ratios equal to 1) in opposite directions. The ≈2.7-Ga greenstone sequence at Kambalda in Western Australia is well known for its volcanics that have interacted with continental crust (Arndt and Jenner 1986; Lesher and Arndt 1995). The ≈4-km-thick sequence comprises komatiites, komatiitic basalts, and tholeiite flows. As discussed by Lesher and Arndt (1995), crustal contamination is minimal in the lower parts of the section but shows a general rise with increasing stratigraphic height. Komatiites that have undergone crustal contamination show large negative Nb–Ta anomalies, relatively enriched incompatible element patterns, and plot on the crustal contamination trend in Figure 7.8(a).

High-pressure experimental studies show that komatiitic melts can be produced by either large degrees of melting of mantle rocks (>50%) at low pressures or by small degrees of melting at high pressures (Nisbet and Walker 1982; Herzberg 1995). Experiments also confirm that at highpressures, majorite garnet and Mg-silicate perovskite

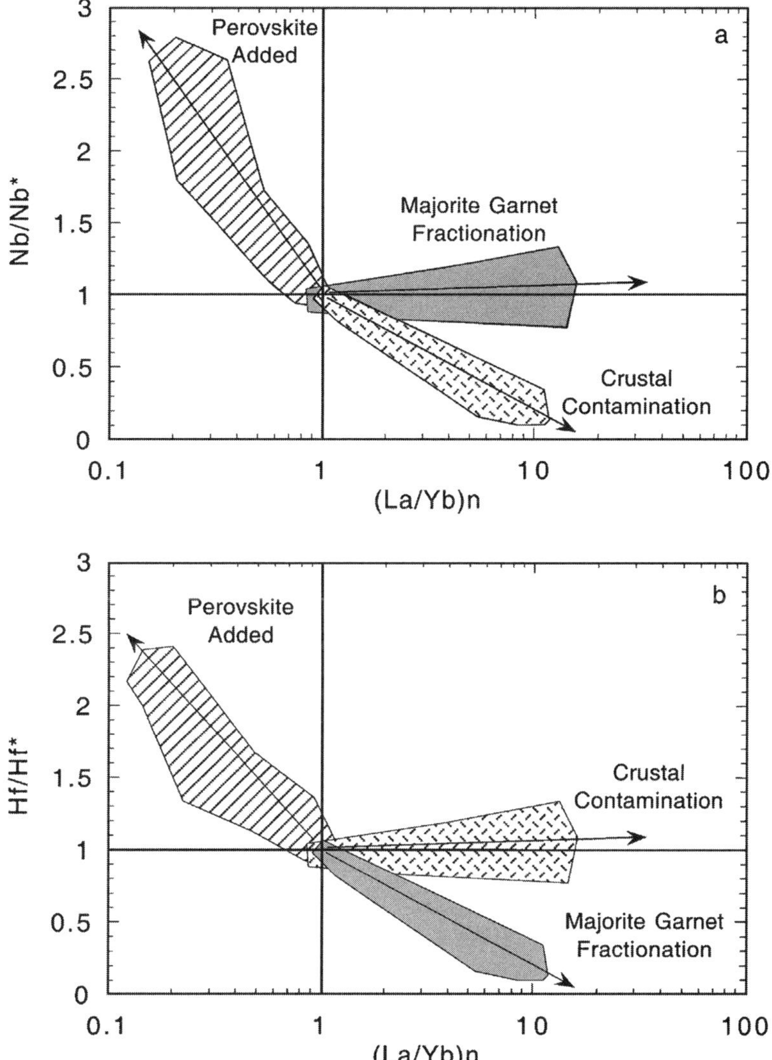

Figure 7.8. Some Nb/Nb* (a) and Hf/ Hf* (b) ratios versus (La/Yb)*n* for Archean komatiites. Nb* and Hf* are interpolated values on primitive-mantle normalized graphs, and (La/Yb)*n* is the La/Yb ratio normalized to primitive mantle. Patterned areas represent komatiite data, and arrows are calculated liquid trends for the processes indicated. Diagrams modified after Xie et al. (1993), Xie and Kerrich (1994), and Fan and Kerrich (1997).

can fractionate high-field-strength elements (HFSE) and rare earth elements (REE) (Drake et al. 1993; McCuaig et al. 1994). Because distribution coefficients of these elements change from low pressure (olivine, pyroxenes), through intermediate pressure (olivine, majorite garnet), to high pressure (magnesiowustite, Mg-silicate perovskite), it is possible to place constraints on the depth that komatiites segregate from their sources in mantle plumes using HFSE and REE distributions (Xie and Kerrich 1994; Fan and Kerrich 1997). Majorite garnet fractionation is recognized by negative Zr and Hf anomalies with no effect on Nb–Ta distributions (Fig. 7.8). These features, which characterize Al-depleted komatiites and basaltic komatiites, reflect magma segregation at depths of 425 to 540 km (Agee 1993).

At pressures exceeding 24 GPa, Mg-silicate perovskite (hereafter perovskite) becomes the major liquidus phase in the mantle (Drake et al. 1993). Because Nb, Ta, Zr and Hf are all compatible in perovskite, whereas light REE are incompatible, komatiitic melts generated at depths below the 660-km seismic discontinuity should be enriched in REE but depleted in Nb, Ta, Zr, and Hf given that perovskite is left in the restite (Drake et al. 1993). The opposite trends result from melting of a perovskite-enriched source or by accumulation of perovskite in komatiitic melts. Positive Nb, Ta, Zr and Hf anomalies in komatiites from Munro township in the Abitibi belt (Figs. 7.7 and 7.8) are interpreted by Xie and Kerrich (1994) to reflect melting in a deep mantle plume source that was previously enriched in perovskite. These rocks plot in the perovskite-added field in Figure 7.8(a) and (b). Komatiite-tholeiite successions within the Abitibi greenstone belt show a wide range of compositions and of HFSE and REE fractionations (Xie and Kerrich 1994; Fan and Kerrich 1997). Unlike the Munro komatiites, Tisdale komatiites, which do not show HFSE anomalies (Fig. 7.7), were likely produced by large degrees of melting of the upper mantle at shallow depths. Komatiites from Boston Township, on the other hand, show incompatible element enrichment and negative Zr–Hf anomalies. These magmas were produced by partial melting at greater depths (300–600 km), leaving majorite garnet in the restite. Because the Munro, Tisdale, and Boston successions occur in different terranes that accreted at 2.7 Ga to make the Abitibi belt, it is possible that each had a different mantle plume source when it formed, and that komatiitic melts segregated at different depths in each plume.

In the Lumby Lake and Wawa greenstone belts in western Ontario, Al-depleted komatiites with an OIB geochemical signature are depleted in heavy REE and display negative Zr and Hf anomalies (Fig. 7.7). These features are explained by Tomlinson et al. (1999) to represent melting with majorite garnet fractionation in the hot axial core of a mantle plume. On the other hand, basalts and Al-undepleted komatiites with an OPB geochemical signature from the same succession represent melting at shallower depths, and these may have come from the plume head (Tomlinson et al. 1999). If so, this indicates tapping of a chemically heterogeneous plume at different depths.

Archean Flood Basalts

The oldest well-documented continental flood basalts are the Ventersdorp (2.7 Ga) and Fortescue (2.77 Ga) successions in South Africa and Western Australia, respectively. The Ventersdorp lavas were erupted on the Early Archean Kaapvaal craton and the Fortescue on the Early Archean Pilbara craton (Arndt et al. 1991; Nelson et al. 1992). The Fortescue Group includes several stratigraphic units composed chiefly of tholeiitic basalts with smaller amounts of basaltic komatiite, mafic and felsic tuffs, and felsic pyroclastics (Barley et al. 1992). Sedimentary rocks are common in the Fortescue Group and include chiefly epiclastic and terrigenous sediments and minor carbonates. Preserved sections are up to 5 km thick, and subaerial tholeiitic flows dominate. Basalts rest directly on Early Archean basement over much of the Pilbara craton. Sedimentary rocks were deposited in lacustrine, deltaic, and various alluvial fan environments.

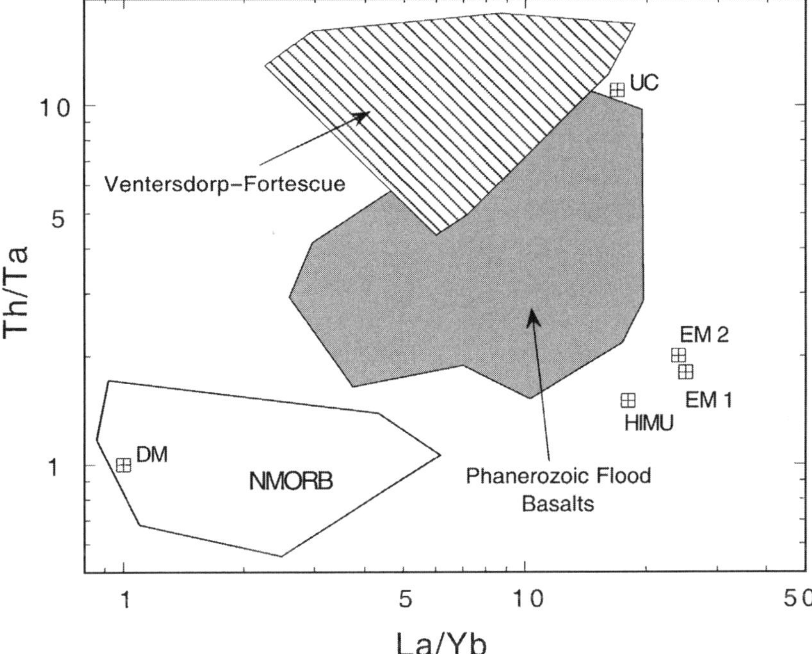

Figure 7.9. A Th/Ta versus La/Yb graph showing the distribution of the Late Archean Ventersdorp and Fortescue flood basalts. Data from Nelson et al. (1992) and Crow and Condie (1988). See Figure 5.8 for other information and comparison with modern basalts.

The Ventersdorp volcanics in southern Africa were erupted on older cratonic sediments that accumulated in an extensional basin (Bickle and Eriksson 1982). The Ventersdorp Supergroup is divided into three groups. The most voluminous volcanics are represented by flood basalts of the lower Klipriviersberg Group, which includes basaltic komatiite and tholeiite flows (Schweitzer and Kroner 1985; Crow and Condie 1988). Following eruption of this group, volcanism was more diverse and included intermediate and felsic units, and eruptions were interspersed with active faulting and sedimentation. Final stages of volcanism are characterized by tholeiitic basalt flows.

Both Fortescue and Ventersdorp basalts are relatively enriched in incompatible elements and show light REE-enriched patterns (Crow and Condie 1988; Nelson et al. 1992). Basalts from both groups have very high Th/Ta and La/Yb ratios with the Th/Ta ratio higher than in most Phanerozoic flood basalts (Fig. 7.9). This suggests considerable contamination of the magmas by the upper continental crust. Both sequences have negative $\epsilon_{Nd}(T)$ values (-1.5 to -4.4) which also are consistent with crustal contamination (Nelson et al. 1992). The intermediate and felsic components appear to be the products of fractional crystallization accompanied in some instances by crustal assimilation. The presence of xenocrystic zircon in Ventersdorp basalts also is strong evidence for crustal contamination of these lavas.

Ventersdorp and Fortescue flood basalts have many features that are consistent with derivation from mantle plumes during decompression melting. In both examples, basaltic komatiites reflect a high temperature source – probably in the plume tail. Geochemical data and subsidence history suggest that both successions were erupted in short periods from a plume source that was 100–150 °C hotter than normal mantle, and that lithospheric rifting allowed the plumes to rise to relatively shallow depths (White 1997). Once melts reached crustal levels, they were able to travel large distances, both as flows and as dykes and sills in the middle crust.

Plume-Head Underplating of the Lithosphere

One of the outstanding problems in Archean history is that of how the very thick Archean lithosphere formed. A recent study of mantle lithosphere xenoliths from the Archean Slave craton in northwestern Canada sheds light on this question. A sudden change in the composition of mantle xenoliths in kimberlites from this area implies a change in the composition of the Archean mantle lithosphere at a depth of about 145 km (Griffin et al. 1998). Garnets from xenoliths derived from depths less than 145 km are extremely depleted in Y, Zr, Ti, and Ga, and the xenoliths have very high MgO contents (Fo_{92-94}) compared with the deeper mantle layer (Fo_{91-92}). The shallow mantle lithosphere layer is similar in composition to mantle xenoliths from convergent plate margins and may have formed in a similar tectonic setting during collision and accretion of Archean arcs that now comprise the Slave craton. The sharp boundary in the lithosphere at 145 km strongly implies a two-stage process in the formation of this deep lithospheric root (Griffin et al. 1998).

Bearing on the origin of the deep layer are eclogite xenoliths from this layer containing diamonds with inclusions of the high-pressure mineral assemblage magnesiowustite + Mg-perovskite, which is stable only at and below the 660-km phase transition in the mantle. This strongly suggests that the material in the deep lithospheric layer has been transported from the deep mantle as a mantle plume. Griffin et al. (1998) proposed a model for the Archean mantle lithosphere beneath the Slave craton where shallow, strongly depleted lithosphere is formed at an active ocean ridge, subduction zone, or both. This lithosphere, with associated arcs as represented by the Yellowknife Supergroup, is accreted to form the Slave craton 2.7 Ga (Fig. 7.10(a) and (b)). At 2.6 Ga, a mantle plume rises to the base of this newly accreted lithosphere, heating the lithosphere and causing remobilization of low-melting fractions, which leads to further depletion and leaves an extremely depleted restite Figure 7.10(c) and (d). The study of Griffin et al. (1998) is important because it shows how mantle xenolith populations can be used to detect underplating of the lithosphere by plume heads. It is possible that the unusually thick lithosphere characteristic of Archean cratons may have resulted from widespread underplating by largely restitic material from plume heads.

Secular Changes in the Mantle

The Appearance of Enriched Mantle

As discussed in Chapter 5, enriched mantle components (EM1, EM2, HIMU) probably represent continental crust (EM2), lithosphere (EM1), and oceanic crust (HIMU) that have been recycled into the mantle during subduction or during delamination and slab break-off accompanying plate collisions. Only a few examples of Archean basalts with

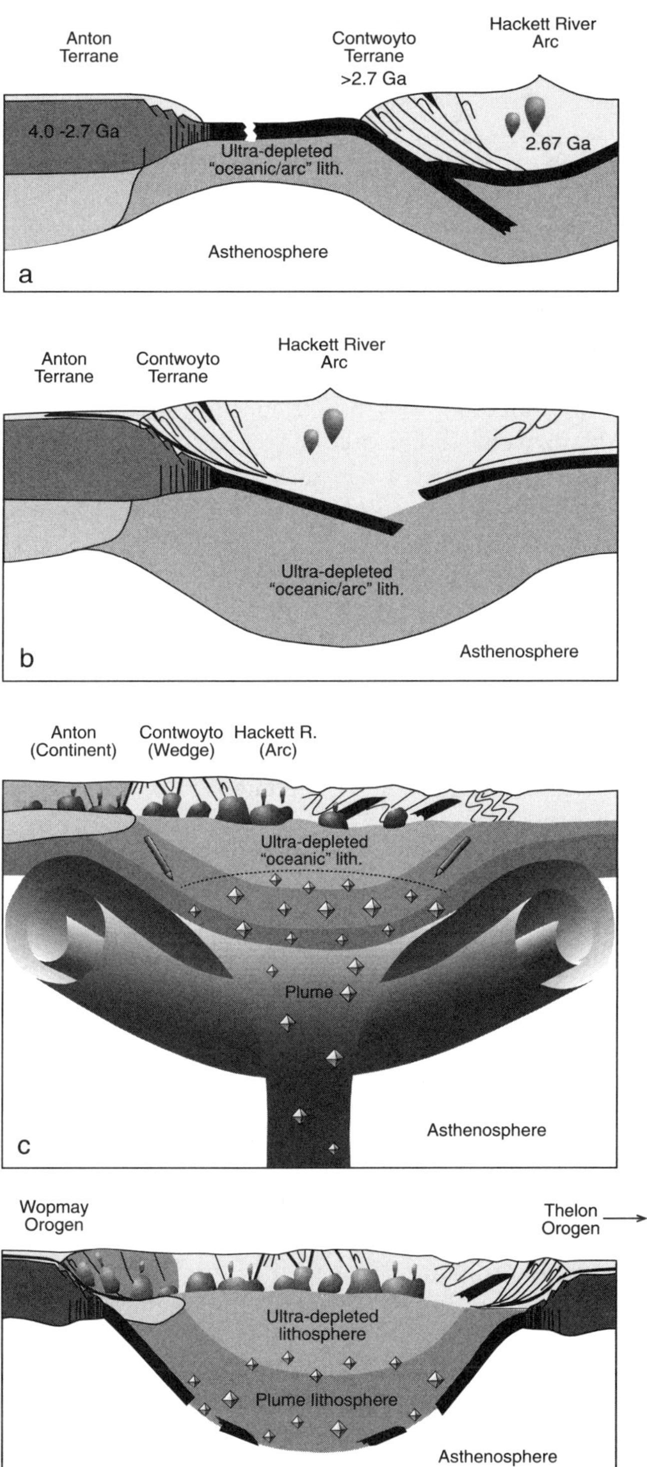

Figure 7.10. Schematic diagram showing development of two mantle lithosphere layers in the Archean Slave craton. The ultra-depleted shallow layer represents oceanic and subarc mantle, whereas the deep layer represents an accreted mantle plume head. Modified after Kusky (1990, 1991) and Griffin et al. (1998), with permission of Oxford University Press.

geochemical evidence for enriched mantle sources have been recognized (Tomlinson et al. 1999; Francis et al. 1999; Tomlinson and Condie 2001), and the first major record for these sources appears in basalts at about 1 Ga (Condie, 1994, 1997b). Mantle isochrons (Nd, Pb, Sr), however, suggest that geochemical domains began to be preserved in the mantle by about 2 Ga (Carlson 1994; Hofmann 1997). Why are these mantle components uncommon in most pre-1-Ga greenstone basalts that were erupted in oceanic tectonic settings? In particular, the HIMU component should be widespread in Archean greenstones because the dominant crust in the Archean was likely oceanic crust and was probably recycled rapidly into the mantle. One possibility is that basalts with enriched mantle sources have been overlooked in Archean greenstones because they are minor. Alternatively, as discussed in Chapter 5 (see Fig. 5.18), the enriched components may not have survived in the Archean mantle owing to more rigorous convection. Also, during probable superplume events at 1.9 and 2.7 Ga (see Chapter 8), the mantle may have been well mixed, losing much of the geochemical evidence for enriched and HIMU mantle domains. The similar Nb/U ratios in MORB and OIB mantle sources (Fig. 5.7) is also consistent with this interpretation.

Komatiites as Geothermometers

Komatiites provide important information on thermal regimes in the early mantle because, as previously mentioned, under dry conditions they imply mantle temperatures up to 400 °C higher than at present. Although Komatiites are generally assumed to form by partial melting of the mantle under dry conditions, recent experimental studies show that if water is present, the liquidus temperature of komatiitic magma is reduced substantially (Asahara et al. 1998). Supporting the existence of hydrous komatiitic melts in the Archean is the presence of a relatively small number of Archean komatiites containing vesicles, magmatic amphibole, or both, and the occurrence of minor pyroclastic komatiites in some greenstones (Stone et al. 1997; Arndt et al. 1998). However, experimental data also show that it is difficult to erupt wet komatiitic lavas because they should erupt explosively and show textural evidence of degassing, or they should solidify before eruption (Arndt et al. 1998). Although pyroclastic komatiites may reflect wet magmatic eruptions, they are rare in the geologic record. Also, the effect of water on liquidus temperature is reduced at the high degrees of melting (30–50%) necessary to produce komatiitic magmas. In addition, trace element distributions and isotopic characteristics of komatiites indicate that their source was depleted in incompatible elements, including water, before magma formation, leaving a dry source to melt. Cr content of olivine in equilibrium with chromite is also very sensitive to melting temperature. Olivines with Cr contents greater than 1550 ppm crystallize at temperatures above 1500 °C, whereas those with Cr contents less than 1000 ppm crystallize at temperatures below 1250 °C (Li et al. 1995). Both komatiite and basaltic komatiite typically have Cr contents of more than 2000 ppm, confirming they are high-rather than low-temperature melts. This is in contrast to low-temperature, high-Mg island arc magmas that have olivine with Cr contents less than 800 ppm.

For these reasons, Arndt et al. (1998) concluded most komatiites are derived from dry mantle sources and thus, that komatiite liquidus temperatures can be used to constrain source temperatures in the Archean mantle and to track the temperature of komatiite sources with time.

How Hot Was the Archean Mantle?

From the liquidus temperatures of dry komatiitic magma it is possible to estimate the potential temperature of the magma source region using the distribution of Fe and Mg between olivine and liquid (Ulmer 1989). Potential temperatures in the magma source can be calculated from the empirical relationship between liquidus and source temperature (McKenzie and Bickle 1988; Abbott, Bugess et al. 1994). Results of these calculations indicate that the Archean mantle sources of komatiite were 200–300 °C hotter than modern MORB sources (Campbell and Griffiths 1992; Nisbet et al. 1993; Abbott et al. 1994b). This, however, does not equate with a similarly elevated temperature for the entire Archean mantle but only for komatiite–picrite sources, which are likely to be mantle plumes. Campbell and Griffiths (1992) have argued that because komatiites are a very minor component in Archean greenstones, it is the tholeiites in these greenstones that more closely reflect the temperature of the upper mantle. Because most Archean tholeiites have MgO contents of 6–10% MgO, which are values similar to those of modern oceanic basalts, it would seem that the Late Archean upper mantle was less than 50 °C hotter than modern upper mantle of about 1400 °C.

Because garnet is stabilized in the mantle relative to pyroxenes, the solubility of garnet in silicate liquids is reduced, resulting in magmas that have very low concentrations of Al_2O_3, which are similar to many Al-depleted komatiites (Herzberg 1995). Because CaO varies much less with residual mineralogy, the CaO/Al_2O_3 ratio is strongly pressure dependent, and a Al_2O_3–CaO/Al_2O_3 plot can be used to estimate the depth of melting of primary picrite and komatiite magmas. As shown on Figure 7.11, high-Mg picrites and komatiites less than 100 Ma have large Al_2O_3 values and CaO/Al_2O_3 ratios near 1. In contrast, Late Archean komatiites have intermediate values of these indices, and Early Archean komatiites have low Al_2O_3 and high CaO/Al_2O_3 ratios. This suggests that young komatiites come from depths of up to 100 km, Late Archean komatiites

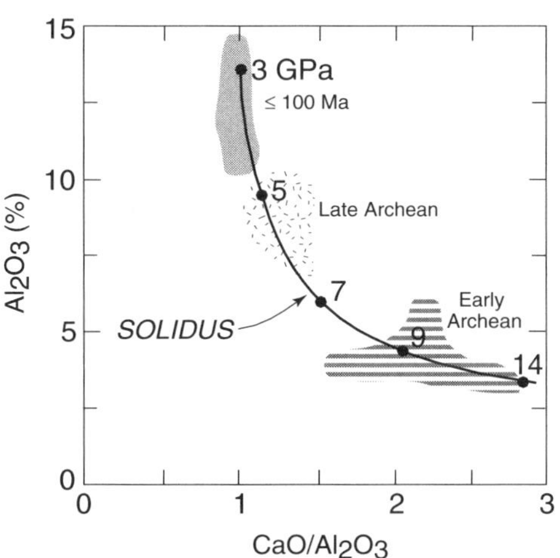

Figure 7.11. An Al_2O_3 versus CaO/Al_2O_3 graph showing the mantle solidus as a function of increasing pressure. Pressure given in GPa. Also shown is the distribution of komatiite and picrite as a function of age (patterned areas). Modified after Herzberg (1995).

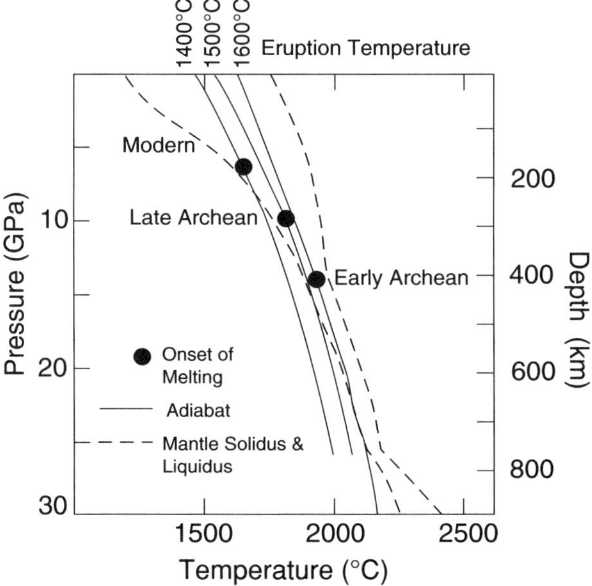

Figure 7.12. Pressure–temperature diagram showing the mantle liquidus and solidus relative to melting depths of high-MgO magmas. After Herzberg (1995).

from depths 150–200 km, and Early Archean komatiites from depths of 300–450 km (Herzberg 1995). Similar depth regimes are apparent from phase equilibria relations (Arndt and Lesher 1992).

The depth of magma segregation in a mantle plume is critically dependent on the temperature of the source. Using the maximum MgO content of mafic and ultramafic lavas, one can calculate source temperatures from experimental data (Campbell and Griffiths 1992). As illustrated in Figure 7.12, young komatiites and picrites have source temperatures on the order of 1500–1600 °C, whereas Late and Early Archean komatiites have source temperatures of 1800–1900 °C and 2000 °C, respectively (Herzberg 1995). Together with the compositional–pressure data in Figure 7.11, this suggests that both temperature and depth of magma segregation in plumes has decreased with time – presumably in response to the cooling of the mantle. The cooling Earth model of Richter (1988) predicts the Earth was hotter by 180 °C at 2.7 Ga and by 300 °C at 3.5 Ga. When compared with this model, it would appear that the average temperature of plumes is 200–300 °C hotter than the temperature beneath ocean ridges throughout geologic time (Fig. 7.13) (Herzberg 1995; Abbott et al., 1994b, Burgess et al. 1994). If this is true, the secular compositional changes in komatiites (Fig. 7.11) may be the result of a hotter bulk Earth rather than just hotter plume sources in the Archean. In any case, the Nb, Ta, Hf, and Zr anomalies previously described in some Archean komatiites, which require fractionation of very high-pressure phases at elevated temperatures, also argue for hotter plume sources in the Archean.

Was the Archean Mantle Iron-Rich?

When normalized to Ti, which is incompatible during high-degree melting of the mantle and also relatively immobile during secondary alteration and metamorphism, Archean

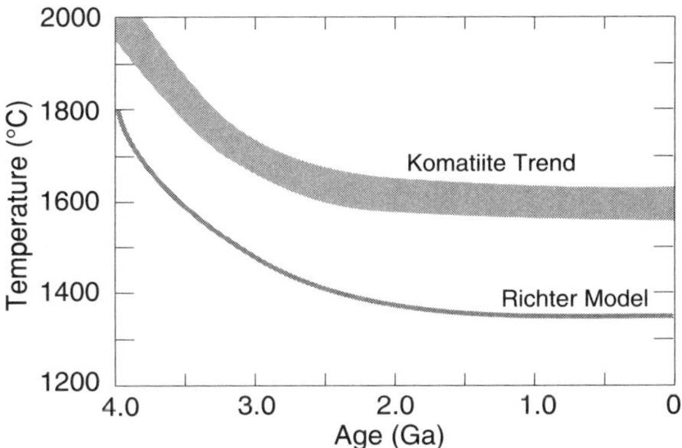

Figure 7.13. Temperature of komatiite sources through time compared with the cooling Earth model of Richter (1988). After Herzberg (1995).

and pre-1-Ga Proterozoic picrites and komatiites show enrichment in Fe compared with Al relative to younger picrites and komatiites (Fig. 7.14) (Francis et al. 1999). Of what significance is this Fe enrichment? It would appear that Fe was somehow enriched in plume sources older than about 1 Ga. Partial melts of mantle lherzolites with greater than 12% MgO have bulk partition coefficients for Fe slightly less than 1, and thus their removal from the upper mantle will leave behind a source that is somewhat enriched in Fe. The predicted enrichment, however, is not nearly enough to account for differences between pre- and post-1-Ga komatiites and picrites (Francis et al. 1999). If the compositions of these picrites are representative of mantle plume sources, it would appear that Fe in these sources must decrease with time, either by loss to the core or sequestering in the lower mantle. However, current estimates of the partitioning of Fe between mantle silicates and metallic Fe in the core would have a negligible effect on Fe content of the lower mantle (Ohtani et al. 1997). It may be that Fe is somehow retained in the deep mantle – perhaps in the D″ layer in one or more high-pressure phases rich in Fe.

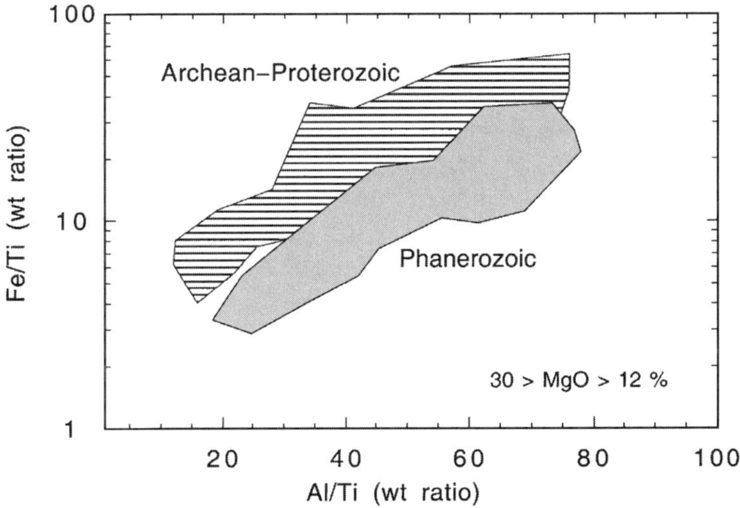

Figure 7.14. Comparison of the composition of Precambrian and Phanerozoic komatiites and picrites on an Fe/Ti versus Al/Ti diagram. After Francis et al. (1999).

Were Mantle Plumes More Widespread in the Archean?

For greenstone belts that have geochemical or lithologic data available, it would appear that about 35% of Late Archean (3.0–2.5 Ga) greenstones have plume affinities (Condie 1994; Tomlinson and Condie 2001). In contrast, about 80% of Early Archean (>3.0 Ga) greenstones have plume affinities, although there are relatively few Early Archean examples upon which to base this estimate. These proportions of Archean, plume-related greenstones are considerably greater than those found in post-Archean greenstones (Condie 1994, 1997c, 1999).

The observation that plume-related basalts and komatiites are a major component of Archean greenstone belts and appear to be more frequent in the Archean than subsequently rests in part on two major assumptions: (1) Archean basalts with oceanic plateau affinities are truly part of oceanic plateaus and are not oceanic crust generated at ocean ridges, and (2) the relative abundance of greenstone belts with plume affinities is not a relict of preservational bias in the Archean. Let us explore each of these assumptions.

1. It is important to distinguish ocean-ridge from oceanic-plateau basalts in Archean greenstone belts because the former do not require a mantle plume component. Because lithologic assemblages formed at ocean ridges and above oceanic mantle plumes may be similar (i.e., submarine basalts and intrusive equivalents together with minor chemical sediments), ocean-ridge and oceanic plateau basalts may be difficult to distinguish from each other – especially in the Archean when the production rate of oceanic crust was probably greater than today (Abbott et al. 1994a). If Archean oceanic crust was on the order of 20-km thick (Sleep and Windley 1982; Galer and Mezger 1998), the oceanic lithosphere would have been more buoyant and could have behaved much like an oceanic plateau. So where do we draw the line between oceanic-plateau and ocean-ridge basalts in greenstone successions? This is a difficult question, and with our present database, three observations favor, but do not prove, plume affinities for Archean mafic-plain greenstones. First of all, komatiites reflect higher mantle temperatures than basalts, and greater melting depths probably occur almost exclusively in these types of greenstone belts. If komatiites require a mantle plume source, as most data suggest (Campbell et al. 1989; Xie and Kerrich 1994), then komatiite-bearing greenstone belts should reflect mantle plume sources. Second, the Th/Ta ratios in many Archean mafic-plain basalts are greater than this ratio in NMORB, and average Th/Ta ratios typically plot in or above the oceanic plateau basalt field on Th/Ta–La/Yb plots (Fig. 7.5). Although some greenstone basalts may come from depleted mantle sources similar to NMORB sources, many if not most of these basalts come from less depleted sources like those of mantle plumes. And finally, the Zr, Hf, Nb, and Th anomalies that have been confirmed to exist in some mafic-plain and platform komatiites and basalts suggest a deep mantle source, one that is characteristic of mantle plumes but not of shallow ocean-ridge sources (Xie and Kerrich 1994; Kerrich et al. 1998; Tomlinson et al. 1999).
2. If plume-related basalts were selectively preserved in the Archean compared with the post-Archean, their higher frequency in Archean greenstone belts may

> reflect preservational bias rather than a greater abundance of Archean compared with post-Archean mantle plumes. What may lead to increased preservation of these greenstone belts in the Archean? The Archean oceanic lithosphere was probably thicker and more buoyant than the post-Archean lithosphere owing to higher mantle temperatures in the Archean (Bickle 1986; Hoffman and Ranalli 1988; Davies 1999). Hence, it would be less subductable, and during plate collisions oceanic plateaus may have accreted to the growing continental crust (Condie 1997a,c), thus selectively preserving volcanic rocks with plume sources compared with volcanic rocks with arc sources. Also supporting this idea is the results of thermal modeling of the mantle by Abbott and Mooney (1995). These investigators showed that that Archean oceanic plateaus, on average, should have been considerably thicker (40 km) and thus should have been less likely to subduct than younger oceanic plateaus. After the Archean, when mantle temperatures fell, oceanic plateaus may have been more subductable and hence, the frequency of preservation of oceanic-plateau-type greenstones decreased. Indeed, the apparent higher frequency of oceanic-plateau-type greenstones in the Archean could be due solely to this factor.

However, the variety of plume-generated assemblages present in the Archean (i.e., mafic-plain, platform, and continental flood basalt assemblages) combined with the high frequency of the mafic plain assemblage suggest that mantle plumes were more widespread in the Archean than subsequently. Also possibly supporting more frequent mantle plumes in the Archean is the thick Archean lithosphere, which, as discussed above, may have been produced by widespread underplating of mantle plumes during the Archean.

A Final Word

If oceanic plateaus were more widespread in the Archean than at present, then several questions arise. What proportion of the Archean ocean floor was plume-generated crust and what proportion was generated at ocean ridges? Although most investigators agree that major changes occurred near the end of the Archean, there is little agreement as to how fast they occurred and what they mean in terms of the Earth's thermal history. If plumes were more widespread in the Archean, did the frequency of plume generation suddenly decrease at about 2.5 Ga, or did it gradually decrease, paralleling the cooling curve of the mantle? To answer this question, we need more precise isotopic ages from plume-generated greenstone assemblages between 3 and 2 Ga. If plume generation suddenly decreased at 2.5 Ga, perhaps a critical thermal threshold was reached in the mantle at that time. For instance, the 660-km discontinuity may have suddenly become less resistant to descending plates and rising plumes as the mantle cooled through a critical temperature threshold. Did plume generation shift from dominantly the 660-km discontinuity to dominantly the D'' boundary zone at the end of the Archean, and, if so, what should we look for in the geologic record that may be sensitive to such a shift? As we will see in Chapter 8, how fast the plume generation rate may have changed at the end of the Archean could be very important in terms of the production rate of continental crust.

8

Superplume Events

Plumes and Supercontinents

Introduction

Supercontinents have aggregated and dispersed several times during geologic history, although our geologic record of supercontinent cycles is only well documented for the last two cycles: Gondwana–Pangea and Rodinia (Hoffman 1989; Rogers 1996). It is generally agreed that the supercontinent cycle is closely tied to mantle processes, including both convection and mantle plumes. Pangea 200 Ma was centered approximately over the African geoid high (Fig. 2.28(a)), and the other continents moved away from this high during breakup of Pangea. Because this geoid high contains many of the Earth's hotspots and is characterized by low seismic-wave velocities in the deep mantle, it is probably hotter than average, as discussed in Chapter 2. Except for Africa, which still sits over the geoid high, continents seem to be moving toward geoid lows, which are also regions with relatively few hotspots and high lower mantle velocities, all of which point to cooler mantle (Anderson 1982). These relationships suggest that supercontinents may affect the thermal state of the mantle as the mantle beneath continents becomes hotter than normal, expands, and produces the geoid highs (Anderson 1982; Gurnis 1988). This is followed by increased mantle plume activity, which may fragment supercontinents or at least contribute to dispersal of cratons.

Mantle Plumes and Supercontinent Breakup

One question not fully understood is the role of mantle plumes in the supercontinent cycle. Are they responsible for fragmenting supercontinents, or do they play a more passive role? Many investigators doubt that mantle convection provides sufficient forces to fragment continental lithosphere and contend instead that mantle plumes play an active role (Richards et al. 1991; Storey 1995). Because plumes have the capacity to generate large quantities of magma, it should be possible to track the role of plumes in continental breakup by the magmas they have left behind as flood basalts and giant dyke swarms.

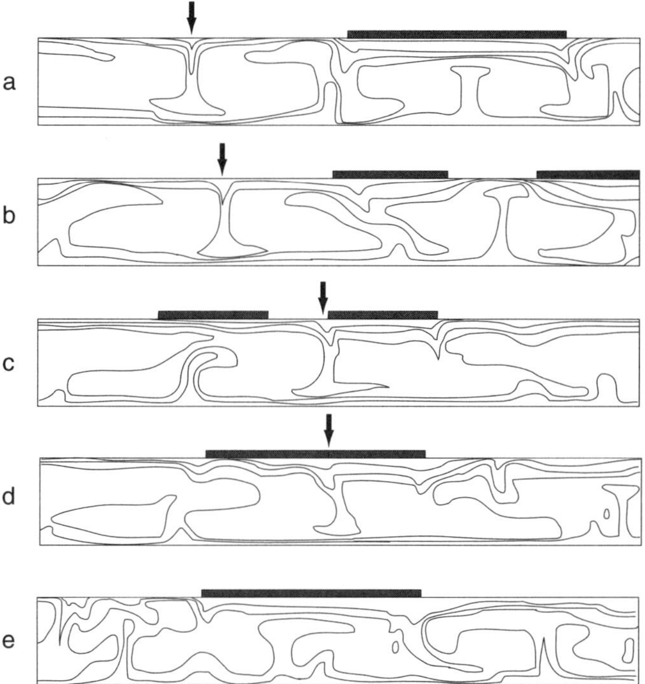

Figure 8.1. Computer-generated model of supercontinent breakup and formation of a new supercontinent. Frame of reference is fixed with respect to the left corner to the diagram, and the right continent moves with respect to the left continent, which is stationary. After Gurnis (1988).

Gurnis (1988) published a numerical model based on feedback between continental plates and mantle convection whereby supercontinents insulate the mantle, causing the temperature to rise beneath a supercontinent. This results in a mantle upwelling that fragments and disperses a supercontinent. Beginning with a supercontinent with cold downwellings along each side, a hot upwelling generated beneath the supercontinent by its insulation fragments the supercontinent (Fig. 8.1(a) and (b)). After breakup, two smaller continental cratons begin to separate rapidly as the hot upwelling extends to the surface between the two plates, producing a thermal boundary layer (b). Both plates rapidly move towards the cool downwellings marked with a vertical arrow in (c). About 150 Myr after breakup, the two continental fragments collide over a downwelling (d). Nearly 450 Myr after breakup, a new thermal upwelling develops beneath the new supercontinent, and the supercontinent cycle starts over (e).

The breakup of Gondwana beginning some 180 Ma provides a means of testing the timing of plume magmatism and supercontinent fragmentation (Storey 1995; Dalziel et al. 2000). The initial rifting stage beginning 180 Ma produced a seaway between West (South America and Africa) and East Gondwana (Antarctica, India, Australia) (Fig. 8.2). Seafloor spreading began in the Somali, Mozambique, and Weddell Sea basins by 156 Ma (b). Approximately 130 Ma, South America separated from Africa–India and Africa–India separated from Antarctica–Australia (c). The breakup was complete by 100 Ma when Australia separated from Antarctica and Madagascar, and the Seychelles separated from India as it migrated northwards on a collision course with Asia (d).

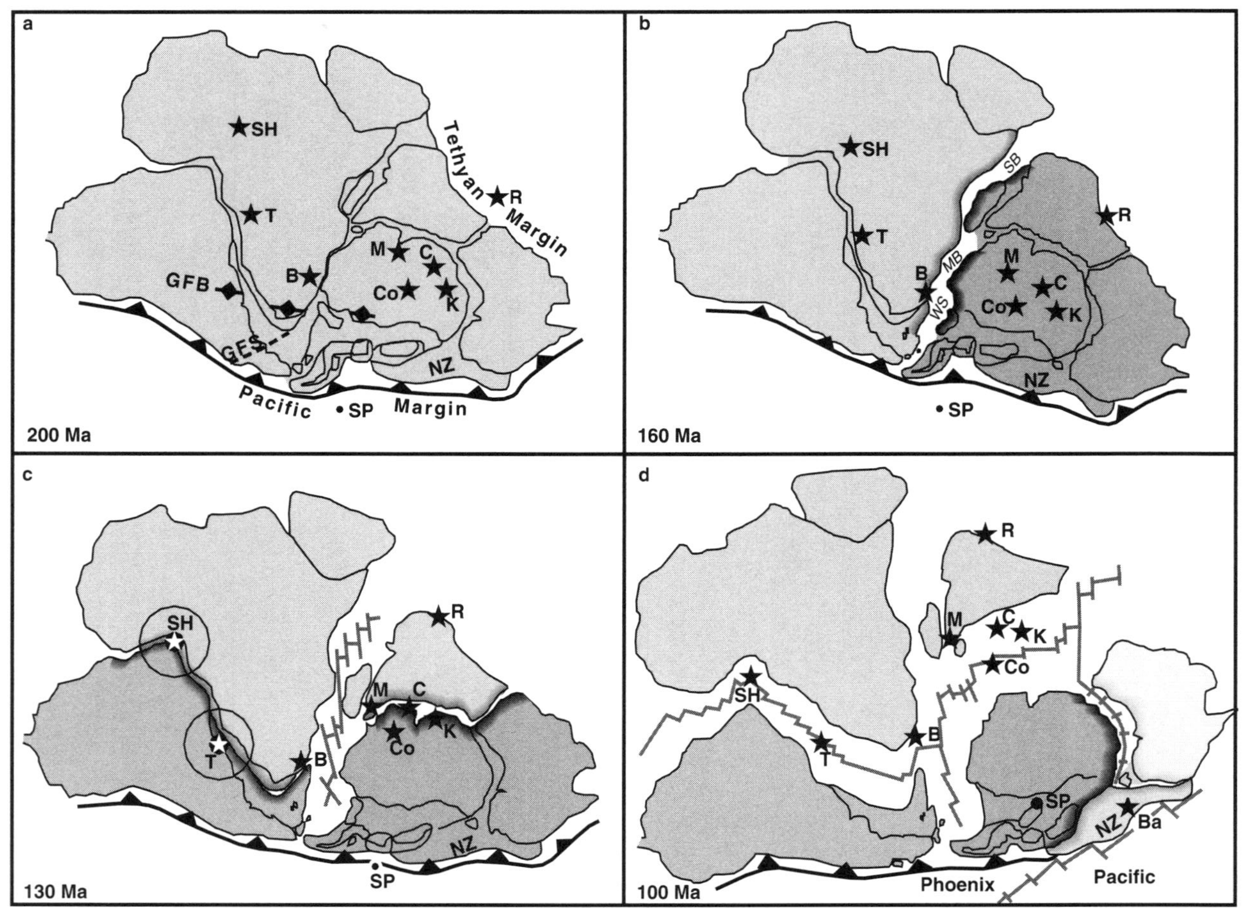

Figure 8.2. Gondwana reconstructions during the last 200 Ma. Shown also are subduction zones (barbed lines), major hotspots (stars), and inferred sizes of plume heads (circles). Ocean ridges are diagrammatic. Symbols: Ba = Balleny; B = Bouvet; C = Crozet; Co = Conrad; K = Kerguelen; M = Marion; R = Réunion; SH = St. Helena; T = Tristan, GFS = Gastre fault system; GFB = Gondwana orogen; MB = Mozambique basin; NZ = New Zealand; SB = Somali basin; SP = South Pole; WS = proto-Weddell Sea. Reprinted from Storey (1995), with permission. Copyright © 1995 by Macmillan Magazines Ltd.

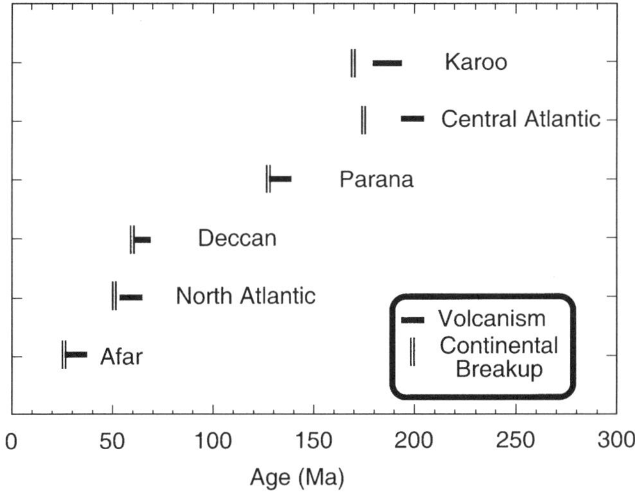

Figure 8.3. Timing of supercontinent breakup and plume volcanism associated with several LIPs. Modified after Courtillot et al. (1999).

Precise isotopic dating suggests that continental separation is closely associated with plume volcanism (Fig. 8.3). In most cases, volcanism begins 3 to 15 Myr before breakup, and in most instances, like the Deccan and Paraná provinces, the most intense volcanism accompanies initial fragmentation of the supercontinent. The onset of major volcanism in the Deccan traps is coeval with continental breakup, and intense volcanism continues for more the 20 Myr. In the case of Iceland, melt production began 60 Ma followed by extensive rifting at 55 Ma, and the first oceanic crust formed about 53 Ma as Greenland and Norway separated (Larsen et al. 1998). In Afar (Ethiopia), oceanic volcanism has not yet begun in the Afar depression. The duration between the onset of flood basalt eruption and the production of oceanic crust ranges from less than 5 Myr in the Paraná and Deccan to 13 Myr for the Karoo and 25 Myr for the Central Atlantic Province (Fig. 8.3). Plume magmatism, of course, can continue after supercontinent fragmentation. In the North Atlantic Province, volcanism continued around the margin of the North Atlantic for at least 27 Myr after the initial rifting of Laurentia from Baltica.

The opening of other basins, such as the Red Sea, Gulf of Aden, Arabian Sea, and the Indian Ocean also appears to be related to plume volcanism. In fact, with the exception of the Siberian and Emeishan traps in eastern Asia, all major flood basalt provinces erupted during the last 200 Myr are associated with the opening a new ocean basin (Wilson 1992; Coffin and Eldholm 1994; Courtillot et al. 1999). The location of plume impacts on the lithosphere may not have been random or uniform in the mantle. In some instances, as illustrated by the breakup of Gondwana (Fig. 8.2), plume impacts were centrally located under supercontinents. In all of the cases cited above, rifting either did not exist prior to flood basalt eruption or it jumped to a new location at or before major eruption began. If the plume head model for flood basalt magma generation is accepted, basalt eruption, uplift (if any), and rifting are all related to rising plume heads, yet they occur in slightly different time sequences in different areas. Most ocean basins not lined with subduction zones may have been shaped by the episodic impact of large plume heads in the interior or at the edges of continents (Courtillot et al. 1999).

Not all investigators, however, accept that mantle plumes are responsible for rupturing continental lithosphere. Ziegler (1993), for instance, has maintained that, although plumes contribute to weakening of the lithosphere, it is stresses within the lithosphere related to plate boundary and drag forces along the base of the lithosphere that are chiefly responsible for rifting of the continental lithosphere.

Large Plates and Mantle Upwelling

The insulating properties of large plates, continental or oceanic, result from the lithosphere's inhibiting mantle convection currents from reaching the surface of the Earth (Ballard and Pollack 1987; Gurnis 1988). An equally, if not more, important effect is that large plates prevent the mantle beneath them from being cooled by subduction. Numerical models show that large plates become increasingly effective as insulators when their width is much greater than the depth of the convecting layer (Guillou and Jaupart 1995; Lowman and Jarvis 1996; Lenardic and Kaula 1996). The net result of this effect is that large mantle upwellings develop beneath large plates. If the large plate happens to carry a supercontinent, the upwelling may promote weakening and eventual breakup of the supercontinent.

The models of Lowman and Jarvis (1999) have been very useful in quantifying the relationships between mantle upwelling, supercontinent fragmentation, and whole-mantle versus layered-mantle convection. Their results indicate that supercontinent rifting varies with the mode of mantle heating (basal heating versus internal radioactive heat sources) and the supercontinent aggregation history. Whether tensile stresses in the interior of model supercontinents exceed the yield stress of the lithosphere of about 80 MPa depends on continental aggregation history, supercontinent size, the Rayleigh number of the convecting mantle, the amount of radioactive heating in the mantle, and the viscosity distribution with depth. In the Lowman–Jarvis models, subduction-related forces are at least as important as mantle upwelling in supercontinent breakup. As observed in the breakup of Gondwana, whole-mantle model results predict that plate velocities should be rapid after supercontinent breakup, reducing in speed thereafter. For layered-mantle convection, the Lowman–Jarvis models require unreasonably long periods of time to generate stresses necessary to rift supercontinents (>600 Myr). Model supercontinents survive for more than 500 Myr when internal heating in the mantle is 40% or less, but less than 250 Myr for models with 80% internal heating of the mantle. Three-dimensional models, which are more applicable to the Earth, show that upwelling beneath large plates results chiefly from the absence of subduction rather than insulation by the lithosphere (Lowman and Gable 1999). Also, three-dimensional models show that, if internal heat sources are abundant in the mantle (≥80%), unusually hot upwellings develop beneath large oceanic plates.

As an example of a numerical mantle upwelling model, let us consider a whole-mantle convection model in which 40% of the heat comes from radioactive sources in the mantle and the remainder from cooling of the core (Lowman and Jarvis 1999). This model is close to an "average" model for Earth's heat budget (Davies 1999). The continental blocks in the model collide to form a supercontinent with a width of 11,600 km (Fig. 8.4). A pair of hot regions is present on each side of the initial cold

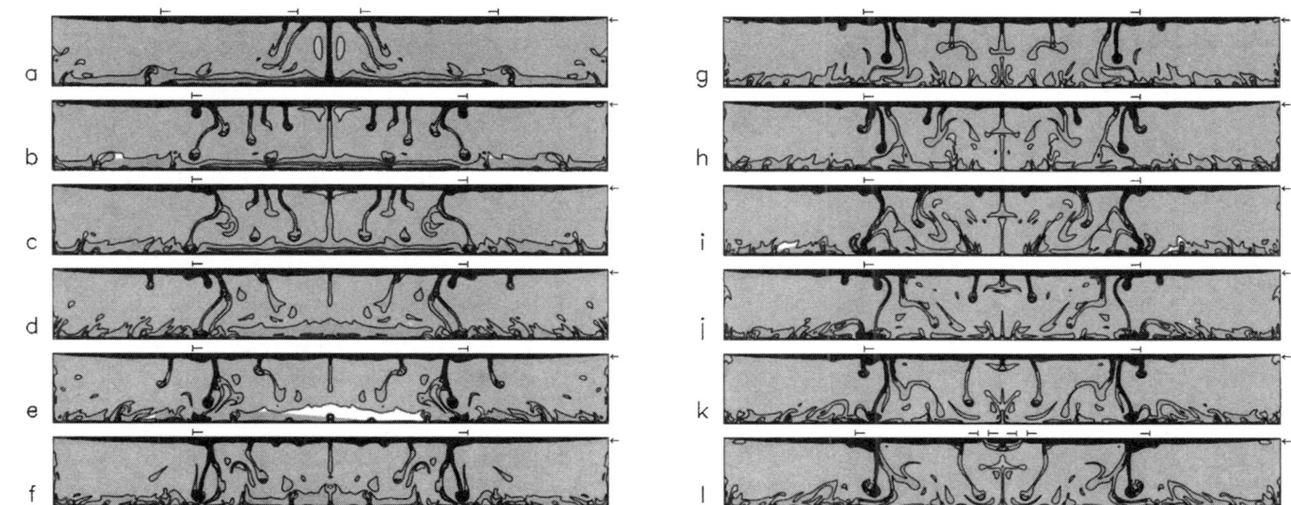

Figure 8.4. Numerical model showing rifting and dispersal of a supercontinent for whole-mantle convection with internal heating (40%). An isothermal boundary layer at the bottom delivers the remainder of the heat. Plate thickness = 99 km; Rayleigh number = 10^7. Model has two continents, each 5800 km wide (a), that collide at the symmetry plane to form a supercontinent 11,600 km across (b). The supercontinent rifts between frames k and l about 600 Myr after its formation. Reproduced from Lowman and Jarvis (1999), with permission. Copyright © 1999 by the American Geophysical Union.

downwelling (subduction zones) where the supercontinent collected (Fig. 8.4(a)). As the downwelling fades, these hot parcels of mantle rise to the base of the lithosphere, which is a feature that facilitates the growth of a thermal instability at the base of the mantle (f,g). Transient plumes develop below the supercontinent and are swept toward the vertical midplane of the model, where they coalesce into a single mantle upwelling (h to j). The yield stress of the lithosphere is exceeded on both sides of the midplane, and thus the supercontinent begins to fragment a short distance from each side of the midplane, leaving a small continental block at the center (k). Final supercontinent breakup occurs after 600 Myr when three upwellings converge simultaneously at the central midplane (1). The only major problem with this model is the long time it takes to break the supercontinent (600 Myr) compared with the observed survival time of supercontinents of 200–400 Myr (Condie 1998).

The Supercontinent Cycle

In the last 1 Gyr, the formation and breakup of three major supercontinents has been documented (Rodinia, Gondwana, and Pangea), with a possible short-lived supercontinent, Pannotia, in the latest Proterozoic (Unrug 1997). Geologic data support the existence of at least two earlier supercontinents, one at the end of the Archean and one in the Early Proterozoic (Hoffman 1989; Barley and Groves 1992; Rogers 1996). As more precise ages for the breakup and aggregation of supercontinents have become available, it seems clear that, in the supercontinent cycle, the breakup of one supercontinent is accompanied in part or entirely by the formation of another supercontinent (Fig. 8.5). For instance, Gondwana was forming at about the same time (650 to 580 Ma) as Rodinia broke up (800 to 600 Ma) (Veevers et al. 1997). Also, the breakup of Pangea in the last 160 Myr was accompanied by numerous major terrane and microcontinent collisions (India, North American Cordillera, Southeast Asia), which may represent the beginnings of a new supercontinent (N in Fig. 8.5). This is not unexpected because supercontinents break up over large mantle upwellings (geoid highs), and the fragmented blocks travel to mantle downwellings (geoid lows), where they collide with other blocks to form a new supercontinent (Anderson 1982). Thus, collisions occur over the geoid

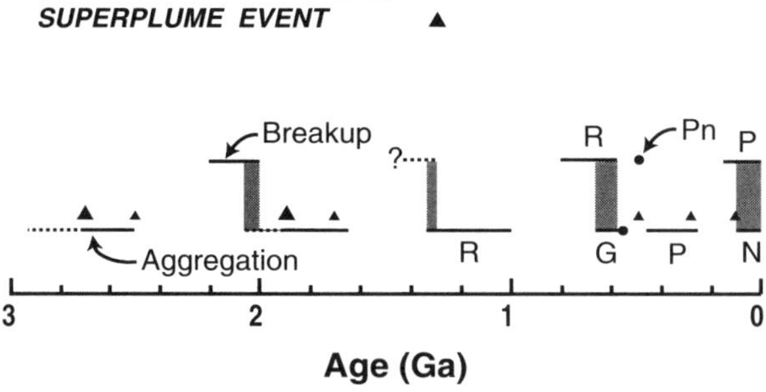

Figure 8.5. The supercontinent cycle. Gray bands refer to overlap between formation of one supercontinent and breakup of another. Black triangles are proposed superplume events; symbol size is proportional to event intensity. After Condie (1998). R = Rodinia; G = Gondwana; Pn = Pannotia; P = Pangea; N = new supercontinent forming today.

lows while supercontinents are still fragmenting over geoid highs. Because there is always an external ocean as supercontinents come and go, the supercontinent cycle can be viewed as the expansion and contraction of this ocean.

A speculative representation of the supercontinent cycle is given in Figure 8.6 modified after Rogers (1996). Locations of older supercontinents and cratons are shown on the Pangea reconstruction in Figure 8.7. Arctica, which includes the Canadian, Greenland, and Siberian shields, was certainly intact by 1.9 Ga, and fusion by 2.5 Ga is possible (Rogers 1996; Aspler and Chiarenzelli 1998). Atlantica includes at least five cratons in West Africa and eastern South America that appear to have come together about 2 Ga (Boher et al. 1992). Other small pre-2-Ga cratons include Baltica, Kaapvaal, India, and Pilbara. Arctica and Baltica (including the Urkrainian shield) collided before 1.8 Ga (perhaps at 1.9 Ga) to form a supercontinent referred to as Nena as constrained by the continuity of 1.8-Ga orogenic belts along the southern margin of both blocks (Gower et al. 1990). If the 1.8–1.7-Ga orogens in the southwestern United States continue into East Antarctica or Australia, one or both of these continents also must have been part of Nena by this time. Nena appears to have remained a coherent plate from 1.8 Ga until its incorporation into the Neoproterozoic supercontinent Rodinia during the Grenville event at 1.3–1.0 Ga. This coherence is supported by the restriction of Grenvillian-age orogens to the margins of Nena with little penetration far into the interior of the continent (Gower and Tucker 1994). During formation of Rodinia, most or all of the remaining continental blocks were accreted to the growing supercontinent, although the positions of many of the small blocks, such as Kalahari, North Africa, Congo, and North China are not known with high degrees of certainty. Rodinia fragmented at 800–600 Ma with the separation of East and West Gondwana and Laurasia (Siberia, Laurentia, Baltica). In the early Paleozoic, East and West Gondwana collided followed by the collision of Laurasia in the mid-Paleozoic to form Pangea.

Episodic Crustal Growth

Although an episodic distribution of isotopic ages has been recognized since the classic article of Gastil (1960), it is only since the early 1990s that the episodic growth of juvenile continental crust has been recognized (Condie 1998). The distribution of U/Pb zircon ages coupled with Nd isotopic data suggest two major peaks in juvenile crust production rate, one at 2.7 Ga and another at 1.9 Ga (Fig. 8.8). In addition, smaller peaks may be present at about 2.8, 2.5, 2.0, and 1.7 Ga, and one or more peaks occur in the Phanerozoic, the most prominent of which is at about 100 Ma.

One of the first quantitative models to explain episodic continental growth was that of McCulloch and Bennett (1994). They proposed a non-recycling model involving three reservoirs: continental crust, depleted mantle, and primitive mantle. The model is based chiefly on Nd isotopic data but also accommodates Sr and Pb isotopes and incompatible element distributions in each reservoir. It is assumed in the model that the volume of depleted mantle increases with time in a stepwise manner, which is linked to major episodes of continental crust formation at 3.6, 2.7, and 1.8 Ga. The isotopic and trace element composition of the upper mantle is buffered by progressive extraction of continental crust and increasing size of the depleted mantle reservoir. The buffered

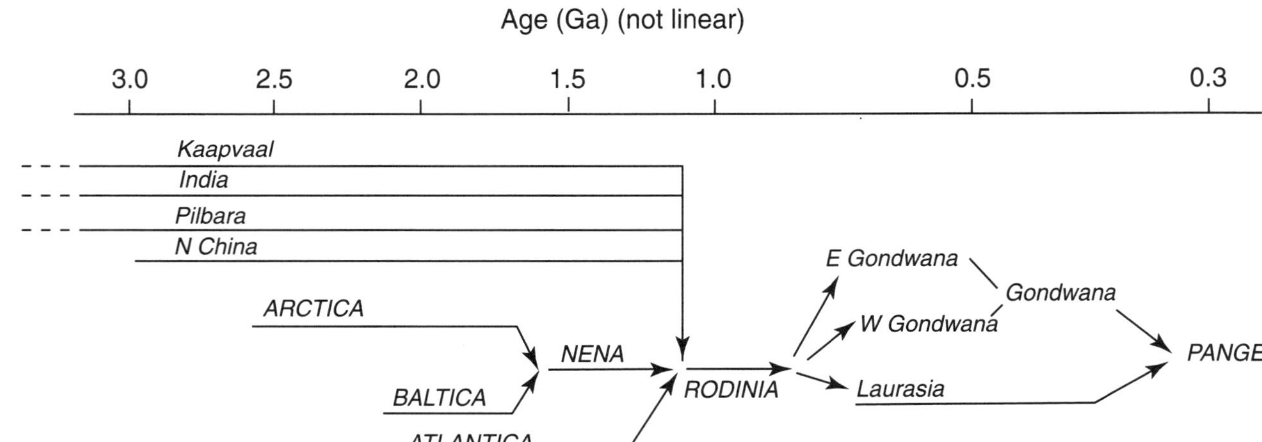

Figure 8.6. Diagrammatic representation of supercontinent formation and breakup. Modified after Rogers (1996).

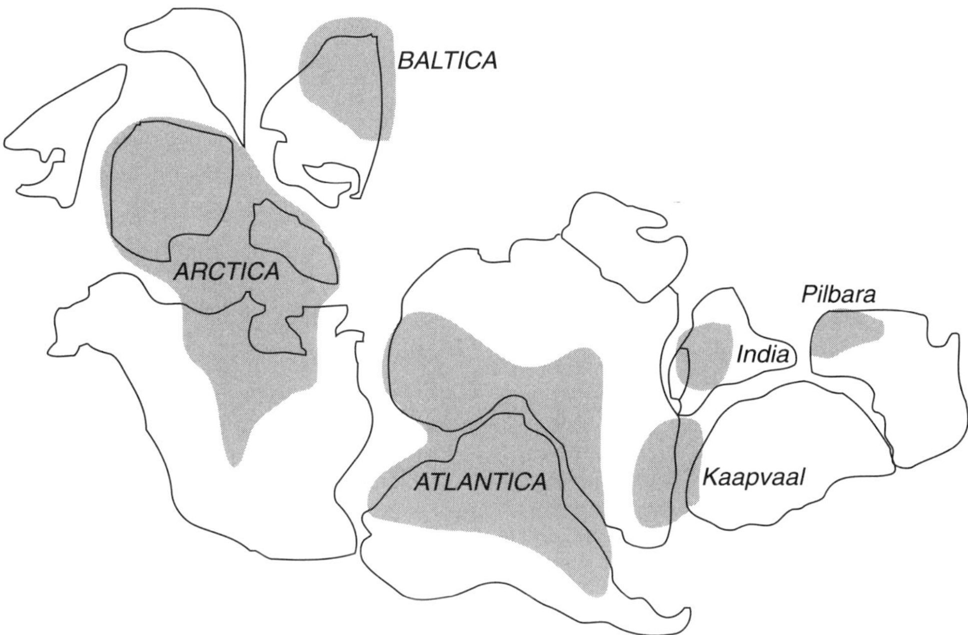

Figure 8.7. Pre-1-Ga supercontinents and cratons shown on a Pangea reconstruction. After Rogers (1996).

composition of the upper mantle remains depleted in incompatible elements through time.

Stein and Hofmann (1994) were among the first to advocate that episodic instability at the 660-km seismic discontinuity controls the growth of continental crust. They suggested that convection patterns changed in the mantle from layered convection (the

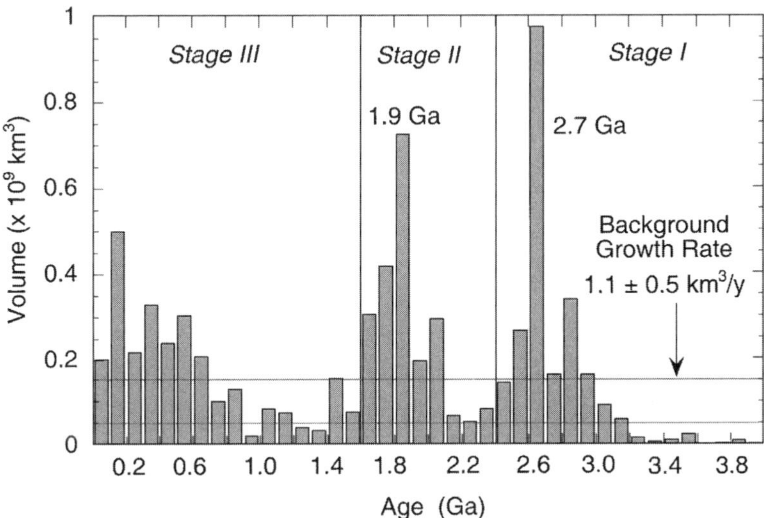

Figure 8.8. Frequency distribution of juvenile continental crust over the last 3.8 Ga. Based on a total volume of continental crust of 7.177×10^9 km^3. Juvenile crust ages are U/Pb zircon ages used in conjunction with Nd isotope data and lithologic associations. Modified after Condie (1998, 2000a).

normal case), when the growth rates of continental crust were relatively low, to whole-mantle convection, when the growth rates were high. Whole-mantle convection occurs in short-lived episodes during which subducted slabs that have accumulated at the 660-km discontinuity catastrophically sink into the lower mantle in a manner similar to that proposed by Tackley et al. (1997). Stein and Hofmann showed that such a model can account for several geochemical observations including the following:

1. ϵ_{Nd} values of the upper mantle are maintained between primitive and depleted mantle values.
2. Trace element models for crustal extraction require that the mantle reservoir source be larger than the upper mantle (above the 660-km discontinuity) but less than the entire mantle.
3. Primordial ^{3}He and radiogenic ^{40}Ar must be present in both the lower and upper mantle.
4. Trace element distributions and Pb isotopes require the upper mantle and continental crust to have evolved as an open system.

One of the important features of the Stein–Hofmann model is that, during periods of whole-mantle convection, plumes rise from the D″ layer above the core and replenish incompatible elements to the upper mantle, which has been depleted by oceanic crust and arc formation.

On the basis of instability at the 660-km discontinuity and using parameterized mantle convection, Davies (1995) proposed catastrophic global magmatic and tectonic events at 1-to-2-Gyr spacings. The favored models show layered convection, which becomes unstable and breaks down episodically to whole-mantle convection, as in the Stein–Hofmann model. The Early Archean overturn cycles may occur on timescales of a few hundred million years. During the catastrophic mantle overturns, hot lower mantle material is transferred to the upper mantle and may be responsible for rapid episodic growth of juvenile crust as well as replenishment of the upper mantle with incompatible elements.

Peltier et al. (1997) extended thermal constraints to evaluate the catastrophic mantle models more thoroughly. These investigators quantified the physical processes that control the Rayleigh number at the 660-km discontinuity, which in turn controls the frequency of slab avalanches at this discontinuity. They also suggested a correlation between avalanche events and the supercontinent cycle. Their results imply that slab avalanches occur at a spacing of 400–600 Myr and that they are brought about by the growth of an instability in the thermal boundary layer at the 660-km discontinuity. During and after slab avalanches, a large mantle downwelling is produced directly above the avalanches, and this downwelling attracts fragments of continental lithosphere, thus leading to the formation of a supercontinent.

On the basis of episodic occurrence of juvenile crust and associated mineral deposits, Barley et al. (1998) proposed a global tectonic cycle beginning in the Late Archean with the breakup of a supercontinent. Enhanced magmatism at 2.8–2.6 Ga results from a global mantle plume event, which is analogous to a superplume event as used in this discussion. Condie (1998) also proposed a model to explain the episodic growth of juvenile crust based on superplume events in the mantle, which is more fully discussed below.

The Mid-Cretaceous Superplume Event

In two important articles, Larson (1991a, 1991b) suggested that one or more superplumes beneath the Pacific basin could explain many geological and geophysical features during the Mid-Cretaceous. The major evidence that Larson presented to support a Mid-Cretaceous superplume event is an enhanced rate in production of juvenile crust (Larson 1991a). Calculated production rates of oceanic crust, which are summarized in different categories in Figure 8.9, show a clear maximum between 120 and 80 Ma. This maximum is particularly striking in the world total, which includes crust produced both at ocean ridges and oceanic plateaus. All three production curves peak at about 120–110 Ma and decline thereafter until 40–30 Ma. Although the average calculated spreading rate of 17 cm/yr during peak crust production is similar to the present-day rate, rates in the Pacific were considerably higher. Many of the largest oceanic plateaus in the modern ocean basins were formed 120–80 Ma (for instance, Ontong Java and Caribbean) (Kerr 1998). The Mid-Cretaceous pulse in production of juvenile crust also correlates closely with the Mid-Cretaceous superchron (long period of normal magnetic polarity), suggesting that the heat source for the crustal pulse is located near the core–mantle boundary and thus supporting the superplume idea.

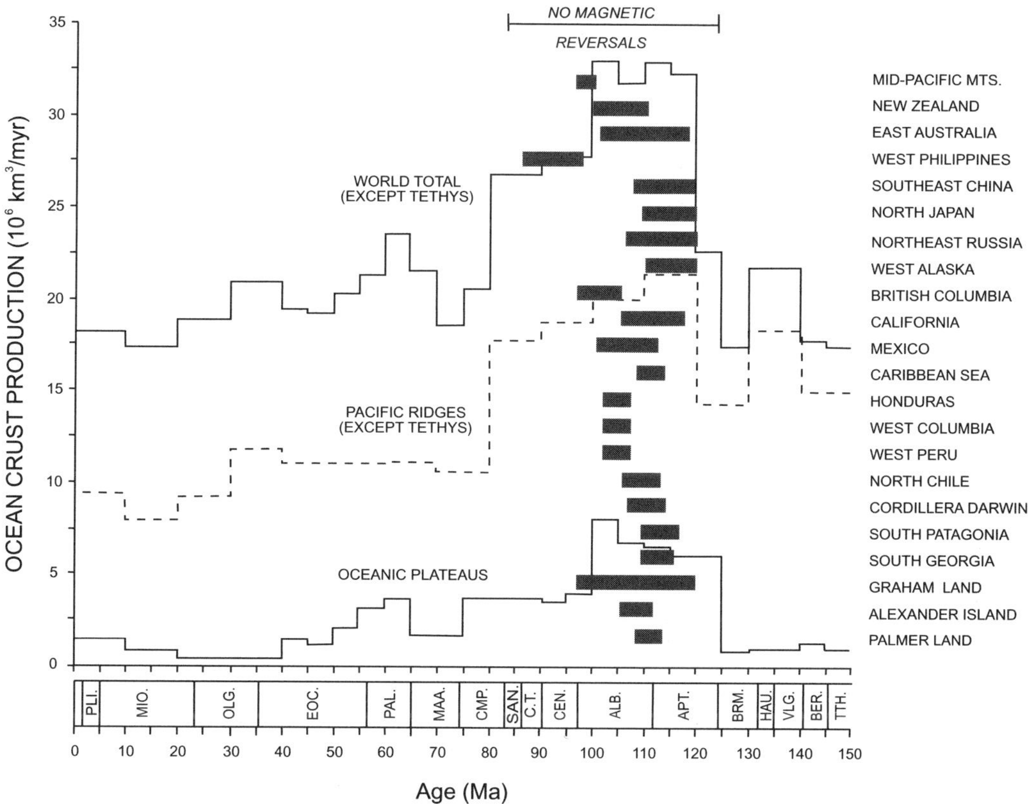

Figure 8.9. Summary of oceanic crust and oceanic plateau production rates in the last 150 Myr. Also shown with the horizontal black bars are major tectonic events around the Pacific basin. Modified after Larson (1991a) and Vaughn (1995).

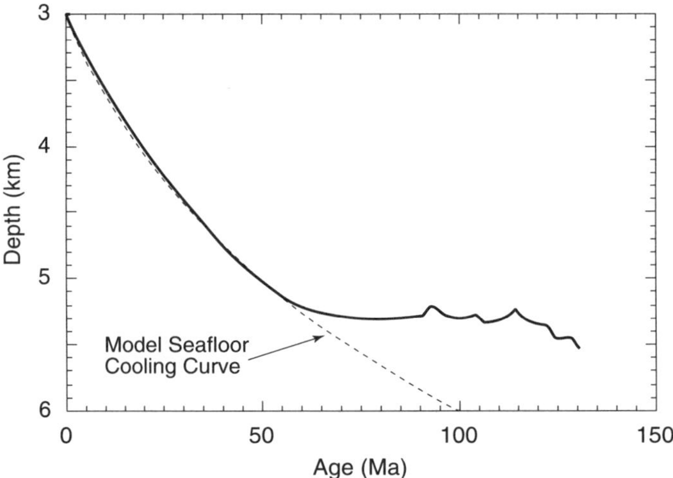

Figure 8.10. Mean variation in depth of seafloor as a function of crustal age. Dashed line shows the predicted depth from cooling of the lithosphere with time. Modified after Humler et al. (1999).

Deformation and orogenesis are also widespread in the continents around the Pacific at this time (Vaughan 1995) (Fig. 8.9). The striking correlation suggests that this deformation is related to the Pacific superplume event. If the Pacific lithosphere were elevated by superplumes, the stress state at continental margins adjacent to subduction zones might switch from tension or strike–slip to compression, leading to enhanced deformation over the new subduction zones (Coblentz et al. 1994). As pointed out by Bott (1993), superplumes may also thermally elevate the Pacific lithosphere, increasing ocean-ridge push forces, which in turn cause increased coupling between subducting and overriding plates. This increased coupling may be manifested at convergent margins as uplift and shortening, as observed around the Pacific margin 120–80 Ma.

Still another factor consistent with a Mid-Cretaceous superplume event in the Pacific is the unusually shallow depth of seafloor of this age. It is well established that depth of the seafloor (away from hotspot effects) is related to crustal thickness and the temperature of the lithosphere. The first-order relationship between depth and age of normal oceanic crust (6–7 km thick) can be modeled by simple cooling and thickening of the lithosphere. However, crust that is 120–80 Ma is shallower than predicted (Fig. 8.10) (Humler et al. 1999). Because the chemical composition of oceanic crust is related to the temperature of the mantle sources, such parameters as sodium, rare earth element distributions, and Zr/Y ratios in oceanic basalts can be used to infer mantle source temperature (Shen and Forsyth 1995). Results from Pacific ocean-ridge basalts 120–80 Ma indicate that the mantle beneath ocean ridges at this time was hotter by about 50 °C and that the depth of ocean ridges was several hundred meters shallower than afterwards (Humler et al. 1999). These results, which are confirmed by seismic data, suggest that Mid-Cretaceous oceanic crust is 1–2 km thicker than normal, reflecting greater magma production rates at Cretaceous oceanic ridges in the Pacific. This means either that the mean mantle temperature was hotter at this time or that Pacific ridges were over hotter parts of the mantle. In the first case, results are consistent with a worldwide superplume event, and in the second case the results imply a Pacific-centered superplume event.

Still another interesting observation comes from paleomagnetic data, which suggest little motion of the Pacific plate between 120 and 80 Ma (Tarduno and Sager 1995). Perhaps the Pacific plate was relatively fixed over the Pacific upwelling at this time, which was also part of the superplume event.

What Is a Superplume Event?

Before describing superplume events, it is necessary to agree on what constitutes a superplume and if there is a need for such a term. Because the term has been used in different ways in the literature, which has led to confusion and misunderstanding, some investigators believe the term should be abandoned. However, as described in Chapter 1, there is a need for a term to describe large mantle plumes that appear to come from the D″ layer as opposed to plumes that may come from shallower boundary layers in the mantle. In that light, a superplume is herein defined as a large mantle plume that spreads at the base of the lithosphere, flattening the plume head to 1500 to 3000 km in diameter. Single superplumes typically give rise to large erupted volumes of mafic magma ($>0.5 \times 10^6$ km^3) in periods of time less than 3 Myr. This differs from the usage of Larson (1991a) and Maruyama (1994), who use the term *superplume* synonymously with *mantle upwelling* as used in this book. I choose to retain the term *mantle upwelling* for the large volumes of mantle that move upward as part of the return flow of mantle convection. Mantle upwellings are not plumes because they do rise from thermal boundary layers as distinct blobs that divide into head and tail components. Rather, they are broad regions of upwelling extending over many thousands of kilometers in diameter. Superplumes, on the other hand, are large plumes with distinct head and tail components that come from deep boundary layers in the mantle – chiefly or entirely the D″ layer.

Condie (1998) and Isley and Abbott (1999) have presented arguments that superplume events have been important throughout Earth history. Although the meaning of the term *superplume event* varies in the scientific literature, we will constrain the term to refer to a short-lived mantle event ($\leq$100 Myr) during which many mantle superplumes as well as smaller plumes bombard the base of the lithosphere. During a superplume event, plume activity may be concentrated in one or more mantle upwellings, as during the Mid-Cretaceous superplume event some 100 Ma, when activity was focused mainly in the Pacific mantle upwelling. However, as pointed out by Condie (1998; 2000a), Precambrian superplume events at 2.7 and 1.9 Ga correlate with maxima in worldwide production rate of juvenile crust and thus may not have been confined to one or two mantle upwellings.

Precambrian Superplume Events

Isley and Abbott (1999) have used the distribution of komatiites, flood basalts, mafic dyke swarms, and layered mafic intrusions in the geologic record to identify superplume events in the Precambrian. A time series analysis of the data shows major superplume

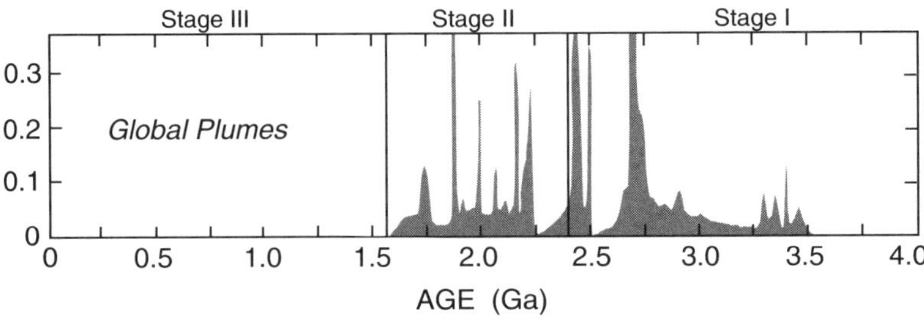

Figure 8.11. Time series distribution of global mantle plumes based on the distribution of plume-related igneous rocks in the geologic record. After Isley and Abbott (1999).

events at 2.75–2.70, 2.45, and 2.0–1.9 Ga and several minor or possible events between 2.5 and 1.75 Ga (Fig. 8.11). The 2.75–2.70 and 2.0–1.9 Ga events correspond well with the peaks in juvenile crust formation at these times (Fig. 8.8). The 2.45 Ga peak corresponds to a juvenile crust formation event recorded in India and the North China craton, and an inferred superplume event at about 2.1–2.15 Ga correlates with the 2.15-Ga crustal formation event in the Guiana shield and in West Africa (Condie 1997a, 1998). One or two peaks at 1.75–1.70 Ga correlate with widespread continental growth in Southern Laurentia and Southern Baltica at this time. A possible superplume event at 2.25 Ga, which is based solely on the ages of dyke swarms in Laurentia, has no counterpart in the juvenile crust record. A small peak at about 3.4 Ga is suggestive of an Early Archean superplume event perhaps associated with the formation of continental nuclei in southern Africa and western Australia.

Kimberlites and Superplumes

Three of the seven major kimberlite-diamond events in the last 500 Myr correlate with three possible superplume events at 480, 280, and 100 Ma, suggesting a cause and effect relationship. Supporting this correlation is the worldwide distribution of kimberlites containing diamonds associated with the superplume events at 480 and 100 Ma (Haggerty 1999). The kimberlites erupted at about 480 Ma (500–450 Ma) include kimberlites in the White Sea area of Russia and examples in China, Canada, South Africa, and Zimbabwe. Those formed at 280 Ma (320–250 Ma) include examples in North America and Siberia, and those formed at 100 Ma (120–80 Ma) include examples from several locations in Africa, Brazil, Canada, India, Siberia, and the United States. The Mid-Cretaceous kimberlite-diamond event at 100 Ma is the most widespread and best documented event. This event, which also correlates with the breakup of Pangea (Figs. 8.5 and 9.2), is not restricted to areas surrounding the Pacific basin but is worldwide in scope.

In addition to these three kimberlite-diamond events, there are kimberlite events that do not correlate with superplume events at 1 Ga (Africa, Brazil, Australia, Siberia, Greenland, India), 410–370 Ma (Siberia, United States), 200 Ma (southern Africa, Canada, Tanzania), and 50 Ma (Canada, Tanzania) (Haggerty 1999). This indicates that not all kimberlite-diamond events require superplume events.

Initiation of Superplume Events

Slab Avalanches

Most models for the episodic growth of continents involve catastrophic sinking of slabs through the 660-km seismic discontinuity in the upper mantle, which, upon arrival at the D″ layer above the core, initiate a superplume event (Stein and Hofmann 1994; Peltier et al. 1997; Brunet and Machetel 1998; Condie 1998). Although seismic tomographic results clearly suggest that descending slabs sink into the lower mantle today (Grand et al. 1997), this may not have been the case in the geologic past when the Earth was hotter. Christensen and Yuen (1985) have shown that, in a hotter mantle with a larger Rayleigh number, such as probably existed in the Archean, the amount of leakage across the 660-km discontinuity is considerably reduced, resulting in layered convection. Computer models of mantle evolution also suggest that increased internal heating of the mantle strongly favors layering in the mantle (Zhao et al. 1992), and this would have been the case during the Archean when heat production by radiogenic isotopes was more pronounced than it is today. Layered convection in the Archean is important because slab avalanches may not occur. It may have been in the Late Archean, when the 660-km discontinuity became less "robust," that slabs occasionally fell through to the lower mantle (Peltier et al. 1997; Condie 1998). Cooling of the Earth also may have been responsible for the shutdown or decrease in intensity of slab avalanches after 1.9 Ga. As the mantle temperature and Rayleigh number decreased with time, slabs should more easily have penetrated the 660-km discontinuity, leading eventually to whole-mantle convection.

Numerical simulations by Yuen et al. (1993) show that at the higher temperatures that existed in the Archean, the mantle would have convected more chaotically, which is a type of convection known as hard turbulence (Chapter 4). Their results show that during hard turbulence with a higher Raleigh number, which is also temperature dependent, phase transitions such a the perovskite transition at 660 km become stronger barriers and result in layered convection. This implies that subducted slabs may accumulate at the 660-km boundary, which eventually leads to gravitational instability and catastrophic collapse of the slabs through the phase boundary. It is this catastrophic collapse that may trigger superplume events in the D″ layer.

Core Rotational Dynamics

Another possible mechanism by which a superplume event could be triggered is by resonance of tidal waves in the fluid outer core (Greff-Lefftz and Legros 1999). When the core rotational frequency and solar tidal waves are in resonance, frictional power may be converted into heat, destablizing the D″ layer above the core and leading to the generation of many mantle plumes. Numerical models predict two major resonances in the past, one at about 3 Ga and another at about 1.8 Ga. These times correspond closely to the observed peaks in juvenile crust production at 2.7 and 1.9 Ga (Fig. 8.8). During the core resonance periods, the temperature near the inner core boundary should increase, which is an effect that could stop inner core growth and produce a new momentum equilibrium for the geodynamo. This, in turn, could lead to a decrease in magnetic reversal frequency, thus accounting for the superchrons associated with the Phanerozoic superplume events.

A Superplume Event Model

Based on the episodic age distribution of continental crust, Condie (1998, 2000a) proposed that continents grew episodically and had major periods of growth at 2.7 and 1.9 Ga (Figs. 8.5 and 8.8). In this model, each maximum in continental growth reflects a superplume event caused by catastrophic slab avalanching at the 660-km discontinuity and also correlates with supercontinent formation. The total duration of each of the Precambrian events is 100 Myr or less.

The following major assumptions are made in developing a superplume model that ties together catastrophic slab avalanches in the mantle, the supercontinent cycle, and episodic production of continental crust (Condie 1998, 2000a):

1. There are two main episodes of production of juvenile continental crust centered at 2.7 and 1.9 Ga, and two minor episodes, one in the late Paleozoic ($\approx$280 Ma) and one in the Mid-Cretaceous ($\approx$100 Ma). In addition, a third episode may occur in the Ordovician (480 Ma).
2. Supercontinents form over geoid lows (mantle downwellings) and break up over geoid highs (mantle upwellings). Mantle upwellings are chiefly responsible for the breakup of supercontinents.
3. Mantle temperature has decreased exponentially with time as Th, U, and K isotopes have decayed, and the Rayleigh number of the mantle has decreased similarly.
4. With time, convection in the mantle has changed from layered in the Precambrian to dominantly whole mantle thereafter. Two episodes of penetrative, whole-mantle convection occurred at 2.7 and 1.9 Ga.
5. The 660-km seismic discontinuity has become a progressively less effective barrier to the descent of lithospheric slabs with time as mantle temperature decreased.
6. During Earth history, there have been two major and five or more minor superplume events. Each event involved avalanching of slabs through the 660-km discontinuity, consequent production of mantle plumes in the D'' layer, and enhanced production of juvenile crust. In addition, the formation of a supercontinent appears to have been associated with some superplume events.

As suggested by Condie (1995, 1998), Earth history can be broadly divided into three stages (Figs. 8.8 and 8.11): Stage I ($>$2.4 Ga) characterized by a hot, chaotically convecting mantle, rapid recycling of juvenile crust into the upper mantle, one major (2.7 Ga) and perhaps one minor superplume event (2.5 Ga), and significant continental growth at 2.7 Ga; Stage II (2.4–1.6 Ga) characterized by one major (1.9 Ga) and several minor superplume events and significant continental growth at 1.9 Ga; and Stage III ($<$1.6 Ga) during which two or three rather minor superplume events occurred with only minor increases in production rate of continental crust.

Stage I. The probable occurrence of coexisting depleted and enriched upper-mantle reservoirs in the same geographic area before 2.4 Ga, as shown by Nd isotopic studies, suggests rapid recycling of continental crust into the mantle (McCulloch and Bennett 1994; Bowring and Housh 1995). The changes in the Nd isotopic composition of clastic sediments with time are also consistent with such recycling (Armstrong 1991). Before the first major superplume event at 2.7 Ga, continental crust survived only as small microcontinents, some of which were "captured" in Late Archean collisional orogens. As the mantle cooled and continental crust became less easy to subduct, collisions between continental blocks led to the formation of the first supercontinent beginning about 3 Ga and culminating with major collisions at 2.7 Ga (Fig. 8.5). Perhaps associated with a second superplume event at 2.5 Ga, continental growth continued in what are now North China and India.

Because it is likely that the upper mantle convected separately from the lower mantle during the Archean, perhaps subducted slabs may have collected at the 660-km discontinuity. If so, in the Late Archean this discontinuity may have destabilized because of slab load and cooling of the mantle, and perhaps it failed at two times, leading to slab avalanches at 2.7 and 2.5 Ga. The first avalanche at 2.7 Ga probably occurred over a wide area on the surface of the 660-km discontinuity, whereas the avalanche at 2.5 Ga may have been localized because it is represented by continental crust only in China and India. It would appear that the avalanche at 2.7 Ga was the most intense. Each slab avalanche produced mantle plumes in the D″ layer, and these bombarded the base of the lithosphere (Larsen and Yuen 1997), where they underwent decompression melting to produce juvenile crust. Plumes may have been temporarily stalled at the 660-km discontinuity owing to a displaced phase boundary within the plumes (Christensen 1995). The plumes should also have heated the upper mantle, resulting in enhanced rates of juvenile crust production at ocean ridges and arcs. Multiple slab avalanching is supported by three-dimensional computer simulations in which unsynchronized and spatially localized flushing is observed to occur at different locations on the 660-km discontinuity surface (Tackley et al. 1994). After the Late Archean slab avalanches, a return to layered convection occurred as the Rayleigh number of the boundary layer at the 660-km discontinuity returned to its previous subcritical value, as the flux of plates crossing the boundary eliminated the temperature gradient across the boundary (Peltier et al. 1997).

When avalanches occur, they may cause "super" subduction zones above them, which attract plates from great distances, thus aiding in the growth of a new supercontinent over the avalanche sites. Also contributing to the growth of supercontinents is juvenile crust produced directly or indirectly by numerous mantle plumes arriving at the base of the lithosphere.

Stage II. After formation of the Late Archean supercontinent, the mantle returned to layered convection, and some fragments of descending slabs again began to accumulate at the 660-km discontinuity. Owing to shielding by the large Late Archean supercontinental plate, a mantle upwelling developed beneath the supercontinent, leading to breakup between 2.2 and 2.0 Ga (Fig. 8.5). As the supercontinent broke up, fragments moved toward geoid lows, where they collided, initiating the growth of a new supercontinent at 1.9 Ga. With the supercontinent breakup, subduction rates began to increase, and perhaps the increase in slabs arriving at the 660-km discontinuity caused another slab avalanche, which in turn initiated a large number of mantle plumes. Again, the slab avalanches are indirectly responsible for increasing the production rate of juvenile crust as well as contributing to the formation of another supercontinent. The width of the 1.9-Ga continental age peak (Fig. 8.8) may reflect several slab avalanches between 1.9 and 1.7 Ga, of which only the 1.9-Ga and perhaps a 1.7-Ga avalanche appear to have had worldwide effects. As the Late Archean supercontinent broke up over a prolonged period of about 200 Myr, fragments traveled to geoid lows and a new supercontinent began to form, the oldest portion of which is the Birimian–Guiana craton, which formed at about 2.1–2.0 Ga. This was followed by continued collision of cratons and arcs from 1.9 to 1.7 Ga with the addition of significant volumes of new continental crust.

Stage III. Between 1.3 and 1.0 Ga, a new supercontinent, Rodinia, formed, although, as discussed in the next section, it is unlikely that a superplume event accompanied growth of this supercontinent. Rodinia survived for about 600 Myr, and it was dispersed by a new mantle upwelling between 800 and 600 Ma (Figs. 8.5 and 8.6). The formation of Gondwana (650–580 Ma) overlapped the breakup of Rodinia, and this was followed by the formation and breakup of the short-lived supercontinent Pannotia between 580 and 545 Ma (Fig. 8.5) (Dalziel 1997). Pangea formed between 450 and 250 Ma, began to fragment about 160 Ma, and broke up in the last 120 Myr. A new supercontinent may have begun to form in the last 100 Myr, as recorded by the India–Tibet collision, collisions between Australia and Indonesia, and numerous terrane collisions in the American Cordillera (N in Fig. 8.5).

Because of the overlap in formation–breakup cycles of Phanerozoic supercontinents, it is not clear how, or if, minor superplume events at 480, 280, and 100 Ma are related to the supercontinent cycle (Fig. 8.5). It is noteworthy, however, that the 480- and 280-Ma events occurred during the formation of Pangea, and the 100 Ma event may correlate with the beginning of a new supercontinent. These superplume events were small because the relatively cool mantle with a low Rayleigh number made it easier for slabs to penetrate the 660-km discontinuity, and only small amounts of slab material collected at this boundary before collapse occurred. The last event, the Mid-Cretaceous superplume event, may have resulted from the last slab avalanche in Earth history. The mantle upwelling beneath the Pacific plate may have developed in the last 100 Myr because the large Pacific plate shielded the underlying mantle from subduction. That the Pacific plate remained in approximately the same location between 130 and 95 Ma supports a shielding origin for the upwelling.

Prevot et al. (2000) have suggested that true polar wander (see Chapter 2) has been intermittent during the last 200 Myr with two long periods of standstill at 80–0 Ma and 200–150 Ma. Corresponding closely to the Cretaceous superplume event is an abrupt tilt of 20° in the Earth's rotational axis, which occurred at 110 ± 10 Ma. These investigators have suggested that this tilt was caused by a major reorganization of mass in the mantle caused perhaps by avalanching of subducted slabs, which, upon arriving at the core–mantle boundary, triggered a superplume event. Furthermore, the stability of Earth's rotation axis both before and after the alleged superplume event implies the existence of steady convection, which does not modify the large-scale mass distribution in the mantle.

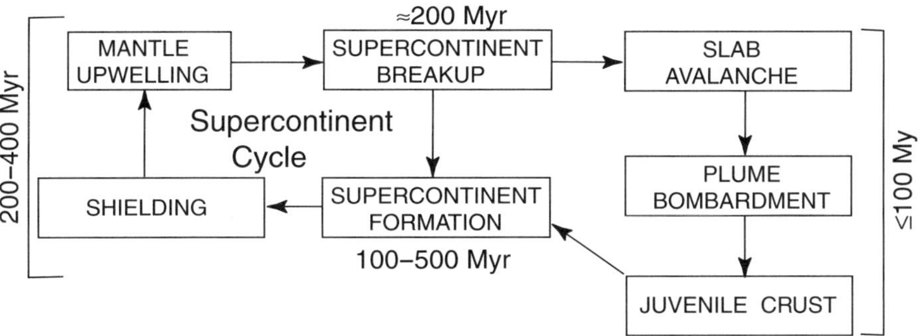

Figure 8.12. Box diagram showing a possible relationship of the supercontinent cycle to superplume events. Modified after Condie (1998).

Superplume Events and Supercontinents

A possible relationship of superplume events to the supercontinent cycle is summarized in Figure 8.12. The timing of various events within this cycle is constrained chiefly by data from two sources: (1) isotopic ages of juvenile continental crust, and (2) results of computer simulations of mantle processes (Tackley et al. 1994; Condie 1998, 2000a). Beginning with a supercontinent, computer models suggest that it takes on average 200–400 Myr for shielding of a large supercontinent to cause a mantle upwelling beneath it (Lowman and Jarvis 1996). The upwelling breaks the supercontinent over approximately a 200-Myr period (160–170 Myr for Rodinia and Pangea). In the case of the 2.7- and 1.9-Ga superplume events, this breakup may be the trigger for the avalanching of slabs through the 660-km discontinuity. Alternatively, the collapse of slabs may result from a threshold for total slab mass at a given location on the 660-km discontinuity, and in this case accumulation could be spread over several hundred million years beginning when a supercontinent fragments. From the time a slab avalanche begins to the time juvenile crust is produced is probably quite short (<100 Myr). This is because slabs can sink to the bottom of the mantle in 100 Myr or less (Larson and Kincaid 1996), and in a mantle in which viscosity increases with depth, mantle plumes can rise to the base of the lithosphere in a few million years (Larsen and Yuen 1997). This scenario suggests a correlation of slab avalanches with supercontinent formation rather than with supercontinent breakup. If supercontinents are really cyclic, it is clear, they are not periodic. Also, supercontinent breakup and aggregation overlap by 50 to 100 Myr, perhaps with an increase in the proportion of overlap in the Phanerozoic (Fig. 8.5).

As suggested by Condie (1998), the duration of supercontinent formation and the total lifespan of supercontinents decrease with age. The duration of supercontinent formation (including enhanced production of juvenile crust) in the first two supercontinents is 500 Myr, or less decreasing to about 300 Myr in Rodinia and then to 100–200 Myr for Gondwana and Pangea. Paralleling this decrease is a decline in the volume of juvenile continental crust produced during each supercontinent cycle. Approximately 30–50% of the present continental crust may have been produced during formation of each of the first two supercontinents at 2.7 and 1.9 Ga, whereas only about 12% was produced

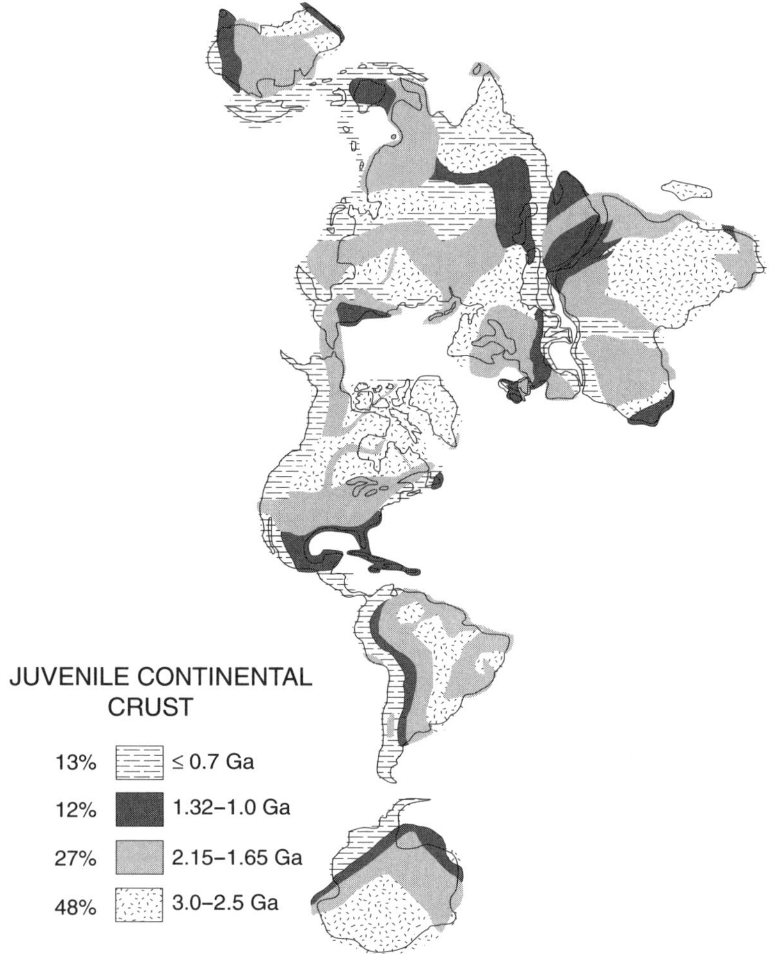

Figure 8.13. Equal-area projection of the continents showing the distribution of juvenile continental crust produced in four time windows. Modified after Condie (1998).

during the formation of Rodinia (Fig. 8.13). The total amount of juvenile continental crust produced during the Phanerozoic appears to have been only 10–15% of the present volume. Only the first two supercontinents, which are both associated with superplume events, contain significant volumes of juvenile crust.

What may have been responsible for decreases in the length of supercontinent cycles and in the volume of juvenile crust produced during each supercontinent cycle? If related to superplume events, the major cause may be the decreasing "strength" of the 660-km discontinuity with time. As the 660-km discontinuity became more permeable to descending slabs with falling mantle temperature, slabs would begin to sink steadily through the boundary, and thus fewer slabs would accumulate at the boundary. Hence, when an avalanche occurred, it would be relatively small. The small volume of plates in an avalanche may have two effects: (1) the duration of a superplume event should be shorter, and perhaps more limited in geographic extent, and (2) because of the smaller mass of sinking slabs, the number of mantle plumes generated would decrease, and thus also would the volume of juvenile crust associated with these plumes. If the proposed superplume model is viable, the decreasing aggregation time of supercontinents with time may reflect cooling of the mantle as radiogenic heat has decreased with time.

One aspect of the catastrophic crustal growth model that is not well understood is that of whether the supercontinent cycle can operate independently of superplume events (Condie 2000a). Peltier et al. (1997) suggested that the supercontinent cycle is caused by slab avalanche events in the mantle. In their model, the avalanches produce mantle downwellings directly over the avalanches, which act as "catchment basins" for an aggregating supercontinent. However, if supercontinents accumulate over geoid lows and break up over mantle upwellings (Anderson 1982; Gurnis 1988; Lowman and Jarvis 1996), both of which are a consequence of the supercontinent cycle, slab avalanches in the mantle may not be a necessary part of supercontinent formation. As discussed earlier, mantle upwellings that eventually break supercontinents result from thermal shielding of a large volume of mantle from subduction and cooling. If so, how are superplume events related to the supercontinent cycle? Perhaps slab avalanches can be considered as an add-on to the supercontinent cycle (Fig. 8.12). A slab avalanche, which initiates plume bombardment of the lithosphere and consequent production of juvenile crust, may occur any time during the supercontinent cycle. In the model of Condie (1998), supercontinent breakup involving increased rates of subduction may trigger a slab avalanche, and this may be the only common denominator to both cycles. This applies, however, only to the 1.9-Ga avalanche because there was no supercontinent to fragment in the Late Archean avalanche. Perhaps superplume events "inject" juvenile crust into the supercontinent cycle, thus increasing the volume of each succeeding supercontinent.

In the proposed model, as the Earth gets older, superplume events will fade out before the supercontinent cycle comes to an end because the 660-km seismic discontinuity becomes more permeable to descending slabs as the mantle cools and the Rayleigh number drops. With the exception of three possible small superplume events in the Ordovician, late Paleozoic, and Cretaceous, superplume events appear to have ended after 1.9 Ga. The supercontinent cycle, however, may continue as long as convection continues in the mantle in response to shielding of parts of the mantle by large lithospheric plates.

The First Supercontinent

One of the intriguing yet puzzling questions of any of the episodic models for production of continental crust is that of just how and why the first supercontinent formed. There are no robust data that support the existence of a supercontinent prior to the Late Archean. For a supercontinent to form requires a significant volume of continental crustal fragments that survive recycling into the mantle. Before the Late Archean, the high mantle temperatures and inferred large mantle convection rates in response to large Rayleigh numbers probably resulted in rapid recycling of continental crust–presumably before continental pieces had time to collide to make a supercontinent (Armstrong 1991; Bowring and Housh 1995). So what happened in the Late Archean that led to formation of the first supercontinent?

One possibility is that the first slab avalanche in the mantle at 2.7 Ga, which liberated mantle plumes from the D″ layer, led to the production of large volumes of continental crust in a relatively short time (≤100 Myr). If this were the case, unlike later supercontinents, the first supercontinent would form in response to the first slab avalanche. The mantle plumes resulting from the avalanche could produce juvenile crust in two ways: directly, by the production of oceanic plateaus, and indirectly, by heating the upper mantle and increasing the production rate of ocean crust due to increased convection rates, or increasing the total length of the ocean ridge system, or both (Larson 1991a). The increased production rates of oceanic crust are accompanied by increased subduction rates, which account for increased production rates of juvenile continental crust in arc systems. As discussed in Chapter 6, a supercontinent may form by collision between Archean oceanic plateaus, surviving fragments of continental crust older than 2.7 Ga, and oceanic arc systems. Also contributing to growth of a Late Archean supercontinent is the thick Archean subcontinental mantle lithosphere, which is relatively buoyant (Griffin et al. 1998), thus resisting subduction during plate collisions. As pointed out in Chapter 7, the frequency of Late Archean greenstones with oceanic plateau geochemical affinities supports the idea that oceanic plateaus were a major contributor to the Late Archean supercontinent.

The Grenville Event at 1 Ga

The last Precambrian supercontinent, Rodinia, formed during the worldwide Grenville orogenic event at 1.3–1.0 Ga. A survey of U/Pb zircon isotopic ages and Nd isotopic data, however, indicates that the volume of juvenile crust formed during the Grenville event was not unusually large (Fig. 8.14) (Condie 2000b). Calculated crustal production rates during the time interval of 1.35–0.9 Ga fall within the background level of crustal production of about 1.1 km^3/yr. Why was so little juvenile continental crust formed during this time period? During formation of the previous two supercontinents at 2.7 and

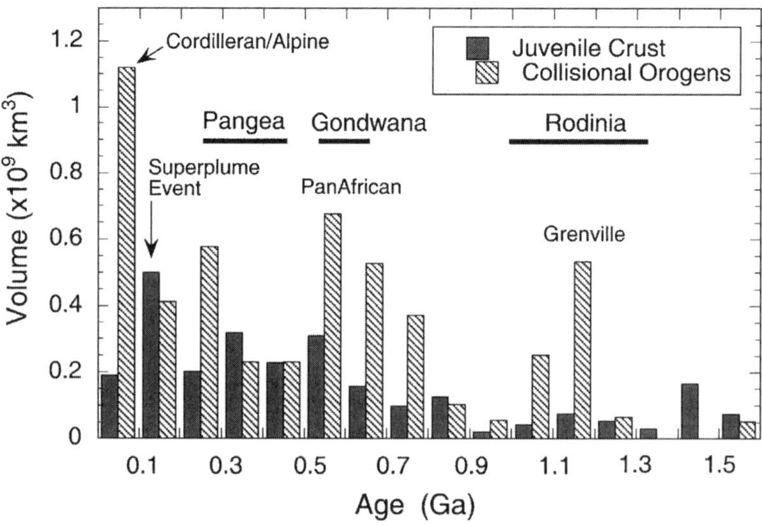

Figure 8.14. Volume distribution of juvenile continental crust and collisional orogens in the last 1.6 Ga. Other information given in Figure 8.8. Modified after Condie (2000b).

1.9 Ga, large volumes of juvenile crust were extracted from the mantle (Condie 1998, 2000a). The first and most obvious reason may be that more juvenile Grenvillian crust is present but that it has not as yet been isotopically dated. The obvious place to look for such crust is in accretionary orogens around the perimeter of Rodinia, for juvenile crust is more frequently preserved in accretionary orogens than in collisional orogens (Windley 1992). However, there are few areas around the margin of Rodinia that are eligible because most of these crustal blocks are known to comprise pre-Grenvillian cratons. The only remaining candidates are perhaps Tibet, the outer edges of the Rio de la Plata and Salvador cratons in South America, and part of the Mauritides in extreme West Africa. Even if minor amounts of Grenvillian crust occur in these areas, it is unlikely that the total amount of juvenile crust of this age exceeds 15% of the total continental crust (Condie 2000b).

If a superplume event is necessary to form large volumes of continental crust, then it follows that a superplume event did not occur in the 1.35–0.9 Ga time window. Only two major superplume events can be verified in the Precambrian at 2.7 and 1.9 Ga. Without a Grenvillian superplume event, a large time gap of the order of 1.5 Gyr exists between the 1.9 Ga and the Phanerozoic events. This indicates that, although superplume events are episodic, they are not periodic.

It appears that the supercontinent cycle and superplume events can operate independently of each other. This conclusion is also supported by the time distribution of juvenile crust and orogenic zircon ages. For instance, if the volume of collisional and accretionary orogenic belts is compared with that of juvenile crust in the last 1.6 Gyr, very different age patterns emerge (Fig. 8.14). A prominent peak in orogenic crust abundance occurs at 75 Ma with other peaks at about 250, 550, 650, and 1150 Ma, whereas the juvenile crust distribution shows a well-defined peak only at about 100 Ma, correlative with the Mid-Cretaceous superplume event. The four oldest orogenic peaks correlate with formation of supercontinents: Rodinia at 1.3–1.0 Ga, Gondwana at 650–580 Ma, and Pangea at 450–250 Ma. The youngest orogenic peak at 75 Ma may correlate with the inception of a modern supercontinent.

As previously discussed, Rogers (1996) suggested that two supercontinents formed during the 1.9 Ga supercontinent (and superplume) event. These are Atlantica and Nena (Figs. 8.6 and 8.7). These two supercontinents appear to have survived breakup of the Paleoproterozoic supercontinent at 1.6–1.4 Ga and were later incorporated intact into the Neoproterozoic supercontinent Rodinia. Why these two supercontinents survived the Mesoproterozoic breakup is unknown. An alternative interpretation is that two supercontinents formed at 1.9 Ga and that neither was large enough to provide adequate lithospheric shielding for the production of mantle upwellings large enough to break the continental lithosphere (Lowman and Jarvis 1996). In either case, however, if supercontinent breakup is necessary to trigger superplume events, survival of two supercontinents could account for the lack of a Grenvillian superplume event. During supercontinent breakup at 1.5 to 1.35 Ga (Fig. 8.5), perhaps only a small portion of the supercontinent (or supercontinents) was fragmented, and the resulting increase in subduction rates was not sufficient to initiate slab collapse at the 660-km seismic discontinuity, which would normally lead to a superplume event.

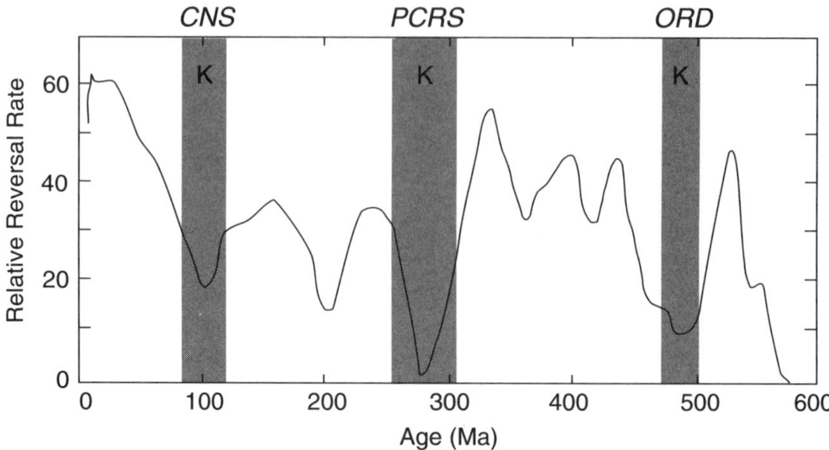

Figure 8.15. Relative magnetic reversal rate during the Phanerozoic. Also shown are three major superchrons: CNS = Cretaceous Normal; PCRS = Permian–Carboniferous Reversed; and ORD = Ordovician. K = major kimberlite-diamond events. After Johnson et al. (1995) and Haggerty (1999).

Superchrons and Superplumes

Also coincident with a Mid-Cretaceous superplume event at 120–80 Ma is the Cretaceous superchron, when very few magnetic reversals occurred for a period of 30–40 My (Figs. 8.15 and 9.5). Larson and Olson (1991) showed that magnetic reversal frequency correlates inversely with mantle plume activity for the past 150 Myr as measured by the production rate of juvenile crust in oceanic areas. To explain this correlation, these authors proposed a model whereby removal of large amounts of heat from the core not only fuels the superplume event but also stops, or greatly retards the frequency of magnetic reversals in the magnetic field. They have suggested that, as mantle plumes rise from the D″ layer, it thins, resulting in core cooling by allowing heat to be conducted more rapidly across the core–mantle boundary. The outer core convection then increases to restore the abnormal heat loss, causing a decrease in frequency of magnetic reversals. In effect, this switches the Earth's geodynamo from a reversing AC state to a nonreversing D C state. When the convective activity increases above some threshold value, a magnetic superchron is initiated.

Although this model has been widely accepted, other explanations for superchrons exist. Loper (1992) suggested that the thermal inertia of the D″ layer may not result in rapid changes in the core's heat flux when superplumes are formed. Cox (1968) proposed a model for magnetic field reversals caused by perturbations of the outer core that feed sufficient reverse flux into the geodynamo to overcome the main dipole. In this model, increased convective vigor generates more frequent instabilities in the fluid outer core, leading to an increase in reversal rate. Courtillot and Besse (1987) suggested still a different model whereby the decrease in reversal rate just before the Mid-Cretaceous superchron is caused by a decrease in core convection due to a buildup of heat at the core–mantle boundary. This excess heat is released with the plumes that rise from D″ near the end of the superchron. The loss of heat results in increased core convection, which in turn causes an increase in reversal rates.

Using the Global Paleomagnetic Database, Johnson et al. (1995) identified two other superchrons in the last 600 Myr. One, the Permo-Carboniferous reversed superchron, is well established, and the other, in the Early Ordovician, may be a previously unrecognized superchron (Fig. 8.15). Each of the Phanerozoic superchrons has a 30–50-Myr duration and a spacing on the order of 200 Myr. If each superchron reflects a superplume event, the spacing between events is considerably less than the spacing between Late Archean and the Early Proterozoic superplume events. The Ordovician and late Paleozoic events need to be further tested against the geologic record as possible superplume events. Data suggest a period of enhanced production of juvenile crust in the late Paleozoic – at least in Asia (Sengor et al. 1993). In the superplume model of Condie (1998), the three Phanerozoic superchrons may reflect superplume events, although less intense than their counterparts in the Precambrian. As previously mentioned, the decrease in intensity results from cooling of the mantle and consequent weakening of the 660-km discontinuity, which would result in more frequent and less intense slab avalanches in agreement with observations. An interesting test of this idea is to check for magnetic reversals in rock successions that formed during the 2.7- and 1.9-Ga superplume events. If the model is feasible, superchrons should occur near each of these peaks. Although single magnetic reversals have been identified in igneous rocks formed at or near the 1.9-Ga peak (Morgan 1985; Zhai et al. 1994; Buchan et al. 1996), multiple reversals in single stratigraphic successions at or near these peaks have not been documented.

Perspective

Although many investigators agree that superplumes exist, whether or not superplume events have occurred in the geologic past still remains debatable and speculative. However, the episodic distribution of continental crustal growth in Earth history seems to presuppose some kind of episodic event in the mantle. One way to test the idea that the episodic growth of continents is due to superplume events is to look in the geologic record for evidence of such events. In the next and final chapter, we will explore the geological record to see if features have been preserved that are consistent with idea of superplume events.

9

Mantle Plumes and Earth Systems

Introduction

One of the most exciting aspects of episodic mantle plume activity is the consequences it may have had in Earth history and especially the effects on near-surface Earth systems (Kerr 1998; Condie et al., 2000; 2001). For instance, the extensive volcanism associated with a superplume event should pump significant amounts methane (CH_4) into the ocean–atmosphere system, where it is rapidly oxidized to carbon dioxide (CO_2). Because CO_2 is an important greenhouse gas, a superplume event should lead to rapid global warming followed by cooling as weathering draws down the CO_2 level. Because new ocean ridges form during supercontinent breakup, the supercontinent cycle may also have important consequences for near-surface Earth systems (Worsley et al. 1986; Veevers 1990). Increased CO_2 levels in the atmosphere should also affect the carbon cycle – perhaps enhancing photosynthesis and the burial rate of carbon. This should be reflected in the sedimentary record by black shales, hydrocarbons, and coal. Increased hydrothermal fluxes on the seafloor may also increase nutrient supplies to oceanic life. The increase in volume of oceanic plateaus as well as elevation of the oceanic lithosphere by superplumes should raise sea level, producing widespread shallow marine environments available for expansion of marine life and for carbonate platform and reef formation. Superplume events may also contribute to mass extinctions. As documented for the Deccan traps 65 Ma and the Siberian traps 250 Ma, even single superplumes may contribute to mass extinctions (Hallam 1987; Renne et al. 1995). During eruption of plume-related magmas, large volumes of sulfur dioxide (SO_2) and halogens are injected into the atmosphere, both of which can produce toxic environments for some organisms. And finally, both radiogenic isotopes, such $^{87}Sr/^{86}Sr$ and $^{143}Nd/^{144}Nd$, and stable isotopes, such as $^{13}C/^{12}C$ and $^{18}O/^{16}O$, should respond to paleoclimatic changes and to input of mantle-derived fluids into the atmosphere and oceans (Veizer et al. 1999).

In this, the final chapter, we will consider potential effects of mantle plume events on the ocean–atmosphere–biosphere system and review evidence in the geologic record that is consistent with such events.

Superplumes, Supercontinents, and the Carbon Cycle

Introduction

The biogeochemical cycle is an integrated system of carbon reservoirs linked by transfer processes (Condie et al. 2000a,b). In Figure 9.1, these reservoirs and processes are depicted as boxes and arrows, respectively. Carbon enters the ocean–atmosphere system by weathering, volcanism, and metamorphism. Part of this carbon is in a reduced state because it includes organic carbon remobilized during destruction of older sediments. In the surface environment, carbon is rapidly cycled through the biosphere (paths labeled a, Fig. 9.1). The amounts of organic carbon and carbonate buried depend upon global rates of erosion and sedimentation (Derry et al. 1992) and upon recycling processes on the seafloor (Betts and Holland 1991). If the average degree of reduction of buried carbon differs from that of the carbon entering the surface environment, then a net transfer of oxidizing or reducing power must occur between the carbon cycle and the cycles of other elements – principally sulfur, oxygen, nitrogen, iron, and manganese.

The operation of the carbon cycle can be monitored by an isotopic mass balance as follows (Des Marais et al. 1992):

$$\delta_{\text{in}} = f_{\text{carb}}\delta_{\text{carb}} + f_{\text{org}}\delta_{\text{org}}, \tag{9.1}$$

where δ_{in} represents the isotopic composition of carbon entering the global surface environment consisting of the atmosphere, hydrosphere, and biosphere. The right side of the equation represents the weighted-average isotopic composition of carbonate ($\delta^{13}C_{\text{carb}}$) and organic ($\delta^{13}C_{\text{org}}$) carbon being buried in sediments, and f_{carb} and f_{org} are the fractions of carbon buried in each form ($f_{\text{carb}} = 1 - f_{\text{org}}$). For timescales longer than 100 Myr, $\delta_{\text{in}} = -5$ per mil, the average value for crustal and mantle carbon (Holser et al. 1988). Thus, where values of sedimentary δ_{carb} and δ_{org} can be measured, it is possible to determine f_{org} for ancient carbon cycles. Note, for example, that higher values of $\delta^{13}C_{\text{carb}}$, $\delta^{13}C_{\text{org}}$, or both indicate a higher value of f_{org}.

The processes that cycle carbon can be affected by crustal tectonics and mantle plume events. A summary of possible effects of superplumes and supercontinents on the carbon cycle is given in Figure 9.1. The feedbacks shown in the figure have been discussed by Worsley et al. (1986), Des Marais (1997), Kerr (1998), and Condie et al. (2000a).

Supercontinent Formation

Supercontinent assembly impacts the carbon cycle in several ways (Kerr 1998; Condie et al. 2001). Continental collisions are initially a net source of CO_2 owing to the burial or thermal destruction of sedimentary organic matter and carbonates within collisional zones, or both (paths b and c, Fig. 9.1) (Bickle 1996). Continued uplift of a supercontinent accelerates erosion of sedimentary rocks and their carbon (paths d and e). Whether this carbon source changes the $\delta^{13}C$ of seawater depends on the ratio of the reduced carbon ($\delta^{13}C = -20$ to -40‰) to oxidized carbon ($\delta^{13}C = 0$‰) recycled back into the oceans (path f). For example, if both carbonate and organic carbon are recycled in

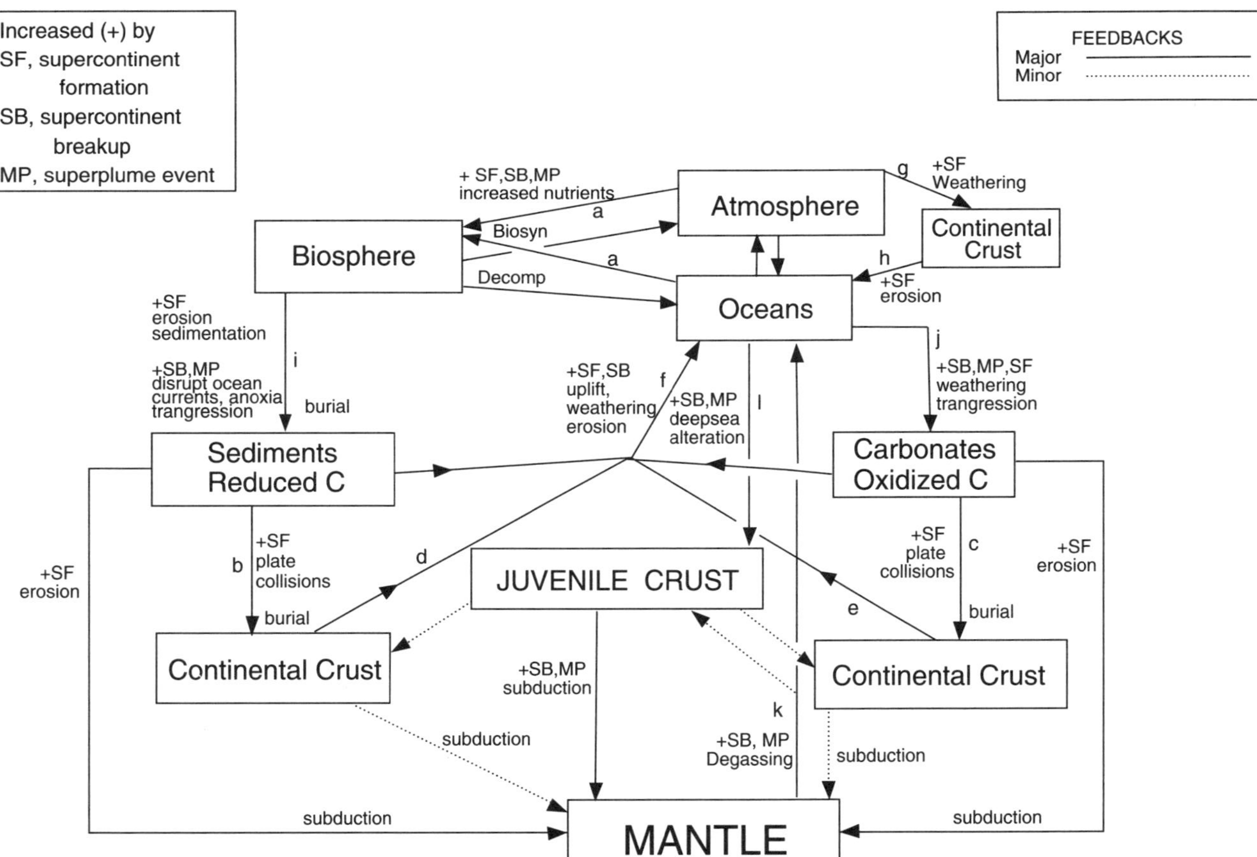

Figure 9.1. Carbon reservoirs in the Earth showing possible effects of supercontinents and superplumes. Each box represents a carbon reservoir. Juvenile crust = oceanic crust + oceanic plateaus + island arcs. Biosyn = biosynthesis; Decomp = decomposition. Lettered paths refer to text discussions. After Condie et al. (2000a).

approximately the same ratio as their ratio prior to supercontinent formation, the $\delta^{13}C$ of seawater will not change (Des Marais et al. 1992). As the surface area of a growing supercontinent increases, weathering of surface rocks withdraws more CO_2 from the atmosphere, transferring it to the continents (path g), where it is eventually returned to the oceans by erosion (path h). Increased erosion also releases more nutrients (e.g., phosphorus), increasing biological productivity (paths h, a) (Worsley and Nance 1989). The nutrient source and CO_2 sinks can draw down atmospheric CO_2 levels, favoring cooler climates that intensify ocean circulation and thus increase nutrient upwelling and marine productivity (Berry and Wilde 1978). Intense drawdown of CO_2 together with increasing albedo caused by the increasing land/ocean ratio can lead to widespread glaciation. These factors collectively promote increased burial rates of organic carbon, relative to carbonates, and thus may raise the $\delta^{13}C$ value of seawater. However, uplift of collisional mountain belts during supercontinent formation can recycle older carbon that is depleted in ^{13}C. For instance, a dramatic drop of $\delta^{13}C$ in marine carbonates about 55 Ma coincides with initial uplift of the Himalayas in response to the India–Tibet collision and may reflect recycling of carbon depleted in ^{13}C (Beck et al. 1995). In addition, the final stages in the formation of Pangea were accompanied by compressive stresses around the margin of most of the supercontinent, leading to significant uplift and erosion. Faure et al. (1995) suggested this enhanced erosion may be responsible for a pronounced minimum in seawater $\delta^{13}C$ at 250 Ma. Hence, it appears the control of $\delta^{13}C$ in seawater during supercontinent formation reflects a delicate balance between carbon burial and carbon recycling.

Two concepts have been proposed for a possible role of gas hydrates (methane–H_2O solids in shallow marine sediments) in global climate change (Kvenvolden 1999). The first one is direct injection of methane, or more likely its oxidized equivalent CO_2, into the ocean–atmosphere system as gas hydrates dissolve during warm climatic regimes. This would provide a strong positive feedback for global warming. The second is that continental-margin gas hydrates release methane during falling sea level, which generally accompanies global cooling. Such cooling, for instance, could occur during glaciation or supercontinent formation. However, with the present-day reserves of gas hydrates, neither of these effects should have significant effects on climate change or on sea level (Kvenvolden 1999; Bratton 1999). If gas hydrates were widespread during the Precambrian, supercontinent formation could lead to gas hydrate evaporation as sea level drops, which would introduce biogenic carbon as CO_2 into the atmosphere, increasing both organic and carbonate burial rates as well as greenhouse warming (Haq 1998). Also, because gas hydrates contain carbon with very negative $\delta^{13}C$ values (averaging about −60‰), they may offset any increase in the $\delta^{13}C$ due to organic carbon burial.

As sea level falls during supercontinent formation, the ensuing regression restricts the deposition of shelf carbonates and mature clastic sediments, and the emerging shelves can accommodate deposition of extensive evaporites. Organic carbon sedimentation occurs either farther offshore or in freshwater basins within the interior of the supercontinent (Berner 1983). Overall, supercontinent formation promotes higher rates of erosion and sedimentation (path i, Fig. 9.1), which correlate with organic carbon burial rates, and platform carbonate deposition becomes more restricted. The net result is that periods of supercontinent formation favor relatively high ratios of organic versus carbonate sedimentation and burial. If this is the case, positive carbon isotope anomalies should develop in seawater during supercontinent formation unless other processes obscure this effect.

Supercontinent Breakup

Supercontinent breakup creates new, narrow ocean basins having restricted circulation and hydrothermally active spreading centers (Kerr 1998; Condie et al. 2001). These features promote anoxia in the deep ocean (path i, Fig. 9.1). The actively eroding escarpments along new rift margins contribute sediments to these basins, and marine transgressions increase the rate of burial of organic and carbonate carbon on stable continental shelves. The amount of shallow marine carbonate deposition (path j), however, critically depends on redox stratification of the oceans, for reducing environments are not conducive to carbonate precipitation. Should anoxic deep-ocean water invade the shelves, it would facilitate organic carbon burial on the shelves, including the deposition of black shale and the accumulation of gas hydrates.

The increase in length of the ocean-ridge network that accompanies supercontinent fragmentation promotes increased degassing of the mantle, including CO_2 (path k). Increasing atmospheric CO_2 levels and rising sea level promote warmer climates, resulting in increased weathering rates (path g) (Berner and Berner 1997) as well as the potential for the marine water column to become stratified and for deep water to become anoxic (path i) (Berry and Wilde 1978). Increasing carbonate in the oceans together with a growing ocean-ridge system would also enhance rates of removal of seawater carbonate by deep-sea alteration (path l). To the extent that these developments enhance the fraction of carbon buried as organic matter, they would also lead to an increase in the $\delta^{13}C$ of seawater because ^{12}C is preferentially incorporated into organic carbon (Des Marais et al. 1992; Melezhik et al. 1999).

Superplume Events

During a superplume event, ascending plumes warm the upper mantle and lithosphere and thereby elevate the seafloor by thermal expansion and create oceanic plateaus by the eruption of large volumes of submarine basalt. Rising sea level triggers marine transgressions (Larson 1991b) (path i, Fig. 9.1). Oceanic plateaus can locally restrict ocean currents (Kerr 1998), thus promoting local stratification of the marine water column which leads to anoxia (path i). Plume volcanism and associated extensive hydrothermal activity exhale both CO_2 and reduced constituents into the atmosphere–ocean system (Larson 1991b; Caldeira and Rampino 1991; Kerr 1998). The increased CO_2 flux warms the climate and enhances weathering rates (path g) (Berner and Berner 1997). During superplume events, when anoxia is widespread in the oceans, gas hydrates could form in large volumes, provided the oceans are not warm enough to dissolve the hydrates.

Biological productivity during a superplume event is enhanced by several factors such as increased concentrations of CO_2, increased nutrient fluxes from both hydrothermal activity (such as P, H_2, sulfides, trace metals, etc.) and enhanced weathering, and elevated temperatures due to CO_2-driven greenhouse warming (paths a).

Carbonate precipitation is enhanced by increased chemical weathering and by marine transgressions (path j). Increased hydrothermal activity on the sea floor should also increase the rate of deep-sea alteration, which in turn should increase the removal rate of carbonate from seawater (path l). Liberation of large amounts of SO_2 into the oceans by increased hydrothermal activity might decrease ocean pH, the net effect of which would be to dissolve marine carbonates – particularly adjacent to high-temperature emanations (Kerr 1998). However, a more acidic ocean would also dissolve more Na and Ca increasing the oceanic pH again. If the oceans are relatively reducing, hydrothermal exhalation of Fe^{2+} promotes siderite deposition, for example, adjacent to banded iron formations (Beukes et al. 1990). This, in turn, promotes carbonate deposition overall because siderite is less soluble than calcite and dolomite. Organic matter burial is enhanced by increased productivity, marine transgression, and the expansion of anoxic waters – in particular onto continental shelves (path i) (Larson 1991b; Kerr 1998). In summary, phenomena associated with superplumes promote the formation and deposition of both organic and carbonate carbon. It has been proposed that the relative deposition of carbonates and organic carbon reflect redox buffering of the crustal and surface environment by the redox state of the upper mantle (Holland 1984). Redox buffering by the mantle should be even stronger during superplume events.

Subduction of carbon may also play a role in determining the response of the carbon cycle to superplume events. During most of Earth history, the relative rates of subduction of carbonate and reduced carbon reflect their relative crustal abundances. If they did not, the mean $\delta^{13}C$ values of crustal versus mantle carbon reservoirs would differ substantially today because the preferential subduction of either oxidized or reduced carbon would have made the crust–mantle exchange of carbon an isotopically selective process. However, the $\delta^{13}C$ values of the total crust and mantle carbon reservoirs are identical within the uncertainties of measurements (Holser et al. 1988). Therefore, subduction has not favored either carbonates or organic carbon. Although Jyotiranjan et al. (1999) suggested that inorganic carbon is subducted and recycled faster than organic carbon during a mantle plume event, their conclusion is based solely on data from Kerguelen, and there is no evidence to support preferential inorganic carbon burial during a worldwide superplume event.

Sea Level

Changes in sea level leave an imprint in the geologic record by the distribution of sediments and biofacies, the aerial extent of shallow seas as manifested by preserved intracratonic sediments, and by unconformity-bounded packages of sediments. A widely used approach to monitor sea level is to map successions of transgressive and regressive facies in marine sediments. To estimate sea level changes, it is necessary to correlate facies and unconformities over large geographic regions on different continents and different tectonic settings using sequence stratigraphy (Vail and Mitchum 1979). Using this technique, onlap and offlap patterns are identified in seismic reflection profiles and are used to construct maps of the landward extent of coastal onlap from which changes in sea level are estimated.

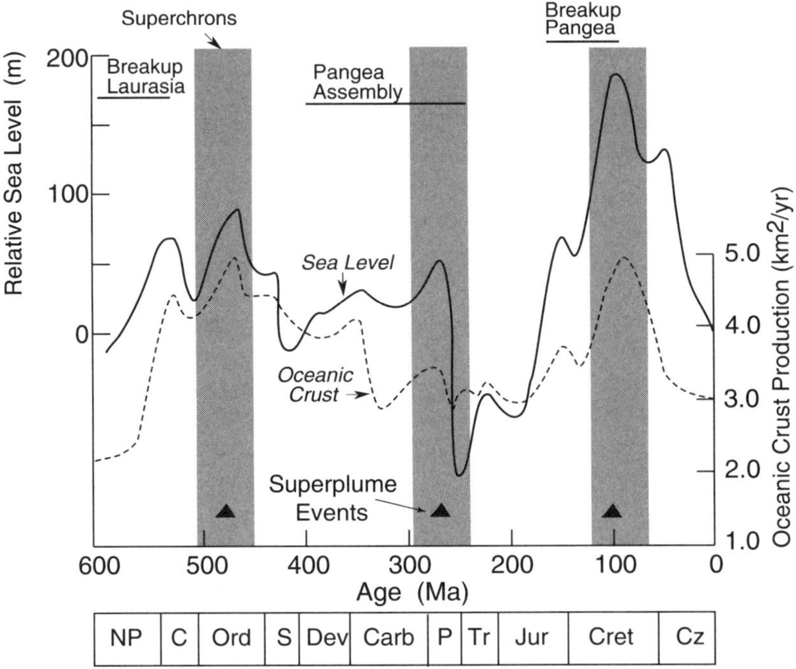

Figure 9.2. Eustatic sea level relative to present sea level over the last 600 Myr. Also shown is calculated production rate of oceanic crust. Data from Gaffin (1987), Haq et al. (1988), and Algeo and Seslavinsky (1995).

Although most investigators seem to agree about a long-term component (200–400 Myr) in sea level change during the Phanerozoic, short-term components and amplitudes of variation remain uncertain. Several factors can cause short-term changes in sea level. Among the most important are tectonic controls, chiefly continental collisions, which operate on timescales of 10^6–10^8 yr, and glacial controls that operate on timescales of 10^4–10^6 yr. Tectonic controls seem to produce sea level changes at rates less than 1 cm/10^3 yr, whereas glacial controls can produce rate changes up to 10 m/10^3 yr.

Major transgressions are recorded in the early and late Paleozoic and in the Late Cretaceous (Fig. 9.2). The long-term changes in sea level may be related to one or more of the following factors: (1) rates of seafloor spreading, (2) the characteristics of subduction, (3) the motion of continents with respect to geoid highs and lows, (4) superplume events, and (5) supercontinent insulation of the mantle (Gurnis 1993; Eriksson 1999). One of the most important causes of long-term sea level changes is the volume of ocean-ridge systems. The cross-sectional area of an ocean ridge is dependent upon spreading rate because the depth of the ridge is a function of its age. For instance, a ridge that spreads 6 cm/yr for 70 Myr has three times the cross-sectional area of a ridge that spreads at 2 cm/yr for the same amount of time. The total ridge volume at any time is dependent on spreading rate (which determines cross-sectional area) and total ridge length.

Subduction can also affect sea level. For instance, if additional convergent and divergent plate margins form in an oceanic plate while spreading rate is maintained at a constant value, the average age of oceanic lithosphere decreases and leads to uplift above subduction zones, causing a drop in sea level (Gurnis 1993). Also, when continental plates move over geoid highs, sea level falls, and conversely, sea level rises when plates move over geoid lows. As the mantle warms up beneath a supercontinent, the continental plate rises isostatically with a corresponding fall in sea level.

Worsley et al. (1984, 1986) proposed a model in which Phanerozoic sea level is directly tied to the supercontinent cycle with a repeat time of about 400 Myr. The rapid increase in sea level observed between 600 and 530 Ma and again at about 100 Ma corresponds to the onset of breakup of Laurasia (Laurentia–Baltica–Siberia) and the extended breakup of Pangea, respectively. The low sea level at about 600 Ma correlates with the assembly of Gondwana and the low at about 250 Ma correlates with the final assembly of Pangea (Fig. 9.2). The low stands of sea level accompanying supercontinents probably reflect the geoid highs developed beneath the supercontinents, whereas the high sea levels accompanying supercontinent dispersal reflect enhanced ocean-ridge activity. Because major changes in sea level in the Phanerozoic appear to be related to changes in volume of the ocean-ridge system, times of high sea level are also times of rapid production of oceanic crust (Fig. 9.2) (Gaffin 1987). The oceanic crust production rate increases by up to a factor of two during supercontinent breakups.

Superplume events should result in a rising sea level (Kerr 1998; Eriksson 1999; Condie et al. 2000a). Sea level rises because of isostatic uplift and thermal erosion of the oceanic lithosphere above plume heads (Kerr 1994; Lithgow-Bertelloni and Silver 1998). Also, oceanic plateaus contribute to a rise in sea level, and during superplume events, when many oceanic plateaus form, the effect could be significant. When supercontinent formation and superplume events coincide, such as at 1.9 and 2.7 Ga, each process affects sea level in opposite directions. Hence, the net effect on sea level, if any, is the weighted average of these competing events.

Global Warming

Carbon dioxide and methane are two important greenhouse gases (Caldeira and Rampino 1991). The primary source for these gases on Earth is volcanism, and especially volcanism related to ocean ridges and to mantle plumes (Fig. 9.3). Because of free oxygen in the atmosphere–ocean system, methane is rapidly oxidized to CO_2, which is the principal terrestrial greenhouse gas. The rate of input of CO_2 into the atmosphere–ocean system is proportional to the rate of seafloor generation and to the rate of mantle plume activity (Berner 1983; Condie et al. 2001). There are three important negative feedbacks that control levels of CO_2 in the atmosphere, as discussed with respect to the carbon cycle earlier in this chapter (Fig. 9.1). The most important is chemical weathering on the continents. Because this effect varies with the surface area of continents exposed to weathering, it is at maximum when supercontinents exist. Photosynthesis and hydrothermal alteration associated with submarine volcanism (both ocean ridge and hotspot) also remove CO_2 from the atmosphere. Hence, the amount of global warming due to CO_2 in the atmosphere at any given time results from a delicate balance of volcanic input and the three processes of extraction (Fig. 9.3).

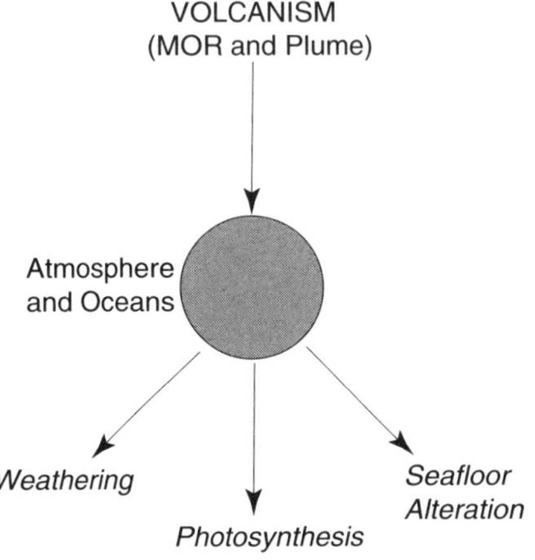

Figure 9.3. Simplified source and sinks for CO_2 in the atmosphere and oceans.

During breakup of supercontinents new ocean ridges form, and the input of CO_2 into the atmosphere–ocean system increases. This could lead to global warming before some combination of the three negative feedbacks reduces the CO_2 level. In contrast, formation of a supercontinent could lead to global cooling in response to enhanced chemical weathering, which, as previously mentioned, could lead to widespread glaciation. A superplume event may inject large quantities of CO_2 into the atmosphere, leading to rapid global warming in a short time (<50 Myr).

Paleoclimatic temperature regimes can be estimated directly in the last few hundred million years from the oxygen isotope record left by shallow marine organisms and, in particular, microfossils (Larson 1991b). The geological sedimentary record can also be useful in that certain sediments, such as laterites, bauxites, and glacial deposits record specific paleoclimates. In addition, black shales and hydrocarbon deposits, both of which are generated from the fossil remains of phytoplankton, are most widespread during warm climatic regimes. The Chemical Index of Alteration (CIA), commonly referred to as the paleoweathering index (Nesbitt and Young 1982), also has been used to estimate the degree of chemical weathering in the source areas of shales. The CIA is calculated from the molecular proportions of oxides, where CaO is the amount in only the silicate fraction, and $CIA = [Al_2O_3/(Al_2O_3 + CaO + Na_2O + K_2O)] \times 100$, the molecular ratio. The higher the CIA, the greater the degree of chemical weathering in sediment sources. For example, CIA values over 85 are characteristic of residual clays in tropical climates (Nesbitt and Young 1982). This index has proved particularly useful in recognizing warm paleoclimatic regimes in the Precambrian geologic record, where other climatic indicators may be rare (Condie et al. 2000a, 2000b).

The Biosphere

Superplume and supercontinent events can affect the biosphere in two major ways: (1) increased CO_2 levels and other nutrients in the atmosphere and oceans can lead to diversification and increases in the biomass, and (2) introduction of toxic gases and trace elements into seawater and climate changes associated with volcanic eruptions can lead to mass extinctions. Also, heat input from submarine eruptions can destabilize the water column, causing large-scale overturn that affects near-surface marine organisms. Increased input of CO_2 into seawater may cause dissolution of carbonates, leading to increased anoxia and episodes of enhanced deposition of black shales and hydrocarbons (Schlanger and Jenkyns 1976). Input of CO_2, CH_4, phosphorus, iron, and trace metals (Sb, As, Se) into seawater is directly tied to the intensity of deep-sea hydrothermal activity, which in turn is related to the intensity of ocean–ridge and plume volcanism (Von Damm 1990; Tunnicliffe and Fowler 1996). Increased nutrients in seawater, such as methane, iron, manganese and phosphorus, can lead to increases in the biomass, especially in microorganisms that abound around submarine hydrothermal vents. Thus, the breakup of supercontinents with increased ocean-ridge activity and superplume events can both lead to increased nutrient supplies for marine organisms. Studies of modern microbial mats show that the rate of carbon fixation in these organisms is higher for greater levels of CO_2 in the atmosphere (Rothschild and Mancinelli 1990). Hence, an increase in hydrothermal venting associated with either supercontinent breakup or superplumes could lead to an increase in the biomass – at least in photosynthesizing microorganisms and in organisms that live around hydrothermal vents on the seafloor.

Eruption of continental flood basalts and subaerial eruptions on oceanic plateaus can transfer immense amounts of CO_2 into the atmosphere in short periods of time (Rampino 1991; Rampino and Caldeira 1993). Eruption of the Deccan traps 65 Ma may have transferred as much as 2×10^{17} moles of CO_2 into the atmosphere in less than one million years, an amount that could have produced a global greenhouse effect of about 2 °C (Caldeira and Rampino, 1991). The Mid-Cretaceous superplume event may have increased atmospheric CO_2 levels by a factor of 5–15 times the modern preindustrial value, resulting in global warming of up to 8 °C (Barron et al. 1995). Volcanic eruptions also transfer toxic components such as As, Se, Sb, as well as volcanic ash into the atmosphere or ocean. If carried into the stratosphere, volcanic gases and aerosols can cause dramatic short-term effects such as acid rain, diminishing sunlight, and global cooling (Officer et al. 1987). Many or all of these effects could lead to mass extinctions of both submarine and terrestrial organisms (Caldeira and Rampino 1991; Courtillot et al. 1996). Given that one superplume, such as that responsible for eruption of the Siberian traps 250 Ma (Campbell et al. 1992), may have had major impact on living organisms, the effect of a superplume event involving many superplumes could be devastating to living systems.

Sedimentary Systems

Strontium Isotopes in Marine Carbonates

Veizer and Compston (1976) were the first to suggest that the growth rate of continental crust can be tracked with Sr isotopes using the $^{87}Sr/^{86}Sr$ ratio in marine carbonates. Because the $^{87}Sr/^{86}Sr$ ratio in marine carbonates reflects chiefly the balance of Sr from continental and deep-sea hydrothermal (mantle) sources entering the oceans (Veizer 1989; Asmerom et al. 1991), relatively high $^{87}Sr/^{86}Sr$ ratios in marine carbonates indicate sources with high Rb/Sr ratios (continental crust), whereas low ratios reflect

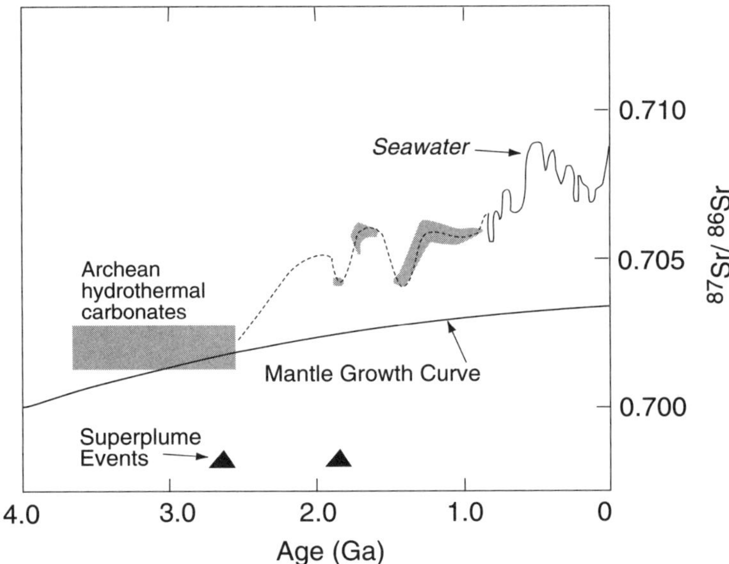

Figure 9.4. The record of $^{87}Sr/^{86}Sr$ ratio in seawater since the Late Archean using marine carbonates as a proxy. Also shown is the mantle growth curve. Gray patterns show distribution of carbonate isotopic ratios. Carbonate strontium isotope data from Mirota and Veizer (1994).

dominant mantle input. Before 2.5 Ga, $^{87}Sr/^{86}Sr$ ratios fall near the mantle growth curve ($\approx$0.702) (Fig. 9.4), indicating that the oceans were largely buffered by mantle input and that little continental crust existed – at least above sea level (Veizer and Compston 1976). After the major period of continental crustal growth at 2.7 Ga, however, the $^{87}Sr/^{86}Sr$ ratio in seawater increased rapidly as increased amounts of continental Sr entered the oceans (dashed line, Fig. 9.4) (Veizer and Compston 1976; Taylor and McLennan 1985; Condie 1998).

Both supercontinent breakup and superplume events should pump more mantle Sr into the oceans via submarine volcanism and hence should decrease the $^{87}Sr/^{86}Sr$ ratio of seawater, provided the effect is intense enough to offset continental inputs of Sr. In contrast, supercontinent formation would lead to uplift and erosion and greater input of continental Sr by rivers, which is an effect that should raise the $^{87}Sr/^{86}Sr$ ratio of seawater. If a significant volume of juvenile continental crust accompanies formation of a supercontinent, the effect should be even more pronounced.

Banded Iron Formation

Banded iron formation, commonly referred to as BIF, is finely laminated, chemical sediment composed chiefly of microcrystalline quartz and iron oxides. The most voluminous BIFs (Superior type) were deposited in intracratonic, passive-margin, or platform basins during stands of high sea level in the Late Archean and Paleoproterozoic (Simonson and Hassler 1996). The iron and silica in BIF appear to have been derived chiefly from hydrothermal vents on the deep seafloor, and the deposition of BIF on shallow continental shelves and intracratonic basins requires one of two processes: (1) upwelling, which brings iron-rich waters from the largely anoxic deep basins into the oxidizing shallow water on the continents, or (2) extensive hydrothermal plumes, depleted in oxygen and enriched in iron, associated with ocean-ridge systems, oceanic plateaus, or both (Klein and Beukes 1992; Isley 1995; Abbott and Isley 2001). In the second process, the hydrothermal plumes transport iron into the upper water column as they rise to a level of neutral buoyancy and then spread outward by ocean currents. In this manner, the iron is carried onto continental shelves without upwelling, and BIF is deposited in various basins along continental margins.

Supercontinent breakup and superplume events can account for several features of BIF deposition. First, the enhanced submarine volcanism and hydrothermal venting associated with ocean-ridge or oceanic-plateau volcanism, or both, during these events may be a source of the iron and silica in BIF. Furthermore, elevated sea level caused by the superplume event or supercontinent breakup provides shallow marine basins on stable continental shelves, which is a feature necessary to preserve BIFs from later subduction. This applies to both the upwelling and hydrothermal plume models of deposition.

Sedimentary Phosphates

In many respects, what we know about the deposition of BIF also applies to many sedimentary marine phosphates. The main difference between the environments of deposition is that phosphates are deposited in biologically productive upwelling zones, whereas BIFs are not. For marine phosphate deposition, we need a source of phosphorus, relatively anoxic seawater to keep the phosphorus in solution, and upwelling along continental margins (or hydrothermal plumes), which brings phosphorus-rich seawater onto continental shelves and basins that contain oxidizing water and biologic activity where phosphates can be precipitated (Cook and Shergold 1984, 1986).

Again, supercontinent breakup and superplume events provide the necessary conditions for enhanced deposition rate of marine phosphates.

Geological Consequences of Superplume Events and Supercontinents

Mid-Cretaceous Event

The last recognized superplume event is the Mid-Cretaceous event (Larson 1991a), and its effects are concentrated chiefly in and around the Pacific basin. The paleotemperature curve for the last 150 Myr shows a broad increase from 150 to 100 Ma and then a steady decrease to the present (Fig. 9.5). The broad temperature peak from 110–90 Ma cannot be explained by supercontinent fragmentation alone but also requires excess CO_2 in the atmosphere (Larson 1991b; Barron et al. 1995). Approximately 2 to 6 times the present content of CO_2 is required to raise Mid-Cretaceous temperatures to the observed levels (Barron 1983; Barron et al. 1995). Although much of this increase may have been caused by increased ocean-ridge volcanism, as reflected by the peak in production of oceanic crust at this time (Fig. 9.5), superplumes in the Pacific basin also may have contributed CO_2. Models by Caldeira and Rampino (1991) show that the Mid-Cretaceous superplume event may have produced atmospheric CO_2 levels of 4 to 15 times the modern preindustrial value. This would have resulted in global warming of 3–8 °C over today's mean temperature. The fragmentation of Pangea at this time

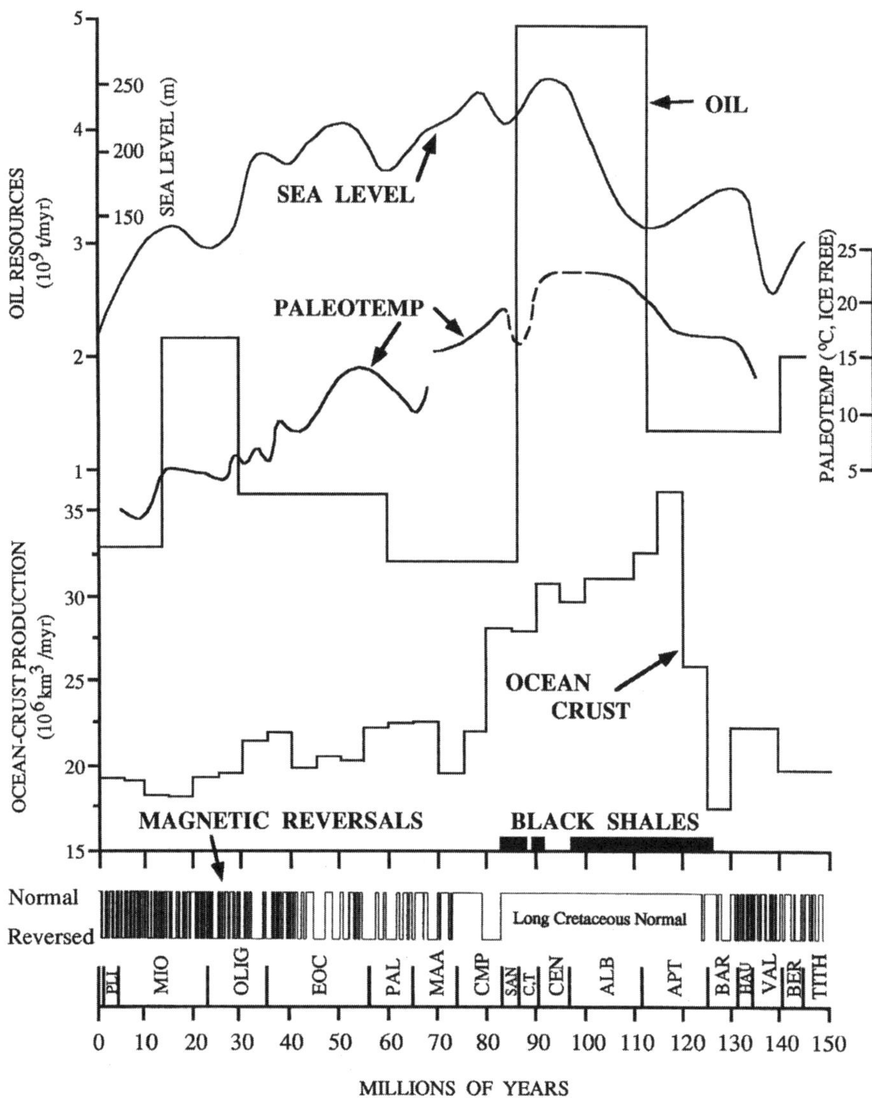

Figure 9.5. Distribution of paleotemperature, oil reserves, black shale, and sea level in the last 150 Myr. Also shown from Figure 8.9 is the production rate of oceanic crust and magnetic reversals. Modified from Larson (1991b), with permission. Copyright © 1991 by the Geological Society of America.

could contribute another 5 °C of warming. Computer models of Barron et al. (1995) show that a combination of increased atmospheric CO_2 and an increased poleward flux of heat is necessary to explain the global distribution of warm climates in the Mid-Cretaceous. The greater poleward heat flux is necessary to prevent the tropical oceans from overheating owing to the increased CO_2 content. The model that best matches observations from Mid-Cretaceous paleoclimatic indicators has a globally averaged surface temperature 6.2 °C higher than at present and atmospheric CO_2 of 4 times the present-day level.

The approximately 125-m increase in eustatic sea level, reaching a maximum at about 100 Ma, can also be related to increased ocean-ridge activity, displacement of seawater by oceanic plateaus, and uplift of the oceanic lithosphere over superplumes (Larson 1991a, 1991b; Kerr 1998). Note that the peak in sea level is near the end of the pulse in ocean-crust production, as predicted by the model of Gaffin (1987) (Fig. 9.5). As pointed out by Hardebeck and Anderson (1996), however, some of this rise in sea level also must have been related to the continuing breakup of Pangea. The effect of the superplume event is superimposed on the supercontinent breakup effect. The gradual decrease in sea level after 100 Ma is not predicted by the superplume model but is probably related to continental collisions associated with the development of a new supercontinent (including India–Tibet, Australia–Southeast Asia, and the Mesozoic terrane collisions in the American Cordillera).

Extensive deposition of black shale is recorded worldwide from about 130 to 85 Ma and may reflect increased CO_2 related to a Mid-Cretaceous superplume event (Jenkyns 1980). Black shales are generally interpreted to result from anoxic events caused by increased organic productivity and poor circulation in basins on continental platforms. As discussed above, a superplume event can supply both of these requirements: directly, by hydrothermal spring input of methane into the oceans, and indirectly, by increasing sea level and the frequency of partially closed basins on the continental shelves (Kerr 1998). The upwelling of trace metals and nutrients from the deep oceans may have increased the habitat of some marine organisms and may have led to the extinction of other organisms – especially those becoming extinct near the Cenomanian–Turonian boundary some 90 Ma (Wilde et al. 1990). Larson (1991b) also pointed out that about 60% of the world's oil reserves were generated between 110 and 88 Ma (Irving et al. 1974), which is consistent with the abundance of black shale at this time. There is also a broad peak in natural gas reserves at about the same time (Bois et al. 1980). A Mid-Cretaceous superplume event may have contributed carbon and other nutrients, such as phosphorus and iron, for the expansion of phytoplankton at this time. The high stand of sea level vastly increased the continental shelf area covered with shallow seas, providing appropriate depositional environments for the hydrocarbon precursors. Also consistent with a superplume event at this time is the peak in seawater $\delta^{13}C$, which is consistent with extensive burial of carbon (Fig. 9.6).

Although there is a small dip in the $^{87}Sr/^{86}Sr$ ratio of seawater at about 100 Ma (Fig. 9.6), the input of mantle Sr at this time from a combination of ocean-ridge and plume-related sources must have been relatively minor (Ingram et al. 1994; Veizer et al. 1999).

Permo–Carboniferous Event

The Permo–Carboniferous reversed superchron centered at about 280 Ma (Fig. 8.15) may also record a superplume event. Many geological features similar to those found associated with the Mid-Cretaceous superplume event occur during the Permo–Carboniferous superchron (Larson 1991b). Paleoclimates at this time, however, were mixed. Swampy, tropical, and wet climates characterized the Northern Hemisphere, whereas in the Southern Hemisphere the supercontinent Gondwana underwent

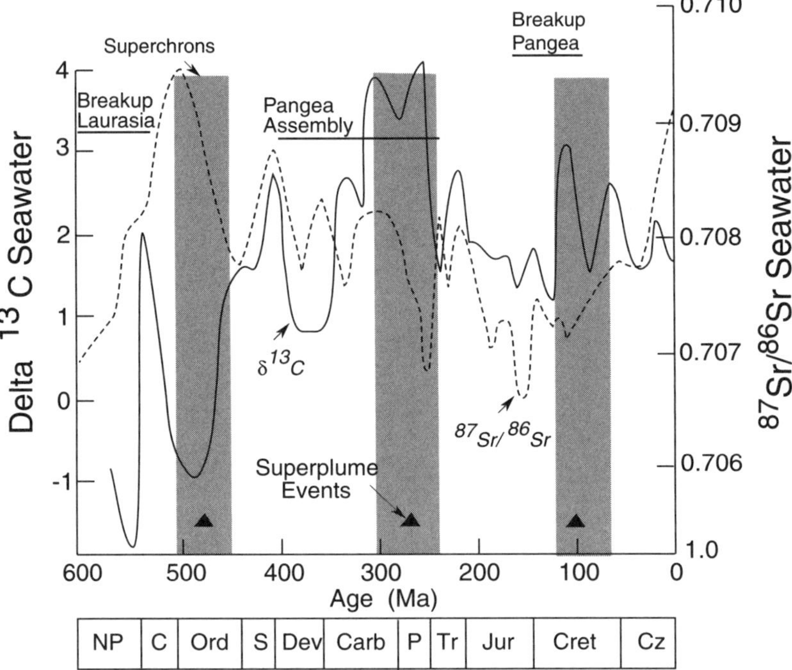

Figure 9.6. Secular changes in $\delta^{13}C$ and $^{87}Sr/^{86}Sr$ ratios of seawater in the last 600 Myr based on data from marine carbonates. Also shown are Phanerozoic superchrons from Figure 8.15. Sr and carbon isotope data from Veizer et al. (1999).

widespread glaciation (Crowell 1999). Also, about 50% of the world's coal reserves formed during the Permo–Carboniferous superchron (Bestougeff 1980), and large reserves of natural gas appear to have formed at this time. The volume of coal is particularly striking because coal contains about nine times the fossil fuel energy of oil and gas combined, and it does not migrate from the source rock. There is also a dramatic minimum in the CO_2 level of the atmosphere at 300–280 Ma, as calculated from buried carbon – chiefly the burial of vascular land plants (Berner 1994). Consistent with the increased burial rate of carbon at this time is the peak doublet in $\delta^{13}C$ of seawater, which rises to a value of about +4 per mil (Veizer et al. 1999) (Fig. 9.6). The rapid decrease in the $^{87}Sr/^{86}Sr$ ratio of seawater between 300 and 250 Ma may represent a response to enhanced input of mantle Sr via plume and ocean-ridge volcanism during a superplume event. The minimum $^{87}Sr/^{86}Sr$ ratio at about 250 Ma could reflect a combination of factors, including waning superplume activity and the final collisions of continental plates forming Pangea, both of which would enhance the continental Sr isotope signature.

There is also a sea level high centered at about 280 Ma, which is also consistent with a superplume event (Fig. 9.2). This high sea level is readily apparent in the Sloss cratonic sequences in North America (Sloss 1972). The Permo–Carboniferous reversed superchron correlates with the Absaroka transgressive sequence, which began at the same time as the superchron. After the peak at about 280 Ma, sea level fell to all-time minimum at the Permian–Triassic boundary (250 Ma) (Fig. 9.2). Although not fully understood, this fall again may have been in response to waning plume activity and the final growth of Pangea, each event providing positive feedback to a drop in sea level.

Ordovician Event

The recent recognition of a superchron in the Ordovician centered at about 480 Ma presents the possibility of still another superplume event in the Phanerozoic (Johnson et al. 1995). Consistent with a superplume event at this time is a peak in eustatic sea level and in calculated production rate of oceanic crust (Fig. 9.2). Black shales are also widespread in the Ordovician. A steep fall in $^{87}Sr/^{86}Sr$ ratio of seawater at this time could have been a response to enhanced input of mantle Sr accompanying a superplume event (Fig. 9.6). Puzzling, however, is the minimum in $\delta^{13}C$ of seawater (about -1) (Fig. 9.6). Perhaps uplift associated with Taconic collisions in Laurentia–Baltica recycled buried organic carbon, thus enriching seawater in ^{12}C. Alternatively, maybe oxidized carbon was preferentially buried on widespread continental shelves at this time. If so, however, why did these environments favor carbonate deposition and not organic carbon? The rapid increase in $\delta^{13}C$ toward the end of the superchron leads into Late Ordovician–Silurian glaciation, and it is interpreted by Kump et al. (1999) as a response to increased weathering of carbonate platforms caused by a glacial–eustatic sea level drop (reaching a minimum in the Silurian, Fig. 9.2) rather than as a response to increased burial of organic carbon.

The 1.9-Ga Event

Sea Level

Although well-preserved shallow marine sedimentary successions are widespread in the Phanerozoic and Neoproterozoic, only small remnants of older Precambrian successions remain in the geologic record. For this reason, it is not possible to use sequence stratigraphy to estimate sea levels in most sedimentary rocks older than about 800 Ma (Eriksson 1999). Remnants of Meso- and Paleoproterozoic marine intracratonic, passive margin, and platform sediments, however, are widespread on the continents, and hence their areal distribution as a function of time yields important insight into the relative elevation of sea level with time. Condie et al. (2000b) estimated the abundance of preserved sediments from these tectonic regimes during the Precambrian as a guide to relative sea level.

The areal distribution of sediments from any particular tectonic setting varies with age owing to the effects of erosion. Veizer and Jansen (1985) and Veizer (1988) have shown that the preservation of sediments decays exponentially with time and varies between tectonic settings. Phanerozoic intracratonic and platform sediments have a half-life of about 350 Myr and an oblivion age (life expectancy) of about 1600 Myr (Veizer 1988). A half-life of about 700 Myr, however, seems more appropriate for Late Archean and Proterozoic sediments, remnants of which are widespread on the continents (Condie et al. 2000b). A histogram of restored areal distribution of these sediments shows prominent peaks at 1.9–1.8 Ga and 800–600 Ma and a smaller peak at

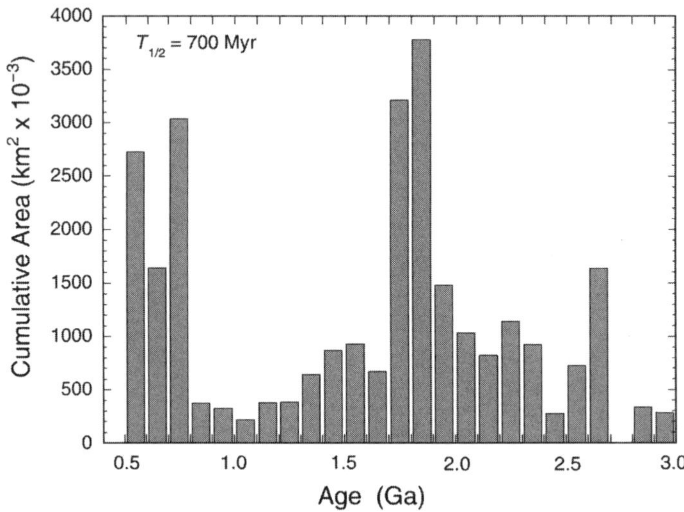

Figure 9.7. Histogram showing the restored areal distribution of intracratonic, passive margin, and platform sediments during the Precambrian. Data from Condie et al. (2000b). Area restored using a sediment half-life of 700 Myr.

about 2.7 Ga (Fig. 9.7). This strongly suggests that shallow marine sediments were more widespread on the continents at these times than at other times during the Precambrian, and by inference, that sea level was also high. It is significant that the highest peak for restored intracratonic sediment occurs at 1.9 Ga, corresponding to one of the two major superplume events identified by Condie (1998) and discussed in the previous chapter.

Also supporting high sea level at about 1.9 Ga is the widespread occurrence of submarine flood basalts of this age erupted on continental platforms. Many examples of such basalts occur in the Ungava orogen in Quebec, in the Birrimian in West Africa, and on the Baltic shield in Scandinavia (Arndt 1999). This suggests that continental shelves were extensively inundated at 1.9 Ga.

The peak in abundance of shallow marine sediments at 1.9 Ga suggests that a 1.9-Ga superplume event overpowered supercontinent formation at this time, resulting in a significant rise in eustatic sea level. This may reflect the relative timing of the two events at 1.9 Ga. Supercontinent formation with many craton and arc collisions at 1.85–1.70 Ga occurred on the tail end of the 1.9-Ga superplume event and may have contributed to the lowering of sea level following the superplume event. This is consistent with the timing of events within the superevent cycle suggested by Condie (1998).

Black Shales

There is a good correlation between a 1.9-Ga superplume event and the distribution of black shales during the Precambrian (Condie et al. 2000, 2001). A cumulative thickness histogram shows a clear maximum in black shale abundance at 1.9–1.7 Ga with smaller peaks at about 2.1 Ga and 600 Ma and the suggestion of a peak at 2.7 Ga (Fig. 9.8(a)). The relatively small cumulative thickness of black shales older than 2 Ga may reflect removal by erosion of older successions. To minimize the effects of different sedimentary environments and facies as well as selective erosion of shale from

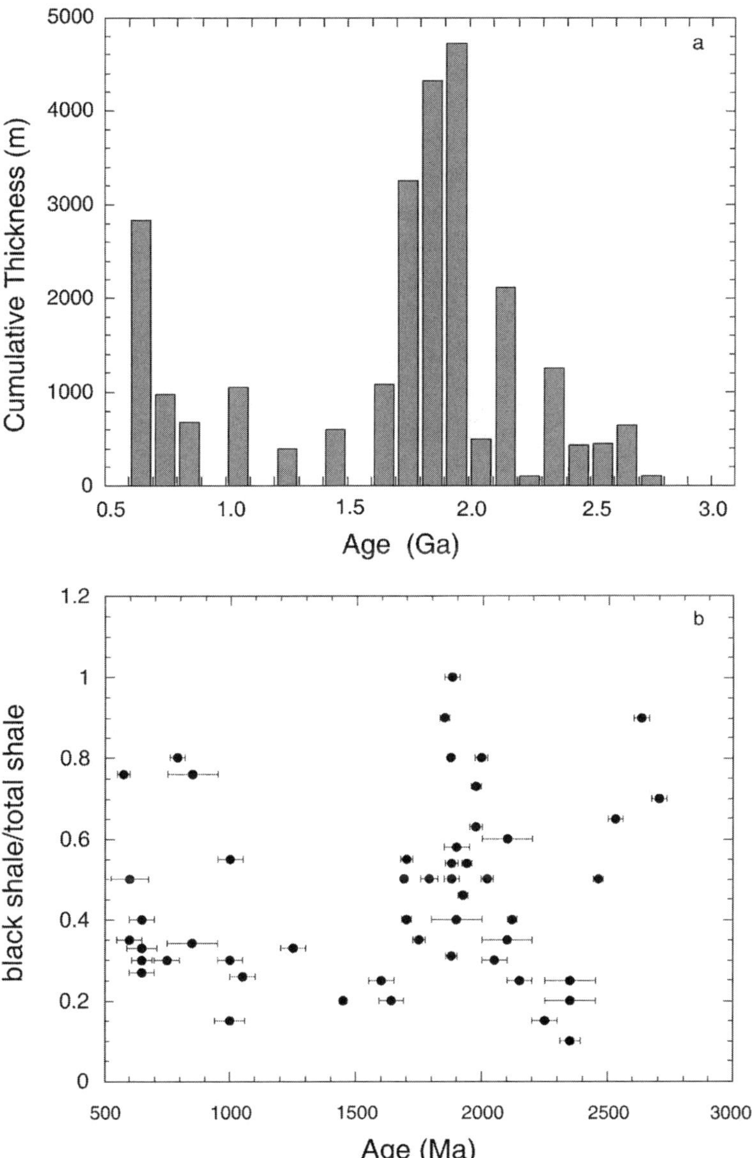

Figure 9.8. Age distribution of black shale in Precambrian passive margin, intracratonic, and platform successions. (a) Frequency distribution of total thickness of black shale. (b) Ratio of black shale to total shale. Error bars are one standard deviation of mean age. From Condie et al. (2000a).

successions, results are also shown for the ratio of black shale to total shale with time (Fig. 9.8(b)). Again, peaks are apparent in the range of 2.0-1.8 Ga as well as in Late Neoproterozoic (800–600 Ma) and in the Late Archean (2.7–2.5 Ga).

When black shale is weighted by preserved thickness, a time series (Isley and Abbott 1999) shows three peaks in the Paleoproterozoic (i.e., at about 2.0, 1.85, and 1.7 Ga) with small peaks at 600 Ma, 1.45 Ga, and near the Archean–Proterozoic boundary at 2.5 Ga (Fig. 9.9). Plotted as a time series weighted by errors in ages and by the ratio of black shale to total shale, a very strong peak occurs at 1.9 Ga with smaller

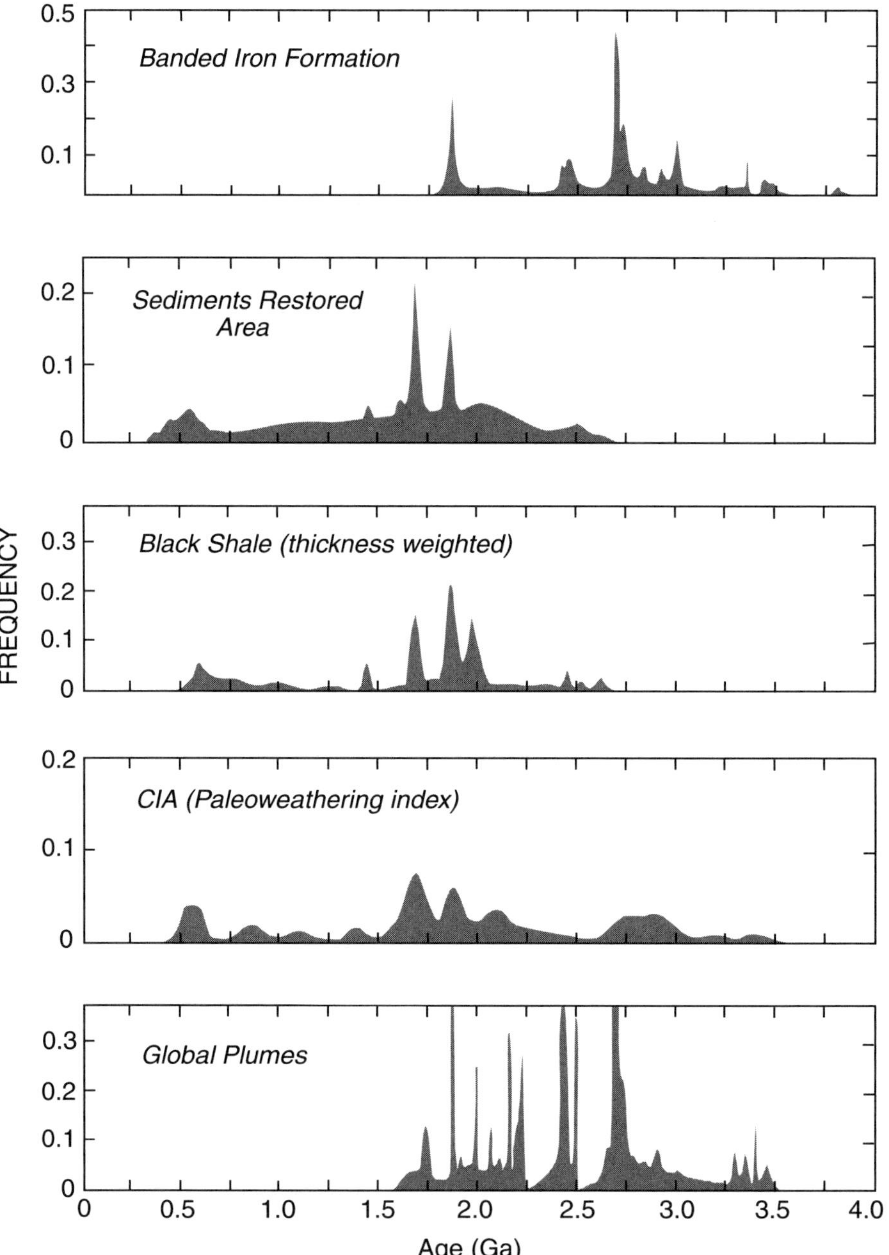

Figure 9.9. Time series of BIF, black shale, CIA in intracratonic shales, and restored intracratonic sediments in the Proterozoic compared with global plumes from Isley and Abbott (1999). Time series generated by summing Gaussian distributions of unit area using mean ages and standard deviations given in Condie et al. (2000a). CIA = $[Al_2O_3/(Al_2O_3 + CaO + Na_2O + K_2O) \times 100]$ molecular ratio with CaO representing the silicate fraction only. Modified after Condie et al. (2000a).

peaks at 1.7, 2.0, and 0.6 Ga and again, three small peaks at the Archean–Proterozoic boundary (Condie et al. 2000). Considered collectively, there is evidence of one-to-three well-defined maxima in black shale deposition at 2.0–1.7 Ga with a broad, less well-defined maximum in the Late Neoproterozoic and one-to-three small maxima in the Late Archean (Fig. 9.9).

If both organic and carbonate carbon were buried during the 1.9-Ga superplume event in approximately the same ratio in which they were buried during most of Earth history, as constrained by carbon isotopes and discussed below, there should also be peaks in carbonate abundance during superplume events. Although a survey of the abundance of marine carbonates in the Precambrian record shows that they are relatively abundant in the Paleoproterozoic (an observation also made by Grotzinger and Kasting 1993), there is only a suggestion of peak abundances at 1.9–1.8 Ga (Fig. 9.10). A possible reason for the absence of strong peaks at 1.9 Ga may be that carbonates weather more rapidly than siliceous sediments and have not survived (Berner and Berner 1997). This, however, requires selectively high rates of carbonate weathering at 1.9 Ga compared with other times in the Proterozoic. Possibly many carbonates of these ages were deposited in deep ocean basins and later subducted. Widespread hydrothermal systems on the seafloor associated with plume magmatism may have produced anoxia in shallow shelf environments where carbonates are normally deposited.

Paleoclimate

Although CIA data from shales show considerable scatter in some stratigraphic sections, owing perhaps to later remobilization of Ca, Na, or K, there is a major peak in CIA

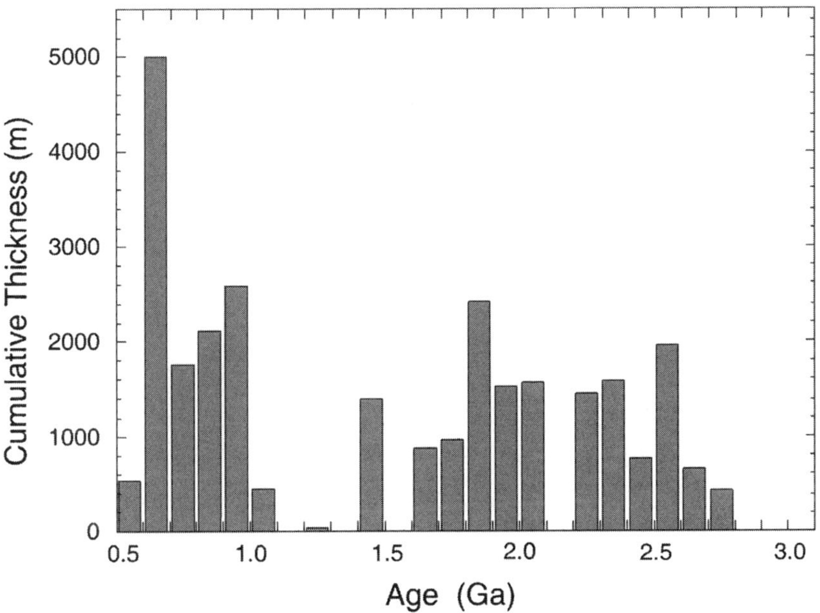

Figure 9.10. Age distribution of shallow marine carbonates in Precambrian passive margin, intracratonic, and platform basins. From Condie et al. (2000a).

at about 1.9 Ga and another at 1.7 Ga (Condie et al. 2000, 2001b) (Fig. 9.9). These peaks in CIA suggest that paleoclimates were unusually warm at these times, which is a feature consistent with increased input of greenhouse gases (principally CO_2) into the atmosphere. This is to be expected during superplume events. The origin and significance of the 1.7 Ga CIA peak is not yet understood. That it seems to correlate with intracratonic sediment and black shale peaks at about the same time indicates a warm climatic regime and relatively high sea level at 1.7 Ga. Isley and Abbott (1999) have proposed a superplume event at about 1.7 Ga (Fig. 9.9) based on the distribution of plume-related igneous rocks in the geologic record. The sediment and CIA peaks at 1.7 Ga seem to support the existence of such an event.

Banded Iron Formation

The last major period of BIF (banded iron formation) deposition was at about 1.9 Ga when the large BIFs of the Labrador Trough in northern Quebec, the Animikie basin in Minnesota, and the Nabberu basin in Western Australia were deposited (Klein and Beukes 1992). As shown by Isley and Abbott (1999), this last peak in BIF deposition correlates well with a 1.9-Ga superplume event and suggests a cause and effect relationship, as discussed previously (Fig. 9.9). A similar correlation with superplume events has been suggested for the voluminous BIFs at 2.7 and 2.5 Ga (Barley et al. 1997; Isley and Abbott 1999).

A superplume event can account for several features of BIF deposition. First, the enhanced submarine volcanism and hydrothermal venting associated both with ocean-ridge and oceanic plateau volcanism during a superplume event may be the source of the iron and silica in BIF. Furthermore, the elevated sea level caused by a superplume event, as discussed previously, provides extensive shallow marine basins along stable continental platforms necessary to preserve BIF against later subduction. This applies to either the upwelling or hydrothermal plume models of deposition. The end of the BIF event at 1.9 Ga may be related to either, or more likely a combination, of the following: (1) a decrease in concentration of ferrous iron in the oceans, which is a feature resulting from decreasing amounts of submarine hydrothermal activity as a superplume event declines in intensity; or (2) increasing oxygenation of deep ocean waters, including, perhaps, introduction of dense plumes of sulfate-rich water. After 1.8 Ga, the effects of oxygenic photosynthesis, together with organic burial and a weaker hydrothermal flux, led to global sulfate deposition and to the complete disappearance of BIF. It would appear that later superplume events, including a possible event at 1.7 Ga, were insufficient to reinitiate deposition of BIF.

Sedimentary Phosphates

Although sedimentary phosphates do not become widespread until after about 800 Ma, some important deposits occur in the Paleoproterozoic (Cook and McElhinny 1979). These are in Australia in the Rum Jungle (1.9 Ga) and Broken Hill (1.9–1.8 Ga) areas, in the Animikie–Gunflint successions in Minnesota ($\approx$1.9 Ga), in the Vayrylankyla area of

Finland (2.0–1.9 Ga), and in the Yenisey Province of Siberia (1.85 Ga) (Cook and McElhinny 1979; Needham et al. 1988; Rosen et al. 1994; Nutman and Ehlers 1998). These phosphates were deposited at or near 1.9 Ga and hence may correlate with a 1.9-Ga superplume event. The source of the increased phosphorus and widespread anoxia in seawater at this time could be from submarine hydrothermal systems associated with a superplume event. This model also explains the association of some phosphates with BIFs, for dissolved phosphate is strongly absorbed on ferric oxides under aerobic conditions (Berner 1999).

Strontium Isotopes in Seawater

In the Paleoproterozoic, the $^{87}Sr/^{86}Sr$ ratio in some of the least radiogenic marine carbonates (such as those from the Albanel Formation in Quebec [1.85 Ga] and the McArthur Group in North Australia [1.75 Ga]) increased to values near 0.706, whereas in slightly older carbonates the ratio remained relatively low (0.704) (for example in the Coronation Supergroup in Canada [1.9 Ga]) near the mantle growth curve (Fig. 9.4) (Mirota and Veizer 1994). Perhaps the relatively low $^{87}Sr/^{86}Sr$ ratios in seawater at 1.9 Ga reflect increased mantle input of Sr from a superplume event at this time, whereas the higher ratios at 1.85–1.75 Ga reflect increased input of continental Sr from a growing supercontinent (Condie 1998). That a superplume event at 1.9 Ga did not decrease seawater Sr isotope ratios to even lower values may be due to the large volume of continental crust formed at 2.7 Ga, which, even during a 1.9-Ga superplume event, continued to supply significant amounts of continental Sr to the oceans. Intermediate $^{87}Sr/^{86}Sr$ ratios in Mesoproterozoic marine carbonates (about 0.705; Fig. 9.4) correlate with supercontinent stasis, perhaps with minor breakup at 1.5–1.4 Ga, and the formation of Rodinia at 1.3–1.0 Ga (Condie 2000).

Stromatolites

Stromatolites, layered structures thought to be deposited by microbial mat communities, are widespread in the Proterozoic with a prominent peak (or peaks) in distribution at about 1.9–1.8 Ga. Maxima at this time are found in the number of stromatolite occurrences, the diversity of stromatolites, and in the number of occurrences of microdigitate stromatolites (Grotzinger and Kasting 1993; Hofmann 1998) (Fig. 9.11). Other investigators have not recognized this peak because data were averaged over long time intervals (Awramik 1992; Semikhatov and Raaben 1996). In addition to stromatolites, there are maxima in the reported occurrences of microfossils, oncoids, and chemofossils (biogenic chemical remains) at about 1.9 Ga (Hofmann 1998). The peaks in abundance and diversity of stromatolites at about 1.9 Ga may reflect a combination of global warming, high sea level stands, and enhanced input of CO_2 into the sedimentary cycle. All of these may be related to a superplume event at 1.9 Ga. Grotzinger and Knoll (1999) suggested that the degree of carbonate saturation in seawater may be very important in controlling stromatolite diversity, and during superplume events, seawater carbonate saturation may increase significantly. Thus, a superplume event may increase both the

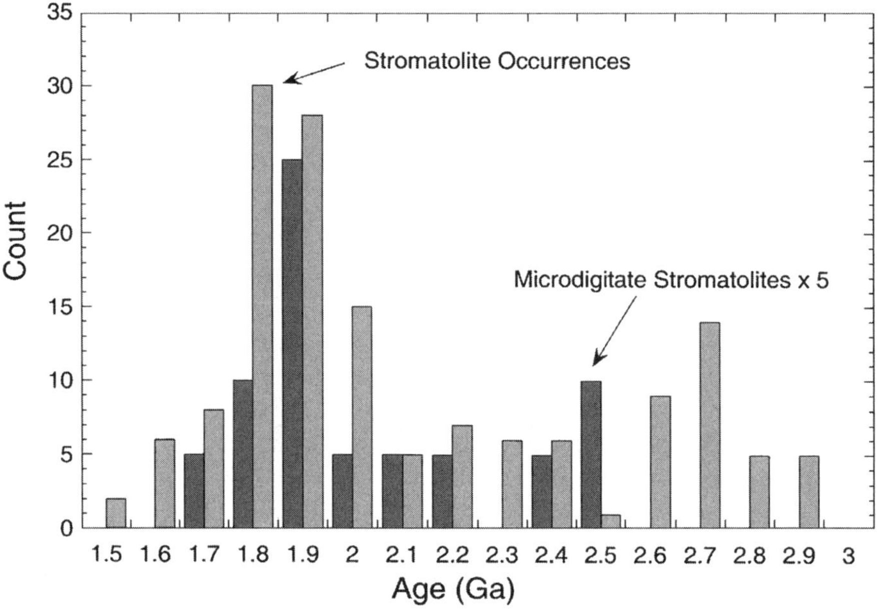

Figure 9.11. Distribution of reported number of occurrences of total stromatolites and of microdigitate stromatolites during the Precambrian. Data from Grotzinger and Kasting (1993) and Hofmann (1998).

availability of carbonate and the proportion of shallow platforms for the deposition and preservation of carbonates.

The distribution of Paleoproterozoic carbonates indicates that cement crusts, and in particular microdigitate stromatolites deposited in tidal flats, were a common mode of deposition of calcium carbonate during the Paleoproterozoic (Grotzinger and Kasting 1993). This feature appears to require Proterozoic seawater that was greatly oversaturated in $CaCO_3$ compared with Phanerozoic seawater. The peak in reported occurrences of microdigitate stromatolites at about 1.9 Ga (Fig. 9.11) is particularly intriguing because it correlates with the suggested superplume event at this time. This is consistent with enhanced CO_2 input into the oceans from submarine volcanism and hydrothermal vents accompanying the superplume event because higher CO_2 means an increase in the HCO_3/SO_4 ratio in seawater, favoring deposition of carbonate over sulfate. Also, high sea level stands create widespread shallow tidal flats, where both Ca^{+2} and HCO_3^{-1} ions increase in concentration in seawater due to evaporation.

Massive Sulfate Evaporites

The first massive sulfate evaporites in the geologic record occur at 1.8–1.6 Ga, following a possible 1.9-Ga superplume event (Fig. 9.12). That widespread sulfate deposition followed a 1.9-Ga superplume event is consistent with the following sequence of events:

1. Large amounts of sulfur were injected into the oceans as sulfide during the superplume event, but only some of the sulfur was deposited as iron sulfides on the deep seafloor (Canfield 1998).

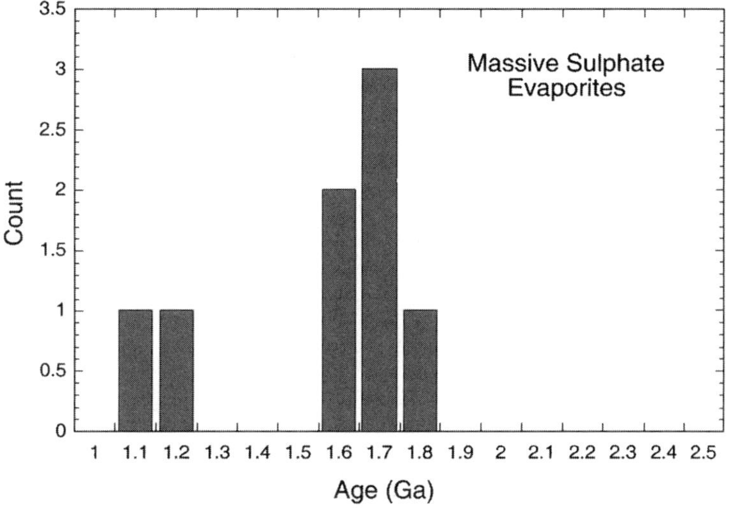

Figure 9.12. Distribution of massive sulfate evaporites during the Precambrian. Data from Grotzinger and Kasting (1993) and miscellaneous sources.

2. The oceans became oxic as submarine volcanic input related to the superplume event subsided.
3. As carbonate levels decreased in seawater owing to falling CO_2 input from the superplumes, marine carbonate deposition became less important and Ca^{+2} ion became available to precipitate as sulfates (Grotzinger and Kasting 1993).

Increasing levels of sulfate in the oceans at this time is supported as well by sulfur isotopes in which the range of $\delta^{34}S$ increases (< -20 to $> +20$) after 2.2 Ga (Canfield 1998). Also, the completion of a supercontinent may have provided numerous partially closed basins for evaporite deposition, as it did on Pangea during the Permian and Triassic.

The sulfate, black shale, and CIA peaks at 1.7 Ga may reflect another superplume event with all of these peaks recording widespread global warming. If so, why is there no evidence of sulfate evaporite deposition corresponding to the 2.7 and 1.9 Ga superplume events? There probably was not enough oxygen available to oxidize sulfur at 2.7 Ga. However, by 1.9 Ga, this should not have been a problem. If there was a 1.7-Ga superplume event, for some reason it must have caused more global warming than the 1.9-Ga event, thus leading to widespread sulfate deposition.

Carbon and Sulfur Isotopes

The impact of a superplume event on the biogeochemical cycles of carbon and sulfur can be explored by considering their stable isotopic records. As discussed previously for carbon, these cycles consist of elemental reservoirs linked by processes that either transport or chemically transform the elements. Both elements are rapidly cycled through the biosphere, which converts a small amount to reduced species by utilizing reducing power provided by weathering, thermal sources, and oxygenic photosynthesis. The amounts of organic carbon, carbonate, sulfides, and sulfate buried depend upon

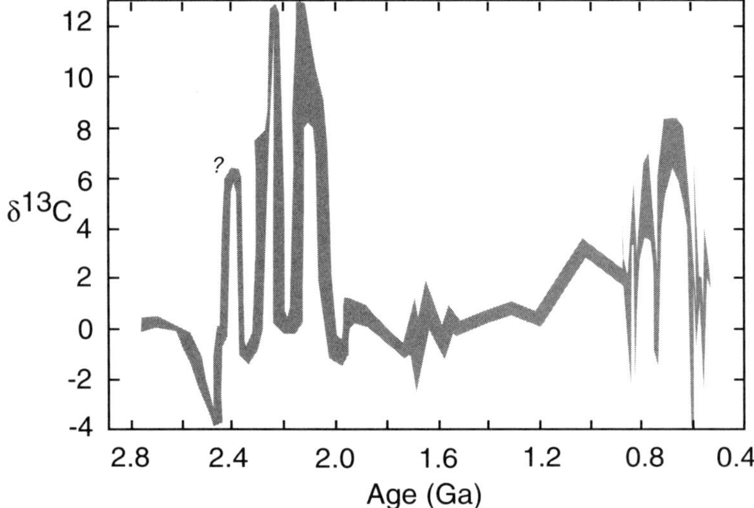

Figure 9.13. Variation in $\delta^{13}C$ in marine carbonates during the Proterozoic. $\delta^{13}C$ data from Karhu and Holland (1996), Kaufman (1997), and Melezhik et al. (1999).

both the elemental fluxes through the surface environment and the burial rates in a range of sedimentary environments that favor sedimentation of either oxidized or reduced species. The burial of both reduced carbon and sulfides is favored by reducing marine environments (Berner 1983).

The net effect of superplumes and supercontinent formation at 1.9 Ga may have been to introduce significant volumes of CO_2 into the atmosphere–ocean system, increasing depositional rates of both organic and carbonate carbon (Condie et al. 2000b). Although data suggest that as many as three positive carbon isotope excursions occur in marine carbonates during the time interval of 2.06 to 2.4 Ga (Karhu and Holland 1996; Melezhik et al. 1999), only a poorly defined and questionable carbon isotope anomaly is observed in marine carbonates at 1.9 Ga (Fig. 9.13). Therefore, the relative burial rates of both reduced and oxidized carbon changed very little at this time, even as their absolute rates increased. The absence of a significant 1.9-Ga carbon isotope anomaly also means that burial of plume-related carbon masked supercontinent-related carbon burial, the latter of which should favor burial of organic carbon. It is worth emphasizing that the absence of a negative $\delta^{13}C$ excursion in carbonates during the 1.9-Ga superplume event indicates that mantle plumes do not preferentially recycle inorganic carbon relative to reduced carbon. After 2.05 Ga, the carbon isotope record indicates that the fraction of carbon buried as organic matter dropped from as much as 50% of the global carbon flux to less than 20% (Karhu and Holland 1996). After 1.92 Ga, f_{org} remained remarkably constant for hundreds of millions of years (Fig. 9.13).

A mass balance equation similar to that for carbon (eq. 9.1) applies also for sulfur, as follows:

$$\delta_{S_{\text{in}}} = f_{SO_4=}\delta_{SO_4=} + f_{S_{\text{red}}}\delta_{S_{\text{red}}}, \tag{9.2}$$

where δ_{Sin} represents the isotopic composition of sulfur entering the surface environment. The right side of the equation represents the weighted-average isotopic composition of sulfate ($\delta^{34}S_{SO_4=}$) and sulfides ($\delta^{34}S_{S_{red}}$) being buried, and $f_{SO_4=}$ and $f_{S_{red}}$ represent the fractions of sulfur buried in each form ($f_{SO_4=} = 1 - f_{S_{red}}$). Over timescales greater than 100 Myr, $\delta_{in} = 0$ per mil, the average value for crustal sulfur (Holser et al. 1988). Hence, lower values of $\delta^{34}S_{S_{red}}$ indicate higher values of $f_{SO_4=}$.

The range of $\delta_{S_{red}}$ values increased between 2.5 and 2.2 Ga and is consistent with an increase in seawater sulfate levels (Knoll and Canfield 1998; Condie et al. 2000b). Because bacterial sulfate reduction utilizes ^{32}S preferentially over ^{34}S (Harrison and Thode 1958), sedimentary sulfides typically have lower $\delta^{34}S$ values than their coeval sulfates. Isotopic mass balance considerations require that the weighted average of the $\delta_{SO_4=}$ and $\delta_{S_{red}}$ values of sulfur species being buried must equal $\delta_{S_{in}}$, which is typically zero. Between 1.9 and 1.8 Ga, $\delta_{S_{red}}$ values become more negative, indicating that a substantial sulfate reservoir existed and that the burial rate of sulfate increased relative to that of sulfide during that time interval. This isotopic trend is consistent with the observed peak in the distribution of sulfate evaporites between 1.8 and 1.6 Ga (Fig. 9.12).

The carbon and sulfur isotope trends in Paleoproterozoic sediments are consistent with a high sea level stand at 1.9 Ga (Condie et al. 2000b). This is followed by a sea level decline and a decline in platform sedimentation in favor of shallow-to-emergent coastal environments that accumulated more oxidized, evaporitic sulfates. This view is corroborated by the decline, from 1.9 to 1.7 Ga, in stromatolite diversity, in the abundance of preserved banded iron formation and black shale, and in declining δ_{org} and $\delta_{S_{red}}$ values. The trend towards greater rates of sulfate deposition between 1.8 and 1.7 Ga is indicated by increased abundance of platform sulfate-rich evaporites and generally lower $\delta_{S_{red}}$ values.

The Case for a 1.9-Ga Superplume Event

Consistent with a superplume event at 1.9 Ga are the following (Condie et al. 2000):

1. High sea level as inferred from the relative abundance of intracratonic, passive margin, and platform sediments.
2. A relatively high abundance of black shale.
3. A peak in CIA (Chemical Index of Alteration) in shale, implying unusually warm paleoclimates.
4. A peak in the abundance of BIF.
5. An abundance of shallow marine phosphate deposits.
6. A peak in the number of reported occurrences and in the diversity of stromatolites.

The peak in distribution of shallow marine sediments at 1.9 Ga (Fig. 9.7) suggests that the superplume event overpowered supercontinent formation, resulting in a significant rise in sea level. The black shale and CIA peaks may reflect a 1.9-Ga superplume event, which introduced massive amounts of CO_2 into the atmosphere–ocean system, increasing depositional rates of carbon and global warming. Increased black shale deposition and preservation at 1.9 Ga are due primarily to anoxia driven onto stable continental shelves. The peak in CIA suggests that paleoclimates were warm at 1.9 Ga, which is a feature consistent with increased input of greenhouse gases (principally CO_2) into the atmosphere.

Increased BIF and marine phosphate deposition at 1.9 Ga reflect increased input of iron and phosphorus into deep, anoxic oceans by submarine hydrothermal activity followed by upwelling (or spreading of hydrothermal plumes) into shallow, oxidizing seas on continental shelves. The slightly low $^{87}Sr/^{86}Sr$ isotope ratios in seawater at 1.9 Ga may reflect increased mantle input of Sr from the superplume event, whereas higher ratios at 1.85–1.75 Ga reflect increased input of continental Sr from the growing supercontinent. A peak in reported occurrences and diversity of stromatolites and in microdigitate stromatolites at about 1.9 Ga requires seawater on continental shelves that was greatly oversaturated in $CaCO_3$ and warm climates, both of which are consistent with high sea level stands and enhanced CO_2 input into the oceans from submarine volcanism and hydrothermal vents accompanying a 1.9-Ga superplume event. The first massive sulfate evaporites in the geologic record at 1.8–1.6 Ga follow the 1.9-Ga superplume event, reflecting oxdizing conditions and greater availability of Ca^{+2} ions as carbonate deposition declined during superplume waning.

The 2.7-Ga Event

On the basis of the distribution of plume-generated mafic igneous rocks, Isley and Abbott (1999) proposed two superplume events in the Late Archean at about 2.5 and 2.7 Ga (Fig. 9.9). A correlation with a superplume event has been suggested for the voluminous BIFs at about 2.5 Ga (Barley et al. 1997; Isley and Abbott 1999). If the production of significant volumes of juvenile continental crust is a necessary consequence of a major superplume event, as suggested by Condie (1998), the 2.5-Ga event probably does not qualify as a major event because the volume of known juvenile crust of this age is relatively minor (Fig. 8.8). Perhaps the 2.5-Ga event is a subsidiary event associated with the 2.7-Ga superplume event, during which large volumes of juvenile continental crust were produced.

There is a good correlation between the superplume event proposed by Condie (1998) and Isley and Abbott (1999) at 2.7 Ga and peaks in the abundance of BIF and the CIA distribution in shale (Fig. 9.9). The peak in BIF abundance may reflect a source of iron from hydrothermal vents associated with plume magmatism. The large volumes of CO_2 introduced into the atmosphere–ocean system during this event may have increased the depositional rates of both organic and carbonate carbon and caused global warming, accounting for the CIA maximum. As at 1.9 Ga, the number of occurrences of marine stromatolites shows a peak at 2.7 Ga (Hofmann 1998) – again perhaps recording enhanced input of CO_2 into seawater. Decreasing amounts of CO_2 pumped into the atmosphere by plume magmas, negative feedback of continental weathering, and increasing albedo caused by the newly formed Late Archean supercontinent appear to have decreased atmospheric CO_2 levels sufficiently to cool worldwide climates after the 2.7-Ga superplume event. This led to widespread glaciation at 2.4–2.3 Ga (Young 1991).

2.0- and 0.6-Ga Events

Although peaks in black shale abundance occur at about 2.0 Ga and 600 Ma (Fig. 9.9), neither of these times has been recognized as a superplume event (Condie 1998; Isley and Abbott 1999). Both of these peaks, however, correlate with breakup of supercontinents, the Late Archean supercontinent at 2.2–2.0 Ga and Rodinia at 800–600 Ma (Condie 1998). It is possible that the increased burial of organic carbon in black shales at these times is the result of supercontinent breakup. As previously discussed, supercontinent breakup might favor the burial of organic carbon compared to carbonate carbon (Fig. 9.1) and thus could result in increases in the depositional rate of black shale. Particularly important may have been the formation of numerous partially closed basins, leading to widespread anoxic environments (Kerr 1998). The supercontinent breakup model is supported by the appearance of positive $\delta^{13}C$ anomalies in marine carbonates at both of these times (Fig. 9.13), reflecting enhanced burial rates of carbon (Karhu and Holland 1996; Des Marais 1997; Kaufman 1997). The occurrence of two or three $\delta^{13}C$ peaks between 2.1 and 2.4 Ga may record stages in the breakup of the Late Archean supercontinent, each stage opening new basins for deposition of organic carbon. Also consistent with increased burial rates of organic carbon at 2.1–2.0 Ga are paleosol data, which indicate that the oxygen level of the atmosphere rose rapidly at this time (Karhu and Holland 1996). The appearance of multicellular organisms in the Late Neoproterozoic again suggests increasing oxygen levels in the atmosphere (Knoll and Canfield 1998). Hence, both the 2.0 and 0.6 Ga black shale peaks may also correlate with increasing diversification of oxygen-dependent biota in response to oxygen liberated into the atmosphere–ocean system.

Why is there no evidence of growth in atmospheric oxygen at 1.9 Ga, when even greater amounts of organic carbon appear to have been buried? Possible factors contributing to minimal atmospheric oxygen input at this time include the following:

1. An increase in total surface area exposed to weathering as the Paleoproterozoic supercontinent grew may have resulted in enhanced removal of oxygen in weathering products (including oxidation of recycled organic carbon).
2. Oxidation of reduced volcanic gases emitted by widespread submarine hydrothermal vents associated with the superplume event could also have consumed free oxygen in the oceans.
3. Because geologic indicators of atmospheric oxygen level are not very sensitive to increases when 10% PAL oxygen levels are reached, increases in oxygen level at 1.9 Ga may not be recognized in the geologic record.

There is a broad peak in CIA at 800–600 Ma (Fig. 9.9), suggesting widespread warm climates at this time. Perhaps increased ocean-ridge volcanism resulting from growth of the worldwide ocean-ridge system as Rodinia broke up resulted in significant CO_2 input into the atmosphere-ocean system, which in turn caused greenhouse warming, leading to an overall increase in the intensity of rock weathering. Marine transgressions during breakup of this supercontinent may also have contributed to global warming.

If the 2.0 and 0.6 Ga black shale peaks correlate with supercontinent fragmentation, why is there no peak in black shale abundance at 1.5–1.4 Ga when a Paleoproterozoic supercontinent allegedly broke up (Condie 1998)? Although there may be a small peak in black shale abundance at this time (Fig. 9.9), it is defined entirely by black shale from one basin, the Belt Basin in western Laurentia. Possibly other basins of this age exist that have not been accurately dated, which would reinforce this peak. Alternatively, the Paleoproterozoic supercontinent may have only partially fragmented. Supporting this idea is increasing evidence that two of the largest pieces of the Mesoproterozoic supercontinent Rodinia remained intact since they formed as part of the Late Archean supercontinent (Rogers 1996) (Fig. 8.7).

Mass Extinctions

A relationship between mass extinctions and flood basalt eruption was first suggested by Vogt (1972), and later the idea was advocated by several other investigators (Officer et al. 1987; Courtillot et al. 1986, 1996). Indeed, the ages of many flood basalt eruptions in the last 250 Myr are close to the ages of mass extinctions (Rampino and Stothers 1988; Stothers 1993) (Fig. 9.14). Although the coincidence of ages is impressive, not all data are of equal reliability (Courtillot et al. 1996). In particular, the isotopic ages of many flood basalt eruptions and extinctions are not precise, as shown by error bars in Figure 9.14. Eight flood basalt provinces have now been dated precisely

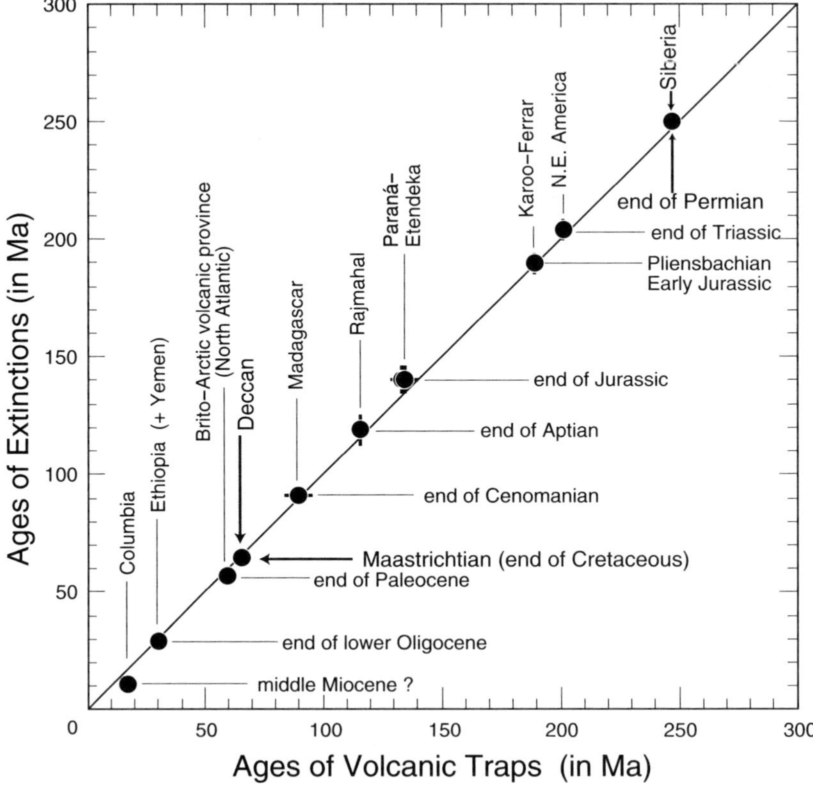

Figure 9.14. Correlation of Phanerozoic flood basalt eruptions with mass extinctions. Modified with permission of Courtillot et al. (1996). Copyright © 1996 by the Geological Society of America.

by the $^{40}Ar/^{39}Ar$ method: the Columbia River, North Atlantic, Central Atlantic, Deccan, Parana–Etendeka, Ferrar, Karoo, and Siberian provinces. Five of these correlate very strongly with mass extinctions: the Cretaceous–Tertiary (K–T), Permian–Triassic (P–T), Triassic–Jurassic, and end-Eocene events. Correlation of the Paraná–Etendeka with the Jurassic–Cretaceous (J–K) extinction is questionable, and two flood basalts do not correlate well with extinctions (the Columbia River and Ferrar–Karoo). The Madagascar and Rajmahal flood basalts with poorly constrained ages may also correlate with mass extinctions. In summary, of the 10 major mass extinctions recognized in the last 250 Myr (Sepkoski 1990), 7 correlate or may correlate with flood basalt eruptions. Of 12 flood basalt events recognized in the same time period, at least 9 may be associated with mass extinctions (Courtillot et al. 1996; Palfy and Smith 2000; Hames et al. 2000). The two largest extinctions at the K–T and P–T boundaries correlate with the two largest flood basalt eruptions. At least six of these extinctions, however, also correlate with large impacts and are associated with evidence diagnostic for an impact origin (Rampino and Haggerty 1994). In particular, the K–T extinction event at 65 Ma is associated with overwhelming evidence for an impact origin, including the probable impact site in Yucatan.

The environmental damage produced by the rapid eruption of large volumes of flood basalt is due chiefly to toxic gases (principally SO_2 and halogens) and sulfate aerosols, and this requires subaerial eruption (Devine et al. 1984; McCartney et al. 1990). In addition, variable amounts of ash may be introduced into the troposphere. Model calculations indicate that tremendous volumes of SO_2 and halogens may be introduced into the atmosphere during single, large flood basalt eruptions. Such eruptions should have severe consequences on global climate and would probably produce acid rain, ozone damage, and increased reflectance of solar radiation, leading to rapid cooling in the hemisphere affected (Handler 1989). The ability of flood basalt eruptions to inject large volumes of toxic aerosols into the stratosphere is also important in changing global climate and in leading to extinction of organisms. The effect of CO_2 emitted during a flood basalt eruption in warming the atmosphere, however, may have been rather minor. Calculations show that, even for a relatively large eruption like the Deccan traps, the surface temperature of the planet would be raised less than 2 °C over a period of 4×10^5 years (Caldeira and Rampino 1990). This indicates that the K–T extinctions were not caused by global warming resulting from volatiles released by eruption of the Deccan traps.

If the effect of single flood basalt province eruptions can lead to mass extinctions, imagine the potential damage of a superplume event. Yet, of the three Phanerozoic superplume events at 100, 280, and 480 Ma, only the 100-Ma event falls at or near a mass extinction (the Cenomanian–Turonian extinction at 90 Ma) (Figs. 9.14 and 9.15). The other two events correlate with minima in extinctions! The Cenomanian–Turonian (C–T) extinction, during which approximately 7% of the families and 26% of the genera became extinct, involved mostly marine invertebrates (Sepkowski 1986). That this extinction affected deep-ocean organisms more than shallow-ocean organisms confirms that enhanced oceanic volcanism at this time played a significant role in the extinction (Kerr 1998). During the 100-Ma superplume event (120–80 Ma), most of the volume of the Ontong Java and Caribbean plateaus was erupted. Perhaps global warming, rising sea level, and anoxia caused by eruption of these and related oceanic plateaus were responsible for the C–T extinctions.

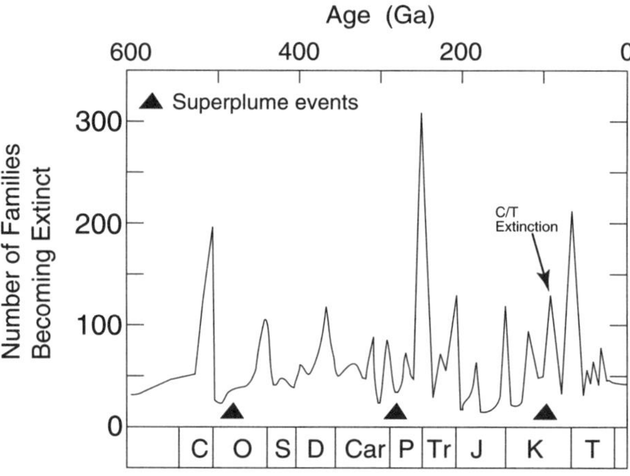

Figure 9.15. Patterns of animal and plant family extinctions during the Phanerozoic. After Benton (1995).

Why mass extinctions are not recognized during possible superplume events at 480- and 280-Ma presents a major problem. Although no major flood basalt eruptions are recognized on the continents during these superplume events, oceanic plateaus (subducted or accreted to the continents) may have been important at these times. If these two superplume events are real, the question of the missing mass extinctions clearly needs to be solved.

Conclusion

So what does the geologic record tell us about the possibility of superplume events in Earth history? With the accuracy of isotopic ages, sediment and fossil distributions in the stratigraphic record allow for superplume events in the geologic past and strongly support major events at 2.7, 1.9, 0.28, and 0.1 Ga and possible events at 2.5, 1.7, and 0.48 Ga. The effects on the atmosphere–ocean–biosphere system predicted to result from superplume events is impressive. As more precise ages become available, it should be possible to distinguish the effects of a superplume event lasting on the order of 50 Myr from supercontinent events that last for 100 Myr or more.

References

Abbott, D. (1996). Plumes and hotspots as sources of greenstone belts. *Lithos* 37:113–27.

Abbott, D., Burgess, L., Longhi, J., and Smith, W. H. F. (1994). An empirical thermal history of the Earth's upper mantle. *J. Geophys. Res.* 99(13):835–50.

Abbott, D., Drury, R., and Smith, W. H. F. (1994). Flat to steep transition in subduction style. *Geol.* 22:937–40.

Abbott, D. and Isley, A. (2001). Oceanic upwelling and mantle plume activity: Paleomagnetic tests of ideas on the source of the iron in Early Precambrian iron formations. Geological Society of America, Spec. Pap. 352 (in press).

Abbott, D. and Mooney, W. (1995). The structural and geochemical evolution of the continental crust: Support for the oceanic plateau model of continental growth. *Rev. Geophys.* 33, Suppl.: 231–42.

Abbott, D., Sparks, D., Herzberg, C., Mooney, W., Nikishin, A., and Zhang, Y. S. (2000). Quantifying Precambrian crustal extraction: The root is the answer. *Tectonophys.* 322:163–190.

Abouchami, W., Bohler, M., Michard, A., and Albarede, F. (1990). A major 2.1-Ga event of mafic magmatism in West Africa: An early stage of crustal accretion. *J. Geophys. Res.* 95(17):605–29.

Acton, G. D. and Gordon, R. G. (1994). Paleomagnetic tests of Pacific plate reconstructions and implications for motion between hotspots. *Science* 263:1246–54.

Agee, C. B. (1993). Petrology of the mantle transition zone. *Annu. Rev. Earth Planet. Sci.* 21:19–41.

Albarede, F. (1998). Time-dependent models of U–Th–He and K–Ar evolution and the layering of mantle convection. *Chem. Geol.* 145:413–29.

Albarede, F. and van der Hilst, R. D. (1999). New mantle convection model may reconcile conflicting evidence. *EOS* 80:535–9.

Algeo, T. J. and Seslavinsky, K. B. (1995). The Paleozoic world: Continental flooding, hypsometry, and sea level. *Amer. J. Sci.* 295:787–822.

Allegre, C. J., Hamelin, B., and Dupre, B. (1984). Statistical analysis of isotopic rations in MORB: The mantle blob cluster model and the convective regime in the mantle. *Earth Planet. Sci. Lett.* 71:71–84.

Anderson, D. L. (1982). Hotspots, polar wander, Mesozoic convection and the geoid. *Nature* 297: 391–3.

———. (1994). The sublithospheric mantle as the source of continental flood basalt; the case against the continental lithosphere and plume head reservoirs. *Earth Planet. Sci. Lett.* 123:269–80.

Anderson, D. L., Zhang, Y., and Tanimoto, T. (1992). Plume heads, continental lithosphere, flood basalts, and tomography. Geological Society of London, Spec. Publ. 68:99–124.

Andrews, J. E. (1985). True polar wander: An analysis of Cenozoic and Mesozoic paleomagnetic poles. *J. Geophys. Res.* 90:7737–50.

Armstrong, R. L. (1991). The persistent myth of crustal growth. *Austral. J. Earth Sci.* 38:613–30.

Arndt, N. T. (1991). High Ni in Archean tholeiites. *Tectonophys.* 187:411–20.

———. (1994). Archean komatiites. In *Archean crustal evolution*, ed. K. C. Condie, 11–44. Amsterdam: Elsevier.

———. (1999). Why was flood volcanism on submerged continental platforms so common in the Precambrian? *Precamb. Res.* 97:155–64.

Arndt, N. T., Albarede, F., and Nisbet, E. G. (1997). Mafic and ultramafic magmatism. In *Greenstone Belts*, eds. M. J. de Wit and L. D. Ashwal, Chap. 3.2. Oxford: Oxford University Press.

Arndt, N. T. and Christensen, U. (1992). The role of lithospheric mantle in continental flood volcanism: Thermal and geochemical constraints. *J. Geophys. Res.* 97:10,967–81.

Arndt, N. T., Czamanske, G. K., Wooden, J. L., and Fedorenko, V. A. (1993). Mantle and crustal contributions to continental flood volcanism. *Tectonophys.* 223:39–52.

Arndt, N. T. and Goldstein, S. L. (1989). An open boundary between lower continental crust and mantle: Its role in crust formation and crustal recycling. *Tectonophys.* 161:201–12.

Arndt, N. T. and Jenner, G. A. (1986). Crustally contaminated komatiites and basalts from Kambalda, Western Australia. *Chem. Geol.* 56:229–55.

Arndt, N. T. and Lesher, C. M. (1992). Fractionation of REEs by olivine and the origin of Kambalda komatiites, Western Australia. *Geochim. Cosmochim. Acta* 56:4191–202.

Arndt, N. T., Nelson, D. R., Compston, W., Trendall, A. F., and Thorne, A. M. (1991). The age of the Fortescue Group, Hamersley Basin, Western Australia, from ion microprobe zircon U/Pb results. *Austral. J. Earth Sci.* 38:261–81.

Arndt, N. T. and Nesbitt, R. W. (1982). Geochemistry of Munro township basalt. In *Komatiites*, eds. N. T. Arndt and E. G. Nesbit, 309–29. London, Allen Unwin.

Arndt, N. T., et al. (1998). Were komatiites wet? *Geol.* 26:739–42.

Asahara, Y., Ohtani, E., and Suzuki, A. (1998). Melting relations of hydrous and fry mantle compositions and the genesis of komatiites. *Geophys. Res. Lett.* 25:2201–4.

Asmerom, Y., et al. (1991). Strontium isotopic variations of Neoproterozoic seawater: Implications for crustal evolution. *Geochim. Cosmochim. Acta* 55:2883–94.

Aspler, L. B. and Chiarenzelli, J. R. (1998). Two Neoarchean supercontinents? *Sed. Geol.* 120:75–104.

Awramik, S. M. (1992). The history and significance of stromatolites. In *Early organic evolution: Implications for mineral and energy resources*, ed. M. Schidlowski, 435–49. New York: Springer-Verlag.

Ballard, S. and Pollack, H. N. (1987). Diversion of heat by Archean cratons: A model for southern Africa. *Earth Planet. Sci. Lett.* 85:253–64.

Banerdt, W. B., Golombek, M. P., and Tanaka, K. L. (1992). Stress and tectonics. In *Mars*, eds. H. H. Kieffer, B. M. Jakosky, and C. W. Snyder, 249–97. Tucson: University of Arizona Press.

Baragar, W. R. A. (1969). The geochemistry of Coppermine River basalts. *Geol. Surv. Can.* Paper 69-44.

Baragar, W. R. A., Ernst, R. E., Hulbert, L., and Peterson, T. (1996). Longitudinal petrochemical variation in the Mackenzie dyke swarm, NE Canadian shield. *J. Petrol.* 37:317–59.

Bargar, K. E. and Jackson, E. D. (1974). Calculated volumes of individual shield volcanoes along the Hawaiian–Emperor chain. *U. S. Geol. Surv., J. Res.* 2:545–50.

Baragar, W. R. A., Lambert, M. B., Baglow, N., and Gibson, I. (1987). Sheeted dykes of the Troodos ophiolite, Cyprus. Geological Association of Canada, Spec. Pap. 34:257–72.

Barker, F., Jones, D. L., Budahn, J. R., and Coney, P. J. (1988). Ocean plateau-seamount origin of basaltic rocks, Angayucham terrane, central Alaska. *J. Geol.* 96:368–74.

Barker, F., et al. (1994). Some accreted volcanic rocks of Alaska and their elemental abundances. *Geology of North America, Geol. Soc. America* G-1:555–87.

Barley, M. E. (1993). Volcanic, sedimentary, and tectonostratigraphic environments of the 3.46-Ga Warrawoona Megasequence: A review. *Precamb. Res.* 60:47–67.

Barley, M. E., Blake, T. S., and Groves, D. I. (1992). The Mount Bruce megasequence set and eastern Yilgarn craton: Examples of Late Archean to Early Proterozoic divergent and convergent craton margins and controls on mineralization. *Precamb. Res.* 58:55–70.

Barley, M. E. and Groves, D. I. (1992). Supercontinent cycles and the distribution of metal deposits through time. *Geol.* 20:291–4.

Barley, M. E., Krapez, B., Groves, D. I., and Kerrich, R. (1998). The Late Archean bonanza: Metallogenic and environmental consequences of the interaction between mantle plumes, lithospheric tectonics and global cyclicity. *Precamb. Res.* 91:65–90.

Barley, M. E., Pickard, A. L., and Sylvester, P. J. (1997). Emplacement of a large igneous province as a possible cause of BIF 2.45 Ga. *Nature* 385:55–8.

Barnes, S. J. (1986). The effect of trapped liquid crystallization on cumulus mineral compositions in layered intrusions. *Contrib. Mineral. Petrol.* 93:524–31.

———. (1989). Are Bushveld U-type parent magmas boninites or contaminated komatiites? *Contrib. Mineral. Petrol.* 101:447–57.

Barnes, S. J. and Francis, D. (1995). The distribution of platinum-group elements, Ni, Cu, and Au in the Muskox layered intrusion, NW Territories, Canada. *Econ. Geol.* 90:135–54.

Barr, S. R., Temperley, S., and Tarney, J. (1999). Lateral growth of the continental crust through deep level subduction–accretion: A reevaluation of central Greek Rhodope. *Lithos* 46:69–94.

Barron, E. J. (1983). A warm, equable Cretaceous: The nature of the problem. *Earth Sci. Rev.* 19: 305–38.

Barron, E. J., Fawcett, P. J., Peterson, W. H., Pollard, D., and Thompson, S. L. (1995). A "simulation" of Mid-Cretaceous climate. *Paleoceanography* 10:953–62.

Basilevsky, A. T. and Head, J. W. (1996). Evidence for rapid and widespread emplacement of volcanic plains on Venus: Stratigraphic studies in the Baltis Vallis region. *Geophys. Res. Lett.* 23:1497–1500.

Batiza, R. (1982). Abundances, distribution, and sizes of volcanoes in the Pacific Ocean and implications for the origin of non-hotspot volcanoes. *Earth Planet. Sci. Lett.* 60:195–206.

———. (1989). Seamounts and seamount chains of the eastern Pacific. In *Geology of North America*, Vol. N, ed. E. L. Winterer et al., 289–306. Boulder, CO: Geological Society of America.

Beck, R. A., Burbank, D. W., Sercombe, W. J., Olson, T. L., and Khan, A. M. (1995). Organic carbon exhumation and global warming during the early Himalayan collision. *Geol.* 23:387–90.

Benton, M. J. (1995). Diversification and extinction in the history of life. *Science* 268:52–8.

Bercovici, D. and Mahoney, J. (1994). Double flood basalts and plume head separation at the 660-km discontinuity. *Science* 266:1367–9.

Berger, W. H. et al. (1992). The record of Ontong Java plateau: Main results of ODP Leg 130. *Geol. Soc. Am. Bull.* 104:954–72.

Berner, R. A. (1983). Burial of organic carbon and pyrite sulfur in sediments over Phanerozoic time: A new theory. *Geochim. Cosmochim. Acta* 47:855–62.

Berner, R. A. (1994). 3Geocarb II: A revised model of atmospheric CO_2 over Phanerozoic time. *Am. J. Sci.* 294:56–91.

———. (1999). A new look at the long-term carbon cycle. *GSA Today* 9(11):1–6.

Berner, R. A. and Berner, E. K. (1997). Silicate weathering and climate. In *Tectonic uplift and climate change*, ed. W.F. Ruddiman, 353–65. New York: Plenum Press.

Berry, W. B. N. and Wilde, P. (1978). Progressive ventilation of the oceans – An explanation for the distribution of the lower Paleozoic black shales. *Am. J. Sci.* 278:257–75.

Bestougeff, M. A. (1980). Summary of world coal resources and reserves, geographic and geologic repartition. *Inst. Francais du Petrole, Revue* 35:353–66.

Betts, J. N. and Holland, H. D. (1991). The oxygen content of ocean bottom waters, the burial efficiency of organic carbon, and the regulation of atmospheric oxygen. *Palaeogeogr. Palaeoclimatol. Palaeoecol.* 97:5–18.

Beukes, J. J., Klein, C., Kaufman, A. J., and Hayes, J. M. (1990). Carbonate petrography, kerogen distribution, and carbon and oxygen isotope variations in an early Proterozoic transition from limestone to iron-formation deposition, Transvaal Supergroup, South Africa. *Bull. Soc. Econ. Geol.* 85:663–90.

Bickle, M. J. (1986). Implications of melting for stabilization of the lithosphere and heat loss in the Archean. *Earth Planet. Sci. Lett.* 80:314–24.

———. (1996). Metamorphic decarbonation, silicate weathering and the long-term carbon cycle. *Terra Nova* 8:270–6.

Bickle, M. J. and Eriksson, K. A. (1982). Evolution and subsidence of early Precambrian sedimentary basins. *Philos. Trans. Roy. Soc. London, Ser. A* 305:235–47.

Bijwaard, H. and Spakman, W. (1999). Tomographic evidence for a narrow whole mantle plume below Iceland. *Earth Planet. Sci. Lett.* 166:121–6.

Birkhold, A. B., Neal, C. R., Mahoney, J. J., and Duncan, R. A. (1999). The Ontong Java plateau: Episodic growth along the SE margin. *Am. Geophys. Union, EOS* 80:F1103.

Bjarnasson, I. T., Menke, W., Flovenz, O. G., and Caress, D. (1993). Tomographic image of the mid-Atlantic plate boundary in SW Iceland. *J. Geophys. Res.* 98:6607–22.

Blichert-Toft, J. and Albarede, F. (1994). Short-lived chemical heterogeneities in the Archean mantle with implications for mantle convection. *Science* 263:1593–6.

Blichert-Toft, J., Frey, F. A., and Albarede, F. (1999). Hf isotope evidence for pelagic sediments in the source of Hawaiian basalts. *Science* 285:879–82.

Bodinier, J. L., Vaseur, G., Vernieres, J., Dupuy, C., and Fabries, J. (1990). Mechanisms of mantle metasomatism: Geochemical evidence from the Lherz orogenic peridotite. *J. Petrol.* 31:597–628.

Boher, M., Abouchami, W., Michard, A., Albarede, F., and Arndt, N. T. (1992). Crustal growth in West Africa at 2.1 Ga. *J. Geophys. Res.* 97:345–69.

Bois, C., Bouche, P., and Pelet, R. (1980). Historie geologique et repartition des reserves d'hydrocarbures dans le Monde. *Inst. Francais du Petrole, Rev.* 35:237–98.

Bott, M. H. P. (1993). Modeling the plate-driving mechanism. *J. Geol. Soc. London* 150:941–51.

Bowring, S. A. and Housh, T. (1995). The earth's early evolution. *Nature* 269:1535–40.

Brandon, A. D., Norman, M. D., Walker, R. J., and Morgan, J. W. (1999). ^{186}Os–^{187}Os systematics of Hawaiian picrites. *Earth Planet. Sci. Lett.* 174:25–42.

Brandon, A. D. and Smith, A. D. (1994). Mesozoic granitoid magmatism in SE British Columbia: Implication for the origin of granitoid belts in the North American Cordillera. *J. Geophys. Res.* 99:11879–96.

Brandon, A. D., Walker, R. J., Morgan, J. W., Norman, M. D., and Prichard, H. M. (1998). Coupled ^{186}Os and ^{187}Os evidence for core-mantle interaction. *Science* 280:1570–3.

Bratton, J. F. (1999). Clathrate eustasy: Methane hydrate melting as a mechanism for geologically rapid sea-level fall. *Geol.* 27:915–18.

Breddam, K., Kurz, M. D., and Storey, M. (2000). Mapping out the conduit of the Iceland mantle plume with helium isotopes. *Earth Planet. Sci. Lett.* 176:45–55.

Brewer, T. S., Hergt, J. M., Hawkesworth, C. J., Rex, D., and Storey, B. C. (1992). Coats Land dolerites and the generation of Antarctic continental flood basalts. Geological Society of London, Spec. Publ. 68, 185–208.

Brunet, D. and Machetel, P. (1998). Large-scale tectonic features induced by mantle avalanches with phase, temperature, and pressure lateral variations of viscosity. *J. Geophys. Res.* 103:4929–45.

Brunet, D. and Yuen, D. A. (2000). Mantle plumes pinched in the transition zone. *Earth Planet. Sci. Lett.* 178:13–27.

Buchan, K. L., Halls, H. C., and Mortensen, J. K. (1996). Paleomagnetism, U–Pb geochronology, and geochemistry of Marathon dykes, Superior Province, and comparison with the Fort Frances swarm. *Can. J. Earth Sci.* 33:1583–95.

Buchanan, P. C. and Reimold, W. U. (1998). Studies of the Rooiberg Group, South Africa: No evidence for an impact origin. *Earth Planet. Sci. Lett.* 155:149–65.

Budahn, J. R. and Schmitt, R. A. (1985). Petrogenetic modeling of Hawaiian tholeiitic basalts: A geochemical approach. *Geochim. Cosmochim. Acta* 49:67–87.

Burke, K. (1988). Tectonic evolution of the Caribbean. *Annu. Rev. Earth Planet. Sci.* 16:201–30.

———. (1996). The African plate. *S. African J. Geol.* 99:341–410.

Burke, K., Fox, P. J., and Sengor, M. C. (1978). Buoyant ocean floor and the origin of the Caribbean. *J. Geophys. Res.* 83:3949–54.

Burke, K. and Wilson, J. T. (1972). Is the African plate stationary? *Nature* 239:387–90.

Caldeira, K. and Rampino, M. R. (1990). Deccan volcanism, greenhouse warming, and the Cretaceous/Tertiary boundary. Geological Society of America, Spec. Pap. 247:117–23.

———. (1991). The mid-Cretaceous superplume, carbon dioxide, and global warming. *Geophys. Res. Lett.* 18:987–90.

Camp, V. E. (1995). Mid-Miocene propagation of the Yellowstone mantle plume head beneath the Columbia River basalt source region. *Geol.* 23:435–8.

Campbell, I. H. (1998). The mantle's chemical structure: Insights from the melting products of mantle plumes. In *The Earth's mantle, composition, structure, and evolution*, ed. Ian Jackson, 259–310. Cambridge: Cambridge University Press.

Campbell, I. H., Czamanske, G. K., Fedorenko, V. A., Hill, R. I., and Stepanov, V. (1992). Synchronism of the Siberian traps and the Permian–Triassic boundary. *Science* 258:1760–3.

Campbell, I. H., Griffiths, R. W., and Hill, R. I. (1989). Melting in an Archean mantle plume: Heads it's basalts, tails it's komatiites. *Nature* 339:697–9.

Campbell, I. H. and Griffiths, R. W. (1990). Implications of mantle plume structure for the evolution of flood basalts. *Earth Planet. Sci. Lett.* 99:79–93.

———. (1992). The changing nature of mantle hotspots through time: implications for the chemical evolution of the mantle. *J. Geol.* 92:497–523.

Canfield, D. E. (1998). A new model for Proterozoic ocean chemistry. *Nature* 396:450–3.

Caress, D. W., McNutt, M. K., Detrick, R. S., and Mutter, J. C. (1995). Seismic imaging of hotspot-related crustal underplating beneath the Marquesas Islands. *Nature* 373:600–3.

Carlson, R. W. (1984). Isotopic constraints on Columbia river flood basalt genesis and the nature of the subcontinental mantle. *Geochim. Cosmochim. Acta* 48:2357–72.

———. (1991). Physical and chemical evidence on the cause and source characteristics of flood basalt volcanism. *Austral. J. Earth Sci.* 38:525–44.

———. (1994). Mechanisms of Earth differentiation: Consequences for the chemical structure of the mantle. *Rev. Geophys. Space Phys.* 38:525–44.

Carlson, R. W. and Hart, W. K. (1987). Crustal genesis on the Oregon plateau. *J. Geophys. Res.* 92:6191–206.

Carlson, R. W. and Irving, A. J. (1994). Depletion and enrichment history of subcontinental lithospheric mantle: And Os, Sr, Nd, and Pb isotopic study of ultramafic xenoliths from the NW Wyoming craton. *Earth Planet. Sci. Lett.* 110:99–119.

Carlson, R. W., Pearson, D. B., Boyd, F. R., Shirey, S. B., Irvine, G., Menzies, A. H., and Gurney, J. J. (1999). Re–Os systematics of lithospheric peridotites: Implications for lithosphere formation and preservation. *7th Inter. Kimberlite Conf. Proc., Cape Town, S. Africa*, pp. 99–108.

Caroff, M., Maury, R. C., Guille, G., and Gotten, J. (1997). Partial melting below Tubuai (Austral Islands, French Polynesia). *Contrib. Mineral. Petrol.* 127:369–82.

Carr, M. H. (1973). Volcanism on Mars. *J. Geophys. Res.* 78, 4049–62.

Case, J. E., MacDonald, W. D., and Fox, P. J. (1990). Caribbean crustal provinces: Seismic and gravity evidence. In *The Caribbean region, geology of North America, decade of North American geology*, eds., G. Dengo and J. E. Case. 15–36. Boulder, CO: Geological Society of America.

Castaing, B. G. et al. (1989). Scaling of hard thermal turbulence in Rayleigh–Bernard convection. *J. Fluid Mech.* 204:1–30.

Castillo, P., Pringle, M. S., and Carlson, R. W. (1994). East Mariana basin tholeiites: Jurassic ocean crust or Cretaceous rift basalts related to the Ontong Java plume? *Earth Planet. Sci. Lett.* 123: 139–54.

Cawthorne, R. G. and Walraven, F. (1998). Emplacement and crystallization time for the Bushveld Complex. *J. Petrol.* 39:1669–87.

Chapel, D. and Small, C. (1996). The distribution of large seamounts in the Pacific. *EOS* 77 (46): F770.

Chesley, J. T. and Ruiz, J. (1998). Crust–mantle interaction in large igneous provinces: Implications from the Re–Os isotope systematics of the Columbia River flood basalts. *Earth Planet. Sci. Lett.* 154:1–11.

Christensen, U. R. (1995). Effects of phase transitions on mantle convection. *Annu. Rev. Earth Planet. Sci.* 23:65–87.

Christensen, U. R. and Yuen, D. A. (1985). Layered convection induced by phase transitions. *J. Geophys. Res.* 90:291–300.

Church, B. N., Dostal, J., Owen, J. V., and Pettipas, A. R. (1995). Late Paleozoic gabbroic rocks of the Bridge River accretionary complex, SW British Columbia: Geology and geochemistry. *Geol. Rund.* 84:710–19.

Clague, D. A. and Dalrymple, G. B. (1989). Tectonics, geochronology, and origin of the Hawaiian–Emperor volcanic chain. In *Geology of North America*, Vol. N, ed. E. L. Winterer et al., 188–217. Boulder, CO: Geological Society of America.

Cloos, M. (1993). Lithospheric buoyancy and collisional orogensis: Subduction of oceanic plateaus, continental margins, island arcs, spreading ridges and seamounts. *Geol. Soc. Am. Bull.* 105:715–37.

Clowes, R. M. et al. (1992). Lithoprobe: New perspectives on crustal evolution. *Can. J. Earth Sci.* 29:1813–64.

Coblentz, D. D., Richardson, R. M., and Sandiford, M. (1994). On the gravitational potential of the Earth's lithosphere. *Tectonics* 13:929–45.

Cocherie, A. (1986). Systematic use of trace element distribution patterns in log–log diagrams for plutonic suites. *Geochim. Cosmochim. Acta* 50:2517–22.

Cordery, M. J., Davies, G. F., and Campbell, I. H. (1997). Genesis of flood basalts from eclogite-bearing mantle plumes. *J. Geophys. Res.* 102 (20):179–97.

Coffin, M. F. and Eldholm, O. (1994). Large igneous provinces: Crustal structure, dimensions, and external consequences. *Rev. Geophys.* 32:1–36.

Comer, R. S., Solomon, S., and Head, J. W. (1985). Mars: Thickness of the lithosphere from the tectonic response to volcanic loads. *J. Geophys. Res.* 23:61–92.

Condie, K. C. (1992). Proterozoic terranes and continental accretion in southwestern North America. In *Proterozoic crustal evolution*, ed. K. C. Condie, 447–80. Amsterdam: Elsevier Scientific Publishers.

———. (1994). Greenstones through time. In *Archean crustal evolution*, Chap. 3, ed. K. C. Condie, 85–120. Amsterdam: Elsevier Scientific Publishers.

———. (1995). Episodic ages of greenstones: A key to mantle dynamics? *Geophys. Res. Lett.* 22:2215–18.

———. (1997a). *Plate tectonics and crustal evolution*, 4th ed. Oxford, UK: Butterworth–Heinemann, 282 pp.

———. (1997b). Sources of Proterozoic mafic dyke swarms: Constraints from Th/Ta and La/Yb ratios. *Precamb. Res.* 81:3–14.

———. (1997c). Contrasting sources for upper and lower continental crust: The greenstone connection. *J. Geol.* 105:729–36.

———. (1998). Episodic continental growth and supercontinents: A mantle avalanche connection? *Earth Planet. Sci. Lett.* 163:97–108.

———. (1999). Mafic crustal xenoliths and the origin of the lower continental crust. *Lithos* 46:95–101.

———. (2000a). Episodic continental growth models: Afterthoughts and extensions. *Tectonophys.* 322:153–62.

———. (2001). Continental growth during formation of Rodinia at 1.35–0.9 Ga. *Gondwana Res.* 4:5–16.

Condie, K. C. and Chomiak, B. (1996). Continental accretion: Contrasting Mesozoic and Early Proterozoic tectonic regimes in North America. *Tectonophys.* 265:101–26.

Condie, K. C., Des Marais, D. J., and Abbott, D. (2001). Precambrian superplumes and supercontinents: A record in black shales, carbon isotopes, and paleoclimates. *Precamb. Res.* 106:239–60.

———. (2000). Geologic evidence for a mantle superplume event at 1.9 Ga. *Geochem. Geophys. Geosyst.* Vol. 1, Paper no. 2000GC000095.

Cook, P. J. and McElhinny, M. W. (1979). A reevaluation of the spatial and temporal distribution of sedimentary phosphate deposits in the light of plate tectonics. *Econ. Geol.* 74:315–30.

Cook, P. J. and Shergold, J. H. (1984). Phosphorus, phosphorites and skeletal evolution at the Precambrian–Cambrian boundary. *Nature* 308:231–6.

———. (1986). Proterozoic and Cambrian phosphorites – nature and origin. In *Phosphate deposits of the world*, eds. P. J. Cook and J. H. Shergold, 369–86. Cambridge: Cambridge University Press.

Courtillot, V. (1999). *The science of mass extinction.* Cambridge: Cambridge University Press.

Courtillot, V. and Besse, J. (1987). Magnetic field reversal, polar wander, and core–mantle coupling. *Science* 237:1140–7.

Courtillot, V., Besse, J., Vandamme, D., Montigny, R., Jaeger, J. J., and Capetta, H. (1986). Deccan flood basalts at the Cretaceous/Tertiary boundary? *Earth Planet. Sci. Lett.* 80:361–74.

Courtillot, V. and Cisowski, C. (1987). The Cretaceous–Tertiary boundary events: External or internal causes? *EOS, Amer. Geophys. Union* 68:193–200.

Courtillot, V., Feraud, G., Maluski, H., Vandamme, D., Moreau, M. G., and Besse, J. (1988). Deccan flood basalts and the Cretaceous/Tertiary boundary. *Nature* 333:843–6.

Courtillot, V., Jaeger, J. J., Yang, Z., Feraud, G., and Hofmann, C. (1996). The influence of continental flood basalts on mass extinctions: Where do we stand? Geological Society of America, Spec. Pap. 307:513–25.

Courtillot, V., Jaupart, C., Manighetti, I., Tapponnier, P., and Besse, J. (1999). On causal links between flood basalts and continental breakup. *Earth Planet. Sci. Lett.* 166:177–95.

Cox. A. (1968). Length of geomagnetic polarity intervals. *J. Geophys. Res.* 73:3247–60.

Cox, K. G. (1988). The Karoo province. In *Continental flood basalts*, ed. J. D. Macdougall, 239–71. Dordrecht: Kluwer Academic Publishers.

Cox, K. G. (1989). The role of mantle plumes in the development of continental drainage patterns. *Nature* 342:873–7.

Crough, T. S. (1983). Hotspot swells. *Annu. Rev. Earth Planet. Sci.* 11:165–94.

Crough, T. S., Morgan, W. J., and Hargraves, R. B. (1980). Kimberlites: Their relation to mantle hotspots. *Earth Planet. Sci. Lett.* 50:260–74.

Crow, C. and Condie, K. C. (1988). Geochemistry and origin of Late Archean volcanics from the Ventersdorp Supergroup, South Africa. *Precamb. Res.* 42:19–37.

Crowell, J. C. (1999). Pre-Mesozoic ice ages: Their bearing on understanding the climate system. Geological Society of America, Mem. 192.

Crumpler, L. S., Head, J. W., and Aubele, J. C. (1993). Relation of major volcanic center concentration on Venus to global tectonic patterns. *Science* 261:591–5.

Cserepes, L., Yuen, D. A., and Shroeder, B. A. (2000). Effect of the mid-mantle viscosity and phase-transition structure on 3D mantle convection. *Phys. Earth Planet. Inter.* 118:135–48.

Czamanske, G. K., et al. (1995). Petrography and geochemical characterization of ore-bearing intrusions of the Norili'sk type, Siberia: With discussion of their origin. *Resource Geology, Spec. Issue* 18:1–48.

Czamanske, G. K., Fedorenk, V. A., and Simonov, V. (1998). Demise of the Siberian plume: Paleogeographic and paleotectonic reconstruction from the pre-volcanic and volcanic record, north-central Siberia. *Int. Geol. Rev.* 40:95–115.

Dalziel, I. W. D. (1997). Neoproterozoic-Paleozoic geography and tectonics: Review, hypothesis, environmental speculation. *Geol. Soc. Amer. Bull.* 109:16–42.

Dalziel, I. W. D., Lawver, L. A., and Murphy, J. B. (2000). Plumes, orogenesis, and supercontinental fragmentation. *Earth Planet. Sci. Lett.* 178:1–11.

———. (1999). Simultaneous generation of hotspots and superswells by convection in a heterogeneous planetary mantle. *Nature* 402:756–60.

Davies, G. F. (1988). Ocean bathymetry and mantle convection 1, large-scale flow and hotspots. *J. Geophys. Res.* 93:10,467–80.

———. (1992). One the emergence of plate tectonics. *Geol.* 20:963–6.

———. (1993). Conjectures on the thermal and tectonic evolution of the earth. *Lithos* 30:281–9.

———. (1994). Thermomechanical erosion of the lithosphere by mantle plumes. *J. Geophys. Res.* 99(15):709–22.

———. (1995). Penetration of plates and plumes through the mantle transition zone. *Earth Planet. Sci. Lett.* 133:507–16.

———. (1999). *Dynamic earth plates, plumes and mantle convection.* Cambridge, U K: Cambridge University Press, 458 pp.

Davies, G. F. and Pribac, F. (1993). Mesozoic seafloor subsidence and the Darwin rise, past and present. American Geophysical Union Mon. 77:39–52.

Davies, G. F. and Richards, M. A. (1992). Mantle convection. *J. Geol.* 100:151–206.

DeBari, S. M. and Sleep, N. H. (1991). High-Mg, low-Al bulk composition of the Talkeetna island arc, Alaska: implications for primary magmas and the nature of the arc crust. *Geol. Soc. Am. Bull.* 103:37–47.

Derry, L. A., Kaufman, A. J., and Jacobsen, S. B. (1992). Sedimentary cycling and environmental change in the Late Proterozoic: Evidence from stable and radiogenic isotopes. *Geochim. Cosmochim. Acta* 56:1317–29.

Des Marais, D. J. (1997). Long-term evolution of the biogeochemical carbon cycle. In *Geomicrobiology*, eds. J. Banfield and K. Nealson, 429–45. Washington, DC: Mineralogical Association of America.

Des Marais, D. J., Strauss, H., Summons, R. E., and Hayes, J. M. (1992). Carbon isotope evidence for the stepwise oxidation of the Proterozoic environment. *Nature* 359:605–9.

Desrochers, J-P., Hubert, C., Ludden, J. N., and Pilote, P. (1993). Accretion of oceanic plateau fragments in the Abitibi greenstone belt, Canada. *Geol.* 21:451–4.

Detrick, R. S. and Crough, S. T. (1978). Island subsidence, hotspots, and lithospheric thinning. *J. Geophys. Res.* 83:1236–44.

Devine, J. D., Sigurdsson, H., and Davis, A. N. (1984). Estimate of sulfur and chlorine yield to the atmosphere from volcanic eruptions and potential climatic effects. *J. Geophys. Res.* 89: 6309–25.

de Wit, M . J. and Ashwal, L. D. (1995). Greenstone belts: What are they? *S. Afr. J. Geol.* 98:505–20.

Dickinson, W. R. (1997). Tectonic implications of Cenozoic volcanism in coastal California. *Geol. Soc. Am. Bull.* 109:936–54.

———. (1998). Geomorphic and geodynamics of the Cook–Austral island-seamount chain in the South Pacific Ocean: Implications for hotspots and plumes. *Int. Geol. Rev.* 40:1039–75.

DiVenere, V. and Kent, D. V. (1999). Are the Pacific and Indo-Atlantic hotspots fixed? Testing the plate circuit through Antarctica. *Earth Planet. Sci. Lett.* 170:105–17.

Dostal, J. and Church, B. N. (1994). Geology and geochemistry of the volcanic rocks of the Pioneer Formation, Bridge River area, SW British Columbia. *Geol. Mag.* 131:243–53.

Douglass, J., Schilling J.-G., and Fontignie, D. (1999). Plume–ridge interactions of the Discovery and Shona mantle plumes with the southern Mid-Atlantic Ridge (40–55°S). *J. Geophys. Res.* 104, B2:2941–62.

Downes, H. (1993). The nature of the lower continental crust of Europe: Petrological and geochemical evidence from xenoliths. *Phys. Earth Planet. Inter.* 79:195–218.

Downes, H., Dupuy, C., and Leyreloup, A. F. (1990). Crustal evolution of the Hercynian belt of western Europe: Evidence from lower-crustal granulitic xenoliths (French Massif Central). *Chem. Geol.* 83:209–31.

Downes, H., Kempton, P. D., Briot, D., Harmon, R. S., and Leyreloup, A. F. (1991). Pb and O isotope systematics in granulite facies xenolith, French Massif, Central: Implications for crustal processes. *Earth Planet. Sci. Lett.* 102:342–57.

Drake, M. J., McFarlane, E. A., Gasparik. T., and Rubie, D. C. (1993). Mg–perovskite/silicate melt and majorite garnet/silicate melt partition coefficients in the system CaO–MgO–SiO_2 at high temperatures and pressures. *J. Geophys. Res.* 98:5427–31.

Duncan, R. A. (1981). Hotspots in the southern oceans – An absolute frame of reference for motion of the Gondwana continents. *Tectonophys.* 74:29–42.

———. (1991). Ocean drilling and the volcanic record of hotspots. *GSA Today* 1(10):214–19.

Duncan, R. A. and Richards, M. A. (1991). Hotspots, mantle plumes, flood basalts, and true polar wander. *Rev. Geophys.* 29:31–50.

Duncan, R. A. and Storey, M. (1992). The life cycle of Indian Ocean hotspots. In *Synthesis of results from scientific drilling in the Indian Ocean*, eds. R. A. Duncan et al., 91–103, Washington, DC: American Geophysical Union.

Durrheim, R. J. and Mooney, W. D. (1991). Archean and Proterozoic crustal evolution: Evidence from crustal seismology. *Geol.* 19:606–9.

Dzurisin, D., Savage, J. C., and Fournier, R. O. (1990). Recent crustal subsidence at Yellowstone Caldera, Wyoming. *Bull. Volcanol.* 52:247–70.

Eales, H. V. and Cawthorn, R. G. (1996). The Bushveld Complex. In *Layered intrusions*, ed. R. G. Cawthorn, 181–229. Amsterdam: Elsevier.

Ebinger, C. J. and Sleep, N. H. (1998). Cenozoic magmatism throughout east Africa resulting from impact of a single plume. *Nature* 395:788–91.

Eldholm, O. (1991). Magmatic–tectonic evolution of a volcanic rifted margin. *Mar. Geol.* 102:43–61.

Eldholm, O. and Grue, K. (1994). North Atlantic volcanic margins: Dimensions and production rates. *J. Geophys. Res.* 99:2955–68.

Eldholm, O., Skogseid, J., Planke, S., and Gladczenko, T. P. (1995). Volcanic margin concepts. In *Rifted ocean–continental boundaries*, eds. E. Banda et al., 1–16. Netherlands: Kluwer Press.

Ellam, R. M., Carlson, R. W., and Shirey, S. B. (1992). Evidence from Re–Os isotopes for plume–lithosphere mixing in Karoo flood basalt genesis. *Nature* 359:718–21.

Elliot, D. H., Fleming, T. H., Kyle, P. R., and Foland, K. A. (1999). Long-distance transport of magmas in the Jurassic Ferrar large igneous province, Antarctica. *Earth Planet. Sci. Lett.* 167:89–104.

Emeleus, C. H. (1991). Tertiary igneous activity. In *The geology of Scotland*, ed. G. Y. Craig, 455–502. London: Geological Society of London.

England, R. W., Butler, R. W. H., and Hutton, D. H. W. (1993). The role of Paleocene magmatism in the Tertiary evolution of basins on the NW seaboard. In *Petroleum Geology of Northwest Europe*, 97–105, Geological Society of London.

Eriksson, P. G. (1999). Sea level changes and the continental freeboard concept: General principles and application to the Precambrian. *Precamb. Res.* 97:143–54.

Ernst, R. E. and Baragar, W. R. A. (1992). Evidence from magnetic fabric for the flow pattern of magma in the Mackenzie giant radiating dyke swarm. *Nature* 356:511–13.

Ernst, R. E. and Buchan, K. L. (1997). Giant radiating dyke swarms: Their use in identifying pre-Mesozoic large igneous provinces and mantle plumes. American Geophysical Union Mon. 100: 297–333.

Ernst, R. E., Head, J. W., Parfitt, E., Grosfils, E., and Wilson, L. (1995). Giant radiating dyke swarms on Earth and Venus. *Earth-Sci. Rev.* 39:1–58.

Fahrig, W. F. (1987). The tectonic settings of continental mafic dyke swarms: Failed arm and early passive margin. Geological Association of Canada Spec. Pap. 34:331–48.

Fan, J. and Kerrich, R. (1997). Geochemical characteristics of Al-depleted and undepleted komatiites and HREE-enriched low-Ti tholeiites, western Abitibi greenstone belt: A heterogeneous mantle plume-convergent margin environment. *Geochim. Cosmochim. Acta* 61:4723–44.

Farnetani, C. G. (1997). Excess temperature of mantle plumes: The role of chemical stratification across D″. *Geophys. Res. Lett.* 24:1583–6.

Farnetani, C. G. and Richards, M. A. (1994). Numerical investigation of the mantle plume initiation model for flood basalt event. *J. Geophys. Res.* 99:13,813–33.

———. (1995). Thermal entrainment and melting in mantle plumes. *Earth Planet. Sci. Lett.* 136: 251–67.

Faure, K., de Wit, M. J., and Willis, J. P. (1995). Late Permian global coal hiatus linked to ^{13}C-depleted CO_2 flux into the atmosphere during the final consolidation of Pangea. *Geol.* 23:507–10.

Feighner, M A., Kellogg, L. H., and Travis, B. J. (1995). Numerical modeling of chemically buoyant mantle plumes a spreading ridges. *Geophys. Res. Lett.* 22:715–18.

Feighner, M. A. and Richards, M A. (1995). The fluid dynamics of plume-ridge and plume-plate interactions: An experimental investigation. *Earth Planet. Sci. Lett.* 129:171–82.

Fitton, J. G., Larsen, L. M., Saunders, A. D., Hardarson, B. S., and Kempton, P. D. (2000). Paleogene continental to oceanic magmatism on the SE Greenland continental margin at 63°N: A review of the results of Ocean Drilling Program Legs 152 and 163. *J. Petrol.* 41:951–66.

Fitton, J. G., Saunders, A. D., Larsen, L. M., Hardarson, B. S., and Norry, M. J. (1998). Volcanic rocks from the SE Greenland margin at 63°N: composition, petrogenesis, and mantle sources. *Proc. Ocean Drill. Program, Sci. Results* 152:331–50.

Fodor, R. V. (1987). Low- and high-TiO_2 flood basalts of southern Brazil: Origin from picritic parentage and a common mantle source. *Earth Planet. Sci. Lett.* 84:423–30.

Francis, D. (1994). Chemical interaction between picritic magmas and upper crust along margins of the Muskox intrusion, NW Territories. Geological Survey of Canada, Paper 92–12.

Francis, D., Ludden, J., Johnstone, R., and Davis, W. (1999). Picrite evidence for more Fe in Archean mantle reservoirs. *Earth Planet. Sci. Lett.* 167:197–213.

Frey, F. A. et al. (2000). Origin and evolution of a submarine large igneous province: The Kerguelen plateau and Broken Ridge, southern Indian Ocean. *Earth Planet. Sci. Lett.* 176:73–89.

Fyfe, W. S. (1978). The evolution of Earth's crust: Modern plate tectonics to ancient hotspot tectonics? *Chem. Geol.* 23:89–114.

Gaffin, S. (1987). Ridge volume dependence on seafloor generation rate and inversion using long term sea level change. *Am. J. Sci.* 287:596–611.

Gaherty, J. B. and Jordan, T. H. (1995). Lehmann discontinuity as the base of an anisotropic layer beneath continents. *Science* 268:1468–71.

Galer, S. J. G. and Mezger, K. (1998). Metamorphism, denudation and sea level in the Archean and cooling of the Earth: *Precamb. Res.* 92:389–412.

Gallagher, K. and Hawkesworth, C. (1992). Dehydration melting and the generation of continental flood basalts. *Nature* 358:57–9.

Garfunkel, Z., Anderson, C. A., and Schubert, B. (1986). Mantle circulation and the lateral migration of subducted slabs. *J. Geophys. Res.* 91:7205–23.

Garland, F. E., Hawkesworth, C. J., and Mantoviani, M. S. M. (1995). Description and petrogenesis of the Paraná rhyolites. *J. Petrol.* 36:1193–227.

Gasperini, D. et al. (2000). Evidence from Sardinian basalt geochemistry for recycling of plume heads into the Earth's mantle. *Nature* 408:701–4.

Gastil, G. (1960). The distribution of mineral dates in time and space. *Am. J. Sci.* 258:1–35.

Geist, D. and Richards, M. (1993). Origin of the Columbia plateau and Snake River plain: Deflection of the Yellowstone plume. *Geol.* 21:789–92.

Gibson, S. A., Thompson, R. N., Dickin, A. P., and Leonardos, O. H. (1995). High-Ti and low-Ti mafic potassic magmas: Key to plume lithosphere interactions and continental flood basalt genesis. *Earth Planet. Sci. Lett.* 136:149–65.

Gladczenko, T. P., Coffin, M. F., and Eldholm, O. (1997). Crustal structure of the Ontong Java plateau: Modeling of new gravity and existing seismic data. *J. Geophys. Res.* 102:22,711–29.

Glen, J. M. G., Renne, P. R., Milner, S. C., and Coe, R. S. (1997). Magma flow inferred from anisotropy of magnetic susceptibility in the coastal Paraná–Etendeka igneous province: Evidence for rifting before flood volcanism. *Geol.* 25:1131–4.

Gorbatschev, R. and Bogdanova, S. (1993). Frontiers in the Baltic shield. *Precamb. Res.* 64:3–22.

Gower, C. F., Ryan, A. B., and Rivers, T. (1990). Mid-Proterozoic Laurentia–Baltica: An overview of its geological evolution and a summary of the contributions made by the volume. Geological Association of Canada Spec. Pap. 38:1–20.

Gower, C. G. and Tucker, R. D. (1994). Distribution of pre-1400 Ma crust in the Grenville province: Implications for rifting in Laurentia Baltica during geon 14. *Geol.* 22:827–30.

Grand, S. P., van der Hilst, R. D., and Widiyantoro, S. (1997). Global seismic tomography: A snapshot of convection in the Earth. *GSA Today* 7(4):1–7.

Green, D. H. (1971). Composition of basaltic magmas as indicators of origin: Applications to oceanic volcanism. *Philos. Trans. Roy. Soc. Lond. Ser. A* 268:707–25.

———. Experimental melting studies on a model upper mantle composition of high pressure under water-saturated and water-undersaturated conditions. *Earth Planet. Sci. Lett.* 19:37–45.

Green, M. G., Sylvester, P. J., and Buick, R. (2000). Crustal growth and the inception of stable continental platforms: Evidence from basalt geochemistry of the 3.5-Ga Coonterunah and Warrawoona Groups, Pilbara craton, Australia. *Tectonophys.* 322:69–88.

Greff-Lefftz, M. and Legros, H. (1999). Core rotational dynamics and geological events. *Science* 286:1707–9.

Gregoire, M., Cottin, J. Y., Giret A., Mattielli, N., and Weis, D. (1998). The meta-igneous granulite xenoliths from Kerguelen Archipelago: Evidence of a continent nucleation in an oceanic setting. *Contrib. Mineral. Petrol.* 133:259–83.

Griffin, W. L., O'Reilly, S. Y., Ryan, C. G., Gaul O., and Ionov, D. A. (1998). Secular variation in the composition of subcontinental lithospheric mantle: Geophysical and geodynamic implications. Geological Society of America, American Geophysical Union, *Geodynamic Ser.* 26:1–25.

Griffiths, R. W. (1986). Particle motion induced by spherical convection elements in Stokes flow. *J. Fluid Mech.* 166:139–59.

Griffiths, R. W. and Campbell, I. H. (1990). Stirring and structure in mantle plumes. *Earth Planet. Sci. Lett.* 99:66–78.

———. (1991). Interaction of mantle plume heads with the Earth's surface and onset of small-scale convection. *J. Geophys. Res.* 96(18):295–310.

Gripp, A. E. and Gordon, R. G. (1990). Current plate velocities related to hotspots incorporating the NUVEL-1 global plate motion model. *Geophys. Res. Lett.* 17:1109–12.

Grosfils, E. B. and Head, J. W. (1994). The global distribution of giant radiating dyke swarms on Venus: Implications for the global stress state. *Geophys. Res. Lett.* 21:701–4.

———. (1996). The timing of giant radiating dyke swarm emplacement on Venus: Implications for resurfacing of the planet and its subsequent evolution. *J. Geophys. Res.* 101:4645–56.

Grotzinger, J. P. and Kasting, J. F. (1993). New constraints on Precambrian ocean composition. *J. Geol.* 101:235–43.

Grotzinger, J. P. and Knoll, A. H. (1999). Stromatolites in Precambrian carbonates: Evolutionary mileposts or environmental dipsticks? *Annu. Rev. Earth Planet. Sci.* 27:313–58.

Grove, T., de Wit, M. J., and Dann, J. (1997). Komatiites from the Komati type section, Barberton, South Africa. In *Greenstone belts*, eds. M. J. de Wit and L D. Ashwal, 438–53. Oxford: Oxford University Press.

Guillou, L. and Jaupart, C. (1995). On the effects of continents on mantle convection. *J. Geophys. Res.* 100:24,217–38.

Gurnis, M. (1988). Large-scale mantle convection and the aggregation and dispersal of supercontinents. *Nature* 332:695–9.

Gurnis, M. (1993). Phanerozoic marine inundation of continents driven by dynamic topography above subducting slabs. *Nature* 364:589–93.

Gurnis, M., Ritsema, J., van Heijst, H. J., and Zhong, S. (2000). Tonga slab deformation: The influence of a lower mantle upwelling on a slab in a young subduction zone. *Geophys. Res. Lett.* 27:2373–6.

Hager, B. H., Clayton, R. W., Richards, M. A., Comer, R. P., and Dziewonski, A. M. (1985). Lower mantle heterogeneity, dynamic topography and the geoid. *Nature* 313:541–5.

Haggerty, S. E. (1994). Superkimberlites: A geodynamic diamond window to the Earth's core. *Earth Planet. Sci. Lett.* 122:57–69.

———. (1999). Diamond formation and kimberlite-clan magmatism in cratonic settings. In *Mantle petrology: Field observations and high-pressure experimentation*, eds., Y. Fei, M. Berka, and B. O. Mysen, 105–23, Geochemical Society Special Publication No. 6. Houston: Geochemical Society.

Haggerty, S. E. and Sautter, V. (1990). Ultradeep ultramafic upper mantle xenoliths. *Science* 248: 993–6.

Hallam, A. (1987). End-Cretaceous mass extinction event: Argument for terrestrial causation. *Science* 29, 1237–42.

Halls, H. C. (1982). The importance and potential of mafic dyke swarms in studies of geodynamic processes. *Geosci. Can.* 9:145–54.

Hames, W. E., Renne, P. R., and Ruppel, C. (2000). New evidence for geologically instantaneous emplacement of earliest Jurassic Central Atlantic magmatic province basalts on the North American margin. *Geol.* 28:859–62.

Hamilton, E. L. (1956). Sunken islands of the Mid-Pacific mountains. Geological Society of America Mem. 64, p. 97.

Hanan, B. B., Blichert-Toft, R., Kingsley, R., and Schilling, J.-G. (2000). Depleted Iceland mantle plume geochemical signature: Aritifact of multicomponent mixing? *Geochem. Geophys. Geosyst.* 1, Paper No. 1999GC000009.

Hanan, B. B. and Graham, D. W. (1996). Lead and helium isotope evidence from oceanic basalts for a common deep source of mantle plumes. *Science* 272:991–5.

Handler, P. (1989). The effect of volcanic aerosols on global climate. *J. Volc. Geotherm. Res.* 37: 233–49.

Hansen, E. C., Newton, R. C., Janardhar, A. S., and Lindenberg, S. (1995). Differentiation of Late Archean crust in the eastern Dharwar craton, Krishagiri–Salem area, South India. *J. Geol.* 103: 629–51.

Hansen, V. L., Banks, B. K., and Ghent, R. R. (1999). Tessera terrain and crustal plateaus, Venus. *Geol.* 27:1071–4.

Hansen, U., Yuen, D. A., and Kroening, S. E. (1990). Transition to hard turbulence in thermal convection at infinite Prandtl number. *Phys. Fluids A* 2(12):2157–63.

Haq, B. U. (1998). Gas hydrates: Greenhouse nightmare? Energy panacea or pipe dream? *GSA Today* 8(11):1–6.

Hardebeck, J. and Anderson, D. L. (1996). Eustasy as a test of a Cretaceous superplume hypothesis. *Earth Planet. Sci. Lett.* 137:101–8.

Harris, C. A., Whittingham, M., Milner, S. C., and Armstrong, R. A. (1990). Oxygen isotope geochemistry of the silica volcanic rocks of the Etendeka–Paraná province: Source constraints. *Geol.* 18, 1119–21.

Hart, S. R. (1984). A large-scale anomaly in the Southern Hemisphere mantle. *Nature* 309:753–7.

———. (1988). Heterogeneous mantle domains: Signatures, genesis and mixing chronologies. *Earth Planet. Sci. Lett.* 90:273–96.

Hart, S. R., Hauri, E. H., Oschmann, L. A., and Whitehead, J. A. (1992). Mantle plumes and entrainment: Isotopic evidence. *Science* 256:517–19.

Hart, S. R. and Zindler, A. (1989). Constraints on the nature and development of chemical heterogeneities in the mantle. In *Mantle Convection*, ed. W. R. Peltier, 262–387. New York: Gordon and Breach Publishers.

Hatton, C. J. (1995). Mantle plume origin for the Bushveld and Ventersdorp magmatic provinces. *J. Afr. Earth Sci.* 21:571–7.

Hauff, F., Hoenle, K., Tilton, G., Graham, D. W., and Kerr, A. C. (2000). Large-volume recycling of oceanic lithosphere over short time scales: Geochemical constraints from the Caribbean large igneous province. *Earth Planet. Sci. Lett.* 174:247–63.

Hauri, E. H. and Hart, S. R. (1993). Re–Os isotope systematics of HIMU and EMII oceanic island basalts from the Pacific Ocean. *Earth Planet. Sci. Lett.* 114:353–71.

Hauri, E. H., Whitehead, J. A., and Hart, S. R. (1994). Fluid dynamic and geochemical aspects of entrainment in mantle plumes. *J. Geophys. Res.* 99(24):275–300.

Hawkesworth, C. J., Gallagher, K., Kelley, S., Mantovani, M., Peate, D. W., Regelous, M., and Rogers, N. W. (1992). Paraná magmatism and the opening of the South Atlantic. Geological Society of London, Spec. Publ. 68:221–40.

Hawkesworth, C. J., Gallaher, K., Kirstein, L., Mantovani, M. S. M., Peate, D. W., and Turner, S. P. (2000). Tectonic controls on magmatism associated with continental break-up: An example from the Paraná–Etendeka Province. *Earth Planet. Sci. Lett.* 179:335–49.

Hawkesworth, C. J., Kelley, S., Turner, S., le Roex, A., and Storey, B. (1999). Mantle processes during Gondwana breakup and dispersal. *J. Afr. Earth Sci.* 28:239–61.

Hawkesworth, C. J., Mantovani, S. M., and Peate, D. W. (1988). Lithosphere remobilization during Paraná CFB magmatism. In *Oceanic and continental lithosphere: Similarities and differences*, eds. M. A. Menzies and K. Cox, 205–23. Oxford: Oxford University Press.

Head, J. W. and Coffin, M. F. (1997). Large igneous province: A planetary perspective. American Geophysical Union, Mon. 100:411–38.

Head, J. W. and Wilson, L. (1986). Volcanic processes and landforms on Venus: Theory, predictions, and observations. *J. Geophys. Res.* 91:9407–46.

Head, J. W. et al. (1992). Venus volcanism: Classification of volcanic features and structures, associations, and global distribution from Magellan data. *J. Geophys. Res.* 97:13,153–97.

Heaman, L. M. and Kjarsgaard, B. A. (2000). Timing of eastern North American kimberlite magmatism: Continental extension of the Great Meteor hotspot track? *Earth Planet. Sci. Lett.* 178: 253–68.

Helmberger, D. V., Wen, L., and Ding, X. (1998). Seismic evidence that the source of the Iceland hotspot lies at the core–mantle boundary. *Nature* 396:251–5.

Hemond, C., Arndt, N. T., Lichenstein, U., and Hofmann, A. W. (1993). The heterogeneous Iceland plume: Nd–Sr–O isotopes and trace element constraints. *J. Geophys. Res.* 98(15):833–50.

Hergt, J. M., Peate, D. W., and Hawkesworth, C. J. (1991). The petrogenesis of Mesozoic Gondwana low-Ti flood basalts. *Earth Planet. Sci. Lett.* 105:134–48.

Herrick, R. R. (1999). Small mantle upwellings are pervasive on Venus and Earth. *Geophys. Res. Lett.* 26:803–6.

Herzberg, C. (1995). Generation of plume magmas through time: An experimental approach. *Chem. Geol.* 126:1–16.

Hieronymus, C. F. and Bercovici, D. (2000). Non-hotspot formation of volcanic chains: Control of tectonic and flexural stresses on magma transport. *Earth Planet. Sci. Lett.* 181:539–54.

Hill, R. I., Campbell, I. H., Davies, G. F., and Griffiths, R. W. (1992). Mantle plumes and continental tectonics. *Science* 256:186–93.

Hoernie, K., et al. (2000). Existence of complex spatial zonation in the Galapagos plume for at least 14 My. *Geol.* 28, 435–8.

Hoffman, H. J. (1998). Synopsis of Precambrian fossil occurrences in North America. Chap. 4, Geol. Soc. America, The Geology of North America, Vol. C-1, p. 273–93.

———. (1989). Speculations on Laurentia's first gigayear (2.0–1.0 Ga). *Geol.* 17:135–8.

Hoffman, P. F. and Ranalli, G. (1988). Archean oceanic flake tectonics. *Geophys. Res. Lett.* 15:1077–80.

Hofmann, A. W. (1986). Nb in Hawaiian magmas: Constraints on source composition and evolution. *Chem. Geol.* 57:17–30.

———. (1997). Mantle geochemistry: The message from oceanic volcanism. *Nature* 385:219–29.

Hofmann, A. W., Jochum, K. P., Seufert, M., and White, W. M. (1986). Nb and Pb in oceanic basalts: New constraints on mantle evolution. *Earth Planet. Sci. Lett.* 79:33–45.

Hofmann, C. et al. (1997). Timing of the Ethiopian flood basalt event and implications for plume birth and global change. *Nature* 38:838–41.

Holbrook, W. S. and Kelemen, P. B. (1993). Large igneous province on the U.S. Atlantic margin and implications for magmatism during continental breakup. *Nature* 364:433–6.

Holland, H. D. (1984). *The chemical evolution of the atmosphere and oceans.* Princeton: Princeton University Press. p. 582.

Hollings, P. N. and Kerrich, R. (1999). Trace element systematics of ultramafic and mafic volcanic rocks from the 3-Ga North Caribou greenstone belt, northwestern Superior Province. *Precamb. Res.* 93:257–79.

Hollings, P. N., Wyman, D., and Kerrich, R. (1999). Komatiite–basalt–rhyolite volcanic associations in Northern Superior province greenstone belts: Significance of plume-arc interaction in the generation of the proto continental Superior province. *Lithos* 46:137–62.

Holser, W. T., Schidlowski, M., Mackenzie, F. T., and Maynard, J. B. (1988). Geochemical cycles of carbon and sulfur. In *Chemical cycles in the evolution of the earth*, eds. C. B. Gregor, R. M. Garrels, F. T. Mackenzie, and J. B. Maynard, 105–173. New York: John Wiley.

Hooper, P. R. (1990). The timing of crustal extension and the eruption of continental flood basalts. *Nature* 345:246–9.

———. (1997). The Columbia River flood basalt province: Current status. American Geophysical Union Mon. 100:1–27.

Horan, M. F., Walker, R. J., Fedorenko, V. A., and Czamanske, G. K. (1995). Os and Nd isotopic constraints on the temporal and spatial evolution of Siberian flood basalt sources. *Geochim. Cosmochim. Acta* 59:5159–68.

Howell, D. G. (1989). *Tectonics of suspect terranes, mountain building and continental growth*. New York: Chapman & Hill, p. 232.

Humler, E., Langmuir, C., and Daux, V. (1999). Depth versus age: New perspectives from the chemical compositions of ancient crust. *Earth Planet. Sci. Lett.* 173:7–23.

Humphreys, E. D., Ducker, K. G., Schutt, D. L., and Smith, R. B. (2000). Beneath Yellowstone: Evaluating plume and nonplume models using teleseismic images of the upper mantle. *GSA Today* 10(12):1–7.

Huang, Y. M., van Calsteren, P., and Hawkesworth, C. J. (1995). The evolution of the lithosphere in southern Africa: A perspective on the basic granulite xenoliths from kimberlites in South Africa. *Geochim. Cosmochim. Acta* 59:4905–20.

Hunter, D. (1978). The Bushveld Complex and its remarkable rocks. *American Scient.* 66:551–9.

Hussong, D. M., Wipperman, L. K., and Kroenke, L. W. (1979). The crustal structure of the Ontong Java and Manihiki oceanic plateaus. *J. Geophys. Res.* 84:6003–10.

Ingram, B. L., Coccioni, R., Montana, A., and Richter, F. M. (1994). Sr isotopic composition of mid-Cretaceous seawater. *Science* 264:546–9.

Irving, E., North, F. K., and Couillard, R. (1974). Oil, climate and tectonics. *Can. J. Earth Sci.* 11: 1–17.

Irvine, T. N. (1970). Crystallization sequences in the Muskox intrusion and other layered intrusion. Geological Society of South Africa, Spec. Publ. 1:441–76.

Ishii, M. and Tromp, J. (1999). Normal-mode and free-air gravity constraints on lateral variations in velocity and density of Earth's mantle. *Science* 285:1231–6.

Isley, A. E. and Abbott, D. H. (1999). Plume-related mafic volcanism and the deposition of banded iron formation. *J. Geophys. Res.* 104(15):461–77.

Isozaki, Y., Maruyama, S., and Furuoka, F. (1990). Accreted oceanic materials in Japan. *Tectonophys.* 181:179–205.

Ita, J. and Stixrude, L. (1992). Petrology, elasticity, and composition of the mantle transition zone. *J. Geophys. Res.* 97:6849–66.

Ito, G., Lin, J., and Gable, C. W. (1997). Interaction of mantle plumes and migrating mid-ocean ridges: Implications for the Galapagos plume–ridge system. *J. Geophys. Res.* 102(15):403–17.

Jackson, E. D., Silver, E. A., and Dalrymple, G. B. (1972). Hawaiian–Emperor chain and its relation to Cenozoic Circum–Pacific tectonics. *Geol. Soc. Am. Bull.* 83:601–18.

James, D. E. et al. (1999). Tomographic and depth phasing imaging of mantle structure beneath the southern African seismic array. *EOS* 80:F12–13.

Janes, D. M. and Squyres, S. W. (1993). Radially fractured domes: A comparison of Venus and the Earth. *Geophys. Res. Lett.* 20:2961–4.

Janes, D. M. et al. (1992). Geophysical models for the formation and evolution of coronae on Venus. *J. Geophys. Res.* 97(16):055–67.

Janney, P. E., Macdougall, J. D., Hatland, J. H., and Lynch, M. A. (2000). Geochemical evidence from the Pukapuka volcanic ridge system for a shallow enriched mantle domain beneath the South Pacific superswell. *Earth Planet. Sci. Lett.* 181:47–60.

Jarrard, R. D. and Clague, D. A. (1977). Implications of Pacific Island and seamount ages for the origin of volcanic chains. *Rev. Geophys. Space Phys.* 15:57–76.

Jeanloz, R. and Morris, S. (1986). Temperature distribution in the crust and in the mantle. *Annu. Rev. Earth Planet. Sci.* 14:377–415.

Jenkyns, H. C. (1980). Cretaceous anoxic events: From continents to oceans. *J. Geol. Soc. Lond.* 137:171–88.

Jochum, K. P., Arndt, N. T., and Hofmann, A. W. (1991). Nb–Th–La in komatiites and basalts: Constraints on komatiite petrogenesis and mantle evolution. *Earth Planet. Sci. Lett.* 107:272–89.

Johnson, H. P., Van Patten, D., Tivey, M., and Sager, W. W. (1995). Geomagnetic polarity reversal rate for the Phanerozoic. *Geophys. Res. Lett.* 22:231–4.

Johnston, S. T. and Thorkelson, D. J. (2000). Continental flood basalts: Episodic magmatism above long-lived hotspots. *Earth Planet. Sci. Lett.* 175:247–56.

Jones, D., Silberling, N. J., and Hillhouse, J. W. (1977). Wrangellia: A displaced terrane in NW North America. *Can. J. Earth Sci.* 14:2565–77.

Jordan, T. H. (1979). Mineralogies, densities, and seismic velocities of garnet lherzolites and their geophysical implications. In *The mantle sample: Inclusions in kimberlites and other volcanic rocks*, eds. F. R. Boyd and H. O. A. Meyer, 1–14, Washington, DC: American Geophysical Union.

Juteau, T. and Maury, R. (1999). *The oceanic crust, from accretion to mantle recycling*. Chichester, UK: Praxis Publishing, p. 390.

Jyotiranjan, S. R., Ramesh, R., and Pande, K. (1999). Carbon isotopes in Kerguelen plume-derived carbonatites: Evidence for recycled inorganic carbon. *Earth Planet. Sci. Lett.* 170:205–14.

Kamber, B. S. and Collerson, K. D. (1999). Origin of ocean island basalts: A new model based on lead and helium isotope systematics. *J. Geophys. Res.* 104(15):479–91.

Karhu, J. A. and Holland, H. D. (1996). Carbon isotopes and the rise of atmospheric oxygen. *Geol.* 24:867–70.

Kato, T., Ringwood, A. E., and Irifune, T. (1988). Experimental determination of element partitioning between silicate perovskites, garnets and liquids: Constraints on early differentiation of the mantle. *Earth Planet. Sci. Lett.* 89:123–45.

Kaufman, A. J. (1997). An ice age in the tropics. *Nature* 386:227–8.

Kay, S. M. and Kay, R. W. (1985). Role of crystal cumulates and the oceanic crust in the formation of the lower crust of the Aleutian arc. *Geol.* 13:461–4.

Keller, R. A., Fisk, M. R., Duncan, R. A., and White, W. M. (1997). 16 My of hotspot and nonhotspot volcanism on the Patton–Murray seamount platform, Gulf of Alaska. *Geol.* 25:511–14.

Kellogg, L. H. (1992). Mixing in the mantle. *Annu. Rev. Earth Planet. Sci.* 20:365–88.

Kellogg, L. H., Hager, B. H., and van der Hilst, R. D. (1999). Compositional stratification in the deep mantle. *Science* 283:1881–4.

Kellogg, L. H. and King, S. D. (1993). Effect of mantle plumes on the growth of D″ by reaction between the core and mantle. *Geophys. Res. Lett.* 20:379–82.

———. (1997). The effect of temperature-dependent viscosity on the structure of new plumes in the mantle: Results of a finite element model in a spherical, axisymmetric shell. *Earth Planet. Sci. Lett.* 148:13–26.

Kempton, P. D., Harmon, R. S., Hawkesworth, C. J., and Moorbath, S. (1990). Petrology and geochemistry of lower crustal granulites from the Geronimo Volcanic Field, SE Arizona. *Geochim. Cosmochim. Acta* 54:3401–26.

Kempton, P. D. et al. (2000). The Iceland plume in space and time: A Sr–Nd–Pb–Hf study of the North Atlantic rifted margin. *Earth Planet. Sci. Lett.* 177:255–71.

Kendall, J. M. and Silver, P. G. (1996). Constraints from seismic anisotropy on the nature of the lowermost mantle. *Nature* 381:409–12.

Kennett, B. L. N. and Widiyantoro, S. (1999). A low seismic wave-speed anomaly beneath NW India: A seismic signature of the Deccan plume? *Earth Planet. Sci. Lett.* 165:145–55.

Kent, R. W. (1991). Lithospheric uplift in eastern Gondwana. *Geol.* 19:19–23.

Keppler, H. (1996). Constraints from partitioning experiments on the composition of subduction-zone fluids. *Nature* 380:237–40.

Kerr, A. C. (1994). Lithospheric thinning during the evolution of continental large igneous provinces. *Geol.* 22:1027–30.

———. (1998). Oceanic plateau formation: A cause of mass extinction and black shale deposition around the Cenomanian–Turonian boundary. *J. Geol. Soc. London* 155:619–26.

Kerr, A. C. et al. (1996a). The geochemistry and tectonic setting of late Cretaceous Caribbean and Colombian volcanism. *J. S. Am. Earth Sci.* 9:111–20.

Kerr, A. C., Iturralde-Vicent, M. A., Saunders, A. D., Babbs, T. L., and Tarney, J. (1999). New plate tectonic model of the Caribbean: Implications from a geochemical reconnaissance of Cuban Mesozoic volcanic rocks. *Geol. Soc. Am. Bull.* 111:1581–99.

Kerr, A. C., Marriner, G. F., Arndt, N. T., Tarney, J., Nivia, A., Saunders, A. D., and Duncan, R. A. (1996b). The petrogenesis of Gorgona komatiites, picrites and basalts: New field, petrographic and geochemical constraints. *Lithos* 37:245–60.

Kerr, A. C., Tarney, J., Nivia, A., Marriner, G. F., and Saunders, A. D. (1998). The internal structure of oceanic plateaus: Inferences from obducted Cretaceous terranes in western Colombia and the Caribbean. *Tectonophys.* 292:173–88.

Kerr, A. C., White, R. V., and Saunders, A. D. (2000). LIP reading: Recognizing oceanaic plateaus in the geological record. *J. Petrol.* 41:1041–56.

Kerrich, R., Polat, A., Wyman, D., and Hollings, P. (1999). Trace element systematics of Mg- to Fe-tholeiitic basalt suites of the Superior Province: Implications for Archean mantle reservoirs and greenstone belt genesis. *Lithos* 46:163–87.

Kerrich, R., Wyman, D., Fan, J., and Bleeker, W. (1998). Boninite series: Low Ti-tholeiite associations from the 2.7-Ga Abitibi greenstone belt. *Earth Planet. Sci. Lett.* 164:303–16.

Khain, V. E. (1985). *Geology of the USSR*. Berlin: Gebruder Borntraeger, p. 272.

Kieffer, H. H., Jakosky, B. M., and Snyder, C. W. (1992). The planet Mars: From antiquity to the present. In *Mars*, H. H. Kieffer, B. M. Jakosky, and C. W. Snyder, eds., 1–33. Tucson: University of Arizona Press.

Kimura, G., Ludden, J. N., Desrochers, J. P., and Hori, P. (1993). A model of ocean-crust accretion for the Superior province, Canada. *Lithos* 30:337–55.

Kimura, G., Sakakibara, M., and Okamura, M. (1994). Plume in central Panthalassa? Deductions from accreted oceanic fragments in Japan. *Tectonics* 13:905–16.

Kincaid, C., Schilling, J. G., and Gable, C. (1996). The dynamics of off-axis plume–ridge interaction in the uppermost mantle. *Earth Planet. Sci. Lett.* 137:29–43.

King, S. D. and Anderson, D. L. (1995). An alternative mechanism of flood basalt formation. *Earth Planet. Sci. Lett.* 136:269–79.

Klein, C. and Beukes, N. J. (1992). Time distribution, stratigraphy, and sedimentologic setting and geochemistry of Precambrian iron formations. In *The proterozoic biosphere: A multidisciplinary study*, 139–46. New York: Cambridge University Press.

Klein, F. W. (1982). Patterns of historical eruptions at Hawaiian volcanoes. *J. Volcan. Geotherm. Res.* 12:1–35.

Klein, F. W. et al. (1987). The seismicity of Kilauea's magma system. U.S. Geological Survey, Prof. Pap. 1350, Vol. 2, pp. 1019–86.

Klein, F. W. and Koyanagi, R. Y. (1989). The seismicity and tectonics of Hawaii. In *Geology of North America*, Vol. N, eds. E. L. Winterer, et al., 238–52. Boulder, CO: Geological Society of America.

Knittle, E. and Jeanloz, R. (1991). Earth's core–mantle boundary: Results of experiments at high pressures and temperatures. *Science* 251:1438–43.

Knoll, A. H. and Canfield, D. E. (1998). Isotopic inferences on early ecosystems. *Paleontol. Soc. Papers* 4:212–43.

Koch, D. M. and Manga, M. (1996). Neutrally buoyant diapirs: A model for Venus coronae. *Geophys. Res. Lett.* 23:225–8.

Kogiso, T., Tatsumi, Y., Shimoda, G., and Barsczus, H. G. (1997). High mu ocean island basalts in southern Polynesia: New evidence for whole mantle scale recycling of subducted oceanic crust. *J. Geophys. Res.* 102:8085–103.

Krapez, B. (1993). Sequence stratigraphy of the Archean supracrustal belts of the Pilbara block, Western Australia. *Precamb. Res.* 60:1–45.

Kumagai, I. and Kurita, K. (2000). On the fate of mantle plumes at density interfaces. *Earth Planet. Sci. Lett.* 179:63–71.

Kump, L. R. et al. (1999). A weathering hypothesis for glaciation at high atmospheric pCO_2 during the Late Ordovician. *Paleogeog. Paleoclimat. Paleoecol.* 152:173–87.

Kurz, M. D. and Geist, D. (1999). Dynamics of the Galapagos hotspot from helium isotope geochemistry. *Geochim. Cosmochim. Acta* 63:4139–56.

Kusky, T. M. (1990). Evidence for Archean ocean opening and closing in southern Slave Province. *Tectonics* 9:1533–66.

———. (1991). Structural development of an Archean orogen, Western Point Lake, NWT. *Tectonics* 10:820–41.

Kusky, T. M. and Kidd, W. S. F. (1992). Remnants of an Archean oceanic plateau, Belingwe greenstone belt, Zimbabwe. *Geol.* 20:43–6.

Kvenvolden, K. A. (1999). Potential effects of gas hydrate on human welfare. *Proc. Natl. Acad. Sci. U.S.A.* 96:3420–6.

Lapierre, H. et al. (1997). Is the lower Duarte complex a remnant of the Caribbean plume-generated oceanic plateau? *J. Geol.* 105:111–20.

Larsen, H. C. and Jakobsdottir, S. (1988). Distribution, crustal properties and significance of seaward-dipping sub-basement reflectors off E. Greenland. Geological Society of London, Spec. Publ. 39: 95–114.

Larsen, T. B. and Yuen, D. A. (1997). Fast plumeheads: Temperature-dependent versus non-Newtonian rheology. *Geophys. Res. Lett.* 24:1995–8.

Larson, R. L. (1991a). Latest pulse of Earth: Evidence for a mid-Cretaceous superplume. *Geol.* 19:547–50.

———. (1991b). Geological consequences of superplumes. *Geol.* 19:963–6.

Larson, R. L. and Kincaid, C. (1996). Onset of mid-Cretaceous volcanism by elevation of the 670 km thermal boundary layer. *Geol.* 24:551–4.

Larson, R. L. and Olson, P. (1991). Mantle plumes control magnetic reversal frequency. *Earth Planet. Sci. Lett.* 107:437–47.

Lassiter, J. C., DePaolo, D. J., and Mahoney, J. J. (1995). Geochemistry of the Wrangellia flood basalt province: Implications for the role of continental and oceanic lithosphere in flood basalt genesis. *J. Petrol.* 36:983–1009.

Lassiter, J. C. and DePaolo, D. J. (1997). Plume/lithosphere interaction in the generation of continental and oceanic flood basalts: Chemical and isotopic constraints. American Geophysical Union, Mon. 100:335–55.

Lawver, L. A. and Muller, R. D. (1994). Iceland hotspot tracks. *Geology* 22:311–14.

Lay, T., Williams, Q., and Garnero, E. J. (1998). The core–mantle boundary layer and deep Earth dynamics. *Nature* 392:461–8.

Le Cheminant, A. N. and Heaman, L. M. (1991). U/Pb ages for the 1.27-Ga Mackenzie igneous events, Canada; support for a plume initiation model. Geological Association of Canada, Waterloo, Ontario, Programs and Abstracts, 16:A73.

Le Dez, A. (1996). Variations petrologiques et geochimiques associees a l'edification des volcans-boucliers de Polynesie Francais: exemples de Nudu Hiva et Hiva Oa et de Moorea. Ph.D. Thesis, University de Bretagne Occidentale, Bres, 407 pp.

Leitch, A. M. and Davies, G. F. (2001). Mantle plumes and flood basalts: Enhanced melting from plume ascent and an eclogite component. *J. Geophys. Res.* 106:2047–60.

Leitch, A. M., Davies, G. F., and Wells, M. (1998). A plume head melting under a rifting margin. *Earth Planet. Sci. Lett.* 161:161–77.

Lenardic, A. and Kaula, W. M. (1996). Near-surface thermal/chemical boundary layer convection at infinite Prandtl number: Two-dimensional numerical experiments. *Geophys. J. Int.* 126:689–711.

Lesher, C. M. and Arndt, N. T. (1995). REE and Nd isotope geochemistry, petrogenesis and volcanic evolution of contaminated komatiites at Kambalda, Western Australia. *Lithos* 34:127–57.

Lewis, J. F. and Jimenez, J. G. (1991). Duarte complex in the La Vega–Jarabacoa–Janico area, central Hispaniola: Geologic and geochemical features of the sea floor during the early stages of arc evolution. Geological Society of America, Spec. Pap. 262:115–41.

Li, J. P., O'Neil, H. S. C., and Seifert, F. (1995). Subsolidus phase relations in the system MgO–SiO_2–Cr–O in equilibrium with metallic Cr, and their significance for petrochemistry of Cr. *J. Petrol.* 36:107–32.

Li, X., Kind, R., Priestley, K., Sobolev, S. V., Tilmann, F., Yuan, X., and Weber, M. (2000). Mapping the Hawaiian plume conduit with converted seismic waves. *Nature* 405:938–41.

Lightfoot, P. C. (1985). Isotope and trace element geochemistry of the South Deccan lavas, India. Ph.D. thesis, Open University, Milton Keyes, U.K.

Lightfoot, P. C. et al. (1990). Geochemistry of the Siberian trap of the Noril'sk area, USSR, with implications for the relative contributions of crust and mantle to flood basalt magmatism. *Contrib. Mineral. Petrol.* 104:631–44.

Lithgow-Bertelloni, C. and Silver, P. G. (1998). Dynamic topography, plate driving forces, and the African superswell. *Nature* 395:269–72.

Loper, D. E. (1992). On the correlation between mantle plume flux and the frequency of reversals of the geomagnetic field. *Geophys. Res. Lett.* 19:25–8.

Loper, D. E. and Lay, T. (1995). The core–mantle boundary region. *J. Geophys. Res.* 100:6397–420.

Lowe, D. R. (1999). Geologic evolution of the Barberton greenstone belt and vicinity. Geological Society of America, Spec. Pap. 329:287–312.

Lowe, D. R. (1999). Shallow-water sedimentation of accretionary lapilli-bearing strata of the Msauli chert: evidence of explosive hydromagmatic komatiitic volcanism. *Geol. Soc. America*, Spec. Pap. 329:213–32.

Lowe, D. R., Byerly, G. R., Ransom, B. L., and Nocita, B. W. (1985). Stratigraphic and sedimentological evidence bearing on structural repetition in Early Archean rocks of the Barberton greenstone belt, South Africa. *Precamb. Res.* 27:165–86.

Lowman, J. P. and Gable, C. W. (1999). Thermal evolution of the mantle following continental aggregation in 3D convection models. *Geophys. Res. Lett.* 26:2649–52.

Lowman, J. P. and Jarvis, G. T. (1996). Continental collisions in wide aspect ratio and high Rayleigh number two-dimensional mantle convection models. *J. Geophys. Res.* 101(25):485–97.

Lowman, J. P. and Jarvis, G. T. (1999). Effects on mantle heat source distribution on supercontinent stability. *J. Geophys. Res.* 104:12,733–46.

Lonsdale, P. (1988). Geography and history of the Louisville hotspot chain in the SW Pacific. *J. Geophys. Res.* 93:3078–104.

Loock, G., Stosch, H. G., and Seck, H. A. (1990). Granulite facies lower crustal xenoliths from the Eifel, West Germany: Petrological and geochemical aspects. *Contrib. Mineral. Petrol.* 105:25–41.

Loper, D. E. and Lay, T. (1995). The core–mantle boundary region. *J. Geophys. Res.* 100:6397–420.

Loper, D. E. and Stacey, F. D. (1983). The dynamical and thermal structure of deep mantle plumes. *Phys. Earth Planet. Inter.* 33:304–17.

MacDonald, G. A. and Katsura, T. (1964). Chemical composition of Hawaiian lavas. *J. Petrol.* 5:82–133.

MacKay, M. E., Moore, G. F., Chochrane, G. R., Moore, J. C., and Kulm, L. D. (1992). Landward vergence and oblique structural trends in the Oregon margin accretionary prism. Implications for fluid flow. *Earth Planet. Sci. Lett.* 109:477–591.

Magee, K. P. and Head, J. W. (2001). Characteristics, distribution and modes of origin of large flow fields on Venus: Implications for plumes, rift associations and resurfacing. Geological Society of America, Mem. (in press).

Mahoney, J. J. (1987). An isotopic survey of Pacific oceanic plateaus: Implications for their nature and origin. American Geophysical Union, Mon. 43:207–20.

———. (1988). Deccan traps. In *Continental flood basalts*, ed. J. D. Macdougall, 151–94. Dordrecht: Kluwer Academic Publishers.

Mahoney, J. J., Frei, R., Tejada, M. L. G., Mo, X. X., Leat, P. T., and Nagler, T. F. (1998). Tracing the Indian Ocean mantle domain through time: Isotopic results from old West Indian, East Tethyan, and South Pacific seafloor, *J. Petrol.* 39:1285–306.

Mahoney, J. J., Jones, W. B., Frey, F. A., Salters, V. J., Pyle, D. G., and Davies, H. L. (1995). Geochemical characteristics of lavas from Broken ridge, the Naturaliste plateau and southernmost Kerguelen plateau: Cretaceous plateau volcanism in the SE Indian Ocean. *Chem. Geol.* 120:315–45.

Mahoney, J. J., Le Roex, A. P., Peng, Z., Fisher, R. L., and Natland J. H. (1992). Southwestern limits of Indian Ocean ridge mantle and the origin of flow – 206Pb/204Pb mid-ocean ridge basalt: Isotope systematics of the central-Southwest Indian Ridge (17°E–50°E). *J. Geophys. Res.* 97(19):771–90.

Mahoney, J. J., Storey, M., Duncan, R. A., Spencer, K. J., and Pringle, M. (1993). Geochemistry and geochronology of Leg 130 basement lavas: Nature and origin of the Ontong–Java plateau. *Proc. Ocean Drill. Project, Scient. Results* 130:3–22.

Malamud, B. D. and Turcotte, D. L. (1999). How many plumes are there? *Earth Planet. Sci. Lett.* 174:113–24.

Marko, G. M. (1980). Velocity and attenuation in partially molten rocks. *J. Geophys. Res.* 85:5173–89.

Marquart, G. and Schmeling, H. (2000). Interaction of small mantle plumes with the spinel–perovskite phase boundary: Implications for chemical mixing. *Earth Planet. Sci. Lett.* 177:241–54.

Marquart, G., Schmeling, H., Ito, G., and Schott, B. (2000). Conditions for plumes to penetrate the mantle phase boundaries. *J. Geophys. Res.* 105:5679–93.

Martin, B. S. (1989). The Roza member, Columbia River Basalt Group: Chemical stratigraphy and flow distribution. Geological Society of America, Spec. Pap. 239:85–104.

Martin, H. (1994). Archean grey gneisses and the genesis of the continental crust, In *Archean crustal evolution*, Chap. 6, ed. K. C. Condie, 205–60. Amsterdam: Elsevier Scientific Publishers.

Maruyama, S. (1994). Plume tectonics. *J. Geol. Soc. Japan* 100:24–49.

———. (1997). Pacific-type orogeny revisited: Miyashiro-type orogeny proposed. *The Island Arc* 6:91–120.

Marzoli, A. et al. (1999). Extensive 200-Ma continental flood basalts of the Central Atlantic magmatic province. *Science* 284:616–18.

Masters, T. G., Johnson, S., Laske, G., and Bolton, H. (1996). A shear-velocity model of the mantle. *Roy. Soc. London, Philos. Trans.* 354:1385–411.

Mattie, P. D., Condie, K. C., Selverstone, J., and Kyle, P. R. (1997). Composition of the lower continental crust in the Colorado Plateau: Geochemical evidence from mafic xenoliths from the Navajo Volcanic Field, SW. United States. *Geochim. Cosmochim. Acta* 61:2007–21.

Mattley, D., Lowry, D., and Macpherson, C. (1994). Oxygen isotope composition of mantle peridotite. *Earth Planet. Sci. Lett.* 128:231–41.

Mauffret, A. and Leroy, S. (1997). Seismic stratigraphy and structure of the Caribbean igneous province. *Tectonophys.* 283:61–104.

McCartney, K. M., Huffman, A. R., and Tredoux, M. (1990). A paradigm for endogenous causation of mass extinctions. Geological Society of America, Spec. Pap. 247:125–38.

McCuaig, T. C., Kerrich, R., and Xie, Q. (1994). Phosphorus and high field strength element anomalies in Archean high-Mg magmas as possible indicators of source mineralogy and depth. *Earth Planet. Sci. Lett.* 124:221–39.

McCulloch, M. T. and Bennett, V. C. (1994). Progressive growth of the Earth's continental crust and depleted mantle: Geochemical constraints. *Geochim. Cosmochim. Acta* 58:4717–38.

McCulloch, M. T. and Gamble, J. A. (1991). Geochemical and geodynamical constraints on subduction zone magmatism. *Earth Planet. Sci. Lett.* 102:358–74.

McDonough, W. F. (1990). Constraints on the composition of the continental lithospheric mantle. *Earth Planet. Sci. Lett.* 101:1–18.

McDougall, I. and Duncan, R. A. (1988). Age progressive volcanism in the Tasmantid seamounts. *Earth Planet. Sci. Lett.* 89:207–20.

McKenzie, D. P. (1984). The generation and compaction of partially molten rock. *J. Petrol.* 25:713–65.

McKenzie, D. P. and Bickle, M. J. (1988). The volume and composition of melt generated by extension of the lithosphere. *J. Petrol.* 29:625–79.

McKenzie, D. P. and O'Nions, R. K. (1991). Partial melt distributions from inversion of REE concentrations. *J. Petrol.* 32:1021–91.

McNutt, M. K. (1998). Superswells. *Rev. Geophys.* 36:211–44.

McNutt, M. K., Caress, D. W., Reynolds, J., Jorhahl, K. A., and Duncan, R. A. (1997). Failure of plume theory to explain midplate volcanism in the southern Austral Islands. *Nature* 389:479–82.

Mege, D. and Ernst, R. E. (2001). Contractional effects of plumes on Earth, Mars, and Venus. Geological Society of America, Spec. Pap. 352 (in press).

Mege, D. and Masson, P. (1996). A plume tectonics model for the Tharsis province, Mars. *Planet. Space Sci.* 44:1499–546.

Melezhik, V. A., Fallick, A. E., Medvedev, P. V., and Makarikhin, V. V. (1999). Extreme $^{13}C_{carb}$ enrichment in ca. 2.0-Ga magnesite–stromatolite–dolomite-'red beds' association in a global context: A case for the world-wide signal enhanced by a local environment. *Earth Sci. Rev.* 48:71–120.

Melluso, L. et al. (1995). Constraints on the mantle sources of the Deccan traps from the petrology and geochemistry of the basalts of Gujarat state, W. India. *J. Petrol.* 36:1393–1432.

Menzies, M. A. (1990). Archean, Proterozoic, and Phanerozoic lithosphere. In *Continental Mantle*, ed., M. A. Menzies, 67–85. Oxford: Clarendon Press.

Miller, D. S. and Smith, R. B. (1999). P and S velocity structure of the Yellowstone volcanic field from local earthquake and controlled source tomography. *J. Geophys. Res.* 104:15,105–21.

Milner, S. C., Le Roex, A. P. and O'Conner, J. M. (1995). Age of Mesozoic igneous rocks in NW Namibia, and their relationship to continental break-up. *J. Geol. Soc. Lond.* 152:97–104.

Mirota, M. D. and Veizer, J. (1994). Geochemistry of Precambrian carbonates: VI. Aphebian Albanel Formations, Quebec, Canada. *Geochim. Cosmochim. Acta* 58:1735–45.

Mitchell, R. H. (1986). *Kimberlites: Mineralogy, geochemistry and petrology*. New York: Plenum, 447 pp.

Molnar, P. and Stock, J. (1987). Relative motions of hotspots in the Pacific, Atlantic, and Indian Oceans since late Cretaceous time. *Nature* 327:587–91.

Monnereau, M., Rabinowicz, M., and Arquis, E. (1993). Mechanical erosion and reheating of the lithosphere: A numerical model for hotspot swells. *J. Geophys. Res.* 98:809–23.

Montague, N. L. and Kellogg, L. H. (2000). Numerical models of a dense layer at the base of the mantle and implications for the geodynamics of D″. *J. Geophys. Res.* 105:11,101–14.

Moore, W. B., Schubert, G., and Tackley, P. J. (1999). The role of rheology in lithospheric thinning by mantle plumes. *Geophys. Res. Lett.* 26:1073–6.

Morgan, G. E. (1985). The paleomagnetism and cooling history of metamorphic and igneous rocks from the Limpopo mobile belt, southern Africa. *Geol. Soc. America Bull.* 96:663–75.

Morgan, J. P. (1999). Isotope topology of individual hotspot basalt arrays: Mixing curves or melt extraction trajectories? *Geochem. Geophys. Geosyst.* Vol. 1, Paper no. 1999GC000004.

Morgan, J. P., Morgan, W. J., and Price, E. (1995). Hotspot melting generates both hotspot volcanism and a hotspot swell? *J. Geophys. Res.* 100:8045–62.

Morgan, W. J. (1971). Convection plumes in the lower mantle. *Nature* 230:42–3.

———. (1972). Deep mantle convection plumes and plate motions. *Amer. Assoc. Petrol. Geol. Bull.* 56:203–13.

———. (1981). Hotspot tracks and the opening of the Atlantic and Indian Oceans. In *The sea*, ed. C. Emiliani, 443–87, New York: John Wiley.

Mortimer, N. and Parkinson, D. (1996). Hikurangi Plateau: A Cretaceous large igneous province in the SW. Pacific Ocean. *J. Geophys. Res.* 101:687–96.

Muller, R. D., Royer, J., and Lawver, L. A. (1993). Revised plate motions relative to the hotspots from combined Atlantic and Indian Ocean hotspot tracks. *Geol.* 21:275–8.

Mysen, B. O. and Boettcher, A. L. (1975). Melting of a hydrous mantle: I. Phase relations of natural peridotite at high pressures and temperatures with controlled activities of water, carbon dioxide, and hydrogen. *J. Petrol.* 16:520–48.

Naldrett, A. J. (1992). A model for the Ni–Cu–PGE ores of the Noril'sk region and its application to other areas of flood basalt. *Econ. Geol.* 87:1945–64.

Neal, C. R. et al. (1997). The Ontong Java plateau. American Geophysical Union, Mon. 100:183–216.

Needham, R. S., Stuart-Smith, P. G., and Page, R. W. (1988). Tectonic evolution of the Pine Creek inlier, Northern Territory. *Precamb. Res.* 40/41:543–64.

Nelson, D. R., Trendall, A. F., de Laeter, J. R., Grobler, N. J. and Fletcher, I. R. (1992). A comparative study of the geochemical and isotopic systematics of Late Archean flood basalts from the Pilbara and Kaapvaal cratons. *Precamb. Res.* 54:231–56.

Nesbitt, H. W. and Young, G. M. (1982). Early Proterozoic climates and plate motions inferred from major element chemistry of lutites. *Nature* 299:715–17.

Nicolaysen, K., Frey, F. A., Hodges, K. V., Weis, D., and Giret, A. (2000). $^{40}Ar/^{39}Ar$ geochronology of flood basalts from the Kerguelen Archipelago, southern Indian Ocean: Implications for Cenozoic eruption rates of the Kerguelen plume. *Earth Planet. Sci. Lett.* 174:313–28.

Nikishin, A. M. et al. (1996). Late Precambrian to Triassic history of the East European craton: Dynamics of sedimentary basin evolution. *Tectonophys.* 268:23–63.

Nisbet, E. G. and Walker, D. (1982). Komatiites and the structure of the Archean mantle. *Earth Planet. Sci. Lett.* 60:105–13.

Niu, F. and Kawakatsu, H. (1997). Depth variation of the mid-mantle seismic discontinuity. *Geophys. Res. Lett.* 24:429–32.

Nur, A. and Ben Avraham, Z. (1982). Oceanic plateaus, the fragmentation of continents, and mountain building. *J. Geophys. Res.* 87:3644–61.

Nutman, A. P. and Ehlers, K. (1998). Evidence for multiple Paleoproterozoic thermal events and magmatism adjacent to the Broken Hill Pb–Zn–Ag orebody, Australia. *Precamb. Res.* 90:203–38.

Nyblade, A. A. et al. (2000). Seismic evidence for a deep upper mantle thermal anomaly beneath east Africa. *Geology* 28:599–602.

O'Connor, J. M. and Duncan, R. A. (1990). Evolution of the Walvis Ridge–Rio Grande rise hotspot system: Implications for African and South American plate motions over plumes. *J. Geophys. Res.* 95(17):475–502.

O'Connor, J. M., Stoffers, P., and McWilliams, M. O. (1995). Time–space mapping of Easter chain volcanism. *Earth Planet. Sci. Lett.* 136:197–212.

O'Connor, J. M., Stoffers, P., van den Bogaard, P., and McWilliams, M. (1999). First seamount age evidence for significantly slower African plate motion since 19 to 30 Ma. *Earth Planet. Sci. Lett.* 171:575–89.

Officer, C. B., Hallam, A., Drake, C. L., and Devine, J. D. (1987). Late Cretaceous and paroxysmal Cretaceous/Tertiary extinctions. *Nature* 326:143–9.

Ohtani, E., Yurimoto, H., and Seto, S. (1997). Element partitioning between metallic liquid, silicate liquid, and lower-mantle minerals: Implications for core formation of the Earth. *Phys. Earth Planet. Inter.* 100:97–114.

Oliver, D. H. and Ghent, R. R. (2000). Superplumes and rotation-induced flow along the CMB. *Geol. Soc. America, Abstracts with Programs* 32(7):A–314.

Olson, P., Schubert, G., and Anderson, C. (1987). Plume formation in the D″ layer and the roughness of the core–mantle boundary. *Nature* 327:409–13.

Operto, S. and Charvis, P. (1996). Deep structure of the southern Kerguelen plateau from ocean bottom seismometer wide-angle seismic data. *J. Geophys. Res.* 101(25):77–103.

Oppliger, G. L., Murphy, J. B., and Brimhall, Jr., G. H. (1997). Is the ancestral Yellowstone hotspot responsible for the Tertiary Carlin mineralization in the Great Basin of Nevada? *Geol.* 25:627–30.

Palfy, J. and Smith, P. L. (2000). Synchrony between Early Jurassic extinction, oceanic anoxic event, and the Karoo–Ferrar flood basalt volcanism. *Geol.* 28:747–50.

Pallister, J. S., Budahn, J. R., and Murchey, B. L. (1989). Pillow basalts of the Angayucham terrane: Oceanic plateau and island crust accreted to the Brooks Range. *J. Geophys. Res.* 94(15):901–23.

Palmason, G. (1986). Model of crustal formation in Iceland, and application to submarine mid-ocean ridges. In *The western North Atlantic region, Geol. North America*, eds. P. R. Vogt and B. E. Tucholke. Boulder, CO: Geological Society of America.

Pankhurst, R. J. and Rapela, C. R. (1995). Production of Jurassic rhyolite by anatexis of the lower crust of Patagonia. *Earth Planet. Sci. Lett.* 134:23–6.

Panuska, B. C. (1990). An overlooked, world class Triassic flood basalt event. *Geol. Soc. America, Abstracts with Programs* 22:168.

Parfitt, E. A. and Head, J. W. (1993). Buffered and unbuffered dyke emplacement of Earth and Venus: Implications for magma reservoir size, depth, and rate of magma replenishment. *Earth, Moon, Planets* 61:249–81.

Patchett, P. J. and Gehrels, G. E. (1998). Continental influence on Canadian Cordilleran terranes from Nd isotopic study, and significance for crustal growth processes. *J. Geol.* 106:269–80.

Pavoni, N. (1997). Geotectonic bipolarity – evidence of bicellular convection on the Earth's mantle. *S. Afr. J. Geol.* 100:291–9.

Peate, D. W. (1997). The Paraná–Etendeka province. American Geophysical Union, Mon. 100:217–45.

Peate, D. W., Hawkesworth, C. J., Mantovani, M. M. S., Rogers, N. W., and Turner, S.P. (1999). Petrogenesis and stratigraphy of the high-Ti/Y Urubici magma type in the Paraná flood basalt province and implications for the nature of "Dupal-type" mantle in the South Atlantic region. *J. Petrol.* 40:451–73.

Pelgram, W. J. and Allegre, C. J. (1992). Osmium isotopic compositions from oceanic basalts. *Earth Planet. Sci. Lett.* 111:59–68.

Peltier, W. R., Butler, S., and Solheim, L. P. (1997). The influence of phase transformations on mantle mixing and plate tectonics. In *Earth's deep interior*, ed. D. J. Crossley, 405–30. Amsterdam: Gordon & Breach.

Peng, Z. X., Mahoney, J., Hooper, P., Harris, C., and Beane, J. (1994). A role for lower continental crust in flood basalt genesis? Isotopic and incompatible element study of the lower six formations of the western Deccan traps. *Geochim. Cosmochim. Acta* 58:267–88.

Percival, J. A. and Card, K. D. (1983). The Archean crust as revealed in the Kapuskasing uplift, Superior province, Canada. *Geol.* 11:323–6.

Percival, J. A., Fountain, D. M., and Salisbury, M. H. (1992). Exposed crustal cross sections as windows of the lower crust. In *Continental lower crust*, eds. D. M. Fountain, R. Arculus, and R. W. Kay, 317–62. Amsterdam: Elsevier Scientific Publishers.

Petterson, M. G. (1995). The geology of north and central Malaita, Solomon Islands. Water & Mineral. Res. Div., Ministry Energy, Honiara, Solomon Islands, Geo. Mem. 1/95.

Petterson, M. G. et al. (1997). Structure and deformation of north and central Malaita, Solomon Islands: Tectonic implications for the Ontong Java Plateau–Solomon arc collision, and for the fate of oceanic plateaus. *Tectonophys.* 283:1–33.

Phinney, E. J., Mann, P., Coffin, M. F., and Shipley, T. H. (1999). Sequence stratigraphy, structure, and tectonic history of the SW Ontong Java Plateau adjacent to the north Solomon Trench and Solomon Islands arc. *J. Geophys. Res.* 104(20):449–66.

Pik, R. et al. (1998). The northwestern Ethiopian plateau flood basalts: Classification and spatial distribution of magma types. *J. Volcan. Geotherm. Res.* 81:91–111.

Pik, R., Deniel, C., Coulon, C., and Yirgu, G. (1999). Isotopic and trace element signatures of Ethiopian flood basalt: Evidence for plume–lithosphere interaction. *Geochim. Cosmochim. Acta* 63: 2263–79.

Plank, T. and Langmuir, C. H. (1998). The chemical composition of subducting sediment and its consequences for the crust and mantle. *Chem. Geol.* 145:325–94.

Polat, A., Kerrich, R., and Wyman, D. A. (1998). The late Archean Schreiber–Hemlo and White River–Dayohessarah greenstone belts, Superior province: Collages of oceanic plateaus, oceanic arcs, and subduction–accretion complexes. *Tectonophys.* 289:295–326.

Pollitz, F. F. (1988). Episodic North American and Pacific plate motions. *Tectonics* 7:711–26.

Prevot, M., Mattern, E., Camps, P., and Daignieres, M. (2000). Evidence for a 20° tilting of the Earth's rotation axis 110 Ma. *Earth Planet. Sci. Lett.* 179:517–28.

Puchtel, I. S., Hofmann, A. H., Mezger, K., Jochum, K. P., Shchipansky, A. A., and Samsonov, A. V. (1998). Oceanic plateau model for continental crustal growth in the Archean: A case study from the Kostomuksha greenstone belt, NW. Baltic Shield. *Earth Planet. Sci. Lett.* 155:57–74.

Puchtel, I. S. et al. (1999). Combined mantle plume-island arc model for the formation of the 2.9 Ga Sumozero–Kenozero greenstone belt, SE. Baltic Shield: Isotope and trace element constraints. *Geochim. Cosmochim. Acta* 63:3579–95.

Rainbird, R. H. (1993). The sedimentary record of mantle plume uplift preceding eruption of the Neoproterozoic Natkusiak flood basalt. *J. Geol.* 101:305–18.

Rainbird, R. H. and Ernst, R. E. (2001). The sedimentary rock record of mantle plume uplift. Geological Society of America, Mem. (in press).

Rainbird, R. H., Jefferson, C. W., and Young, G. M. (1996). The early Neoproterozoic sedimentary succession B of NW. Laurentia: Correlations and paleogeographic significance. *Geol. Soc. Am. Bull.* 108:454–70.

Raja Rao, C. S., Sahasrabudhe, Y. S., Deshmukh, S. S., and Raman, R. (1978). Distribution, structure and petrography of the Deccan trap, India. *Rept. Geol. Survey India*, 43 pp.

Rampino, M. R. (1991). Volcanism, climatic change, and the geologic record. Society of Economic Paleontalogists and Mineralogists, Spec. Publ. 45:9–18.

Rampino, M. R. and Caldeira, K. (1993). Major episodes of geologic change: Correlations, time structure and possible causes. *Earth Planet. Sci. Lett.* 114:215–27.

Rampino, M. R. and Haggerty, B. M. (1994). Extraterrestrial impacts and extinction of life. In *Hazards due to comets and asteroids*, ed. T. Gehrels, 827–57. Tucson: University of Arizona Press.

Reidel, S. P. (1983). Stratigraphy and petrogenesis of the Grande Ronde basalt from the deep canyon country of Washington, Oregon, and Idaho. *Geol. Soc. Am. Bull.* 94:519–42.

Renne, P. R. and Basu, A. R. (1991). Rapid eruption of the Siberian traps flood basalts at the Permo–Triassic boundary. *Science* 253:175–8.

Renne, P. R., Zichao, Z., Richards, M. A., Black, M. T., and Basu, A. R. (1995). Synchrony and causal relations between Permian–Triassic boundary crises and Siberian flood volcanism. *Science* 269:1413–16.

Revillon, S., Arndt, N. T., Chauvel, C., and Hallot, E. (2000). Geochemical study of ultramafic volcanic and plutonic rocks from Gorgona Island, Colombia: Plumbing system of an oceanic plateau. *J. Petrol.* 41:1127–54.

Reynaud, C. et al. (1999). Oceanic plateau and island arcs of SW Ecuador: Their place in the geodynamic evolution of NW. South America. *Tectonophys.* 307:235–54.

Ribe, N. M. and Christensen, U. R. (1999). The dynamical origin of Hawaiian volcanism. *Earth Planet. Sci. Lett.* 171:517–31.

Ribe, N. M., Christensen, U. R., and Theissing, J. (1995). The dynamics of plume–Ridge interaction, 1: Ridge-centered plumes. *Earth Planet. Sci. Lett.* 134:155–68.

Ribe, N. M. and de Valpine, D. P. (1994). The global hotspot distribution and instability of D″. *Geophys. Res. Lett.* 21:1507–10.

Richards, M. A., Duncan, R. A., and Courtillot, V. E. (1989). Flood basalts and hotspot tracks: Plume heads and tails. *Science* 246:103–7.

Richards, M. A. and Griffiths, R. W. (1988). Deflection of plumes by mantle shear flow: Experimental results and a simple theory. *Geophys. J. Roy. Astron. Soc.* 94:367–76.

Richards, M. A., Hager, B. H., and Sleep, N. H. (1988). Dynamically supported geoid heights over hotspots: Observation and theory. *J. Geophys. Res.* 93:7690–708.

Richards, M. A., Jones, D. L., Duncan, R. A., and DePaolo, D. J. (1991). A mantle plume initiation model for the Wrangellia flood basalt and other oceanic plateaus. *Science* 254:263–7.

Richardson, W. P., Okal, E. A., and van der Lee, S. (2000). Raleigh wave tomography of the Ontong Java plateau. *Phys. Earth Planet. Inte.* 118:29–51.

Richter, F. M. (1988). A major change in the thermal state of the Earth at the Archean–Proterozoic boundary: Consequences for the nature and preservation of continental lithosphere. *J. Petrol., Spec. Lithosphere Issue* 39–52.

Ringwood, A. E., Kesson, S. E., Hibberson, W., and Ware, N. (1992). Origin of kimberlites and related magmas. *Earth Planet. Sci. Lett.* 113:521–38.

Ritsema, J. and van Heijst, H. (2000). New seismic model of the upper mantle beneath Africa. *Geology* 28:63–6.

Ritsema, J., van Heijst, H., and Woodhouse, J. H. (1999). Complex shear wave velocity structure imaged beneath Africa and Iceland. *Science* 286:1925–8.

Roach, T. A., Roeder, P. L., and Hulbert, L. J. (1998). Composition of chromite in the upper chromitite, Muskox layered intrusion, NW. Territories. *Canad. Mineral.* 36:117–35.

Roberts, D. G., Backman, J., Morton, A. C., Murray, J. W., and Keene, J. B. (1984). Evolution of volcanic rifted margins: Synthesis of Leg 81 results on the west margin of the Rockall plateau. *Initial Repts. Deep Sea Drill. Proj.* 81:883–911.

Rogers, J. J. W. (1996). A history of continents in the past three billion years. *J. Geol.* 104:91–107.

Rogers, N., Macdonald, R., Fitton, J. G., George, R., Smith, M., and Barreiro, B. (2000). Two mantle plumes beneath the East African rift system: Sr, Nd and Pb isotope evidence from Kenya rift basalts. *Earth Planet. Sci. Lett.* 176:387–400.

Rollinson, H. (1993). *Using geochemical data*. New York: Longman Scientific & Technical, 353 pp.

Rosen, O. M., Condie, K. C., Natapov, L. M., and Nozhkin, A. D. (1994). Archean and early proterozoic evolution of the Siberian craton: A preliminary assessment. In *Archean crustal evolution*, Chap. 10, ed. K. C. Condie, 411–59. Amsterdam: Elsevier Scientific Publishers.

Rothschild, L. J. and Mancinellli, R. L. (1990). Model of carbon fixation in microbial mats from 3500 Myr ago to the present. *Nature* 345:710–12.

Rudnick, R. L. (1990). Nd and Sr isotopic compositions of lower crust xenoliths from north Queensland, Australia: Implications for Nd model ages and crustal growth processes. *Chem. Geol.* 83:195–208.

Rudnick, R. L. (1992a). Xenoliths–samples of the lower continental crust, In *Continental lower crust*, eds. D. Fountain, R. Arculus, and R. Kay, 269–316. Amsterdam: Elsevier Scientific Publishers.

———. (1995). Making continental crust. *Nature* 378:571–8.

Rudnick, R. L. and Fountain, D. M. (1995). Nature and composition of the continental crust: A lower crustal perspective. *Rev. Geophys.* 33:267–309.

Rudnick, R. L. and Goldstein, S. L. (1990). The Pb isotopic composition of lower crustal xenoliths and the evolution of lower crustal Pb. *Earth Planet. Sci. Lett.* 98:192–207.

Rudnick, R. L. and Jackson, I. (1995). Measured and calculated elastic wave speeds in partially equilibrated mafic granulite xenoliths: Implications for the properties of an underplated lower continental crust. *J. Geophys. Res.* 100:10211–18.

Russell, S. A., Lay, T., and Garnero, E. J. (1998). Seismic evidence for small-scale dynamics in the lowermost mantle at the root of the Hawaiian hotspot. *Nature* 396:255–8.

Saemundsson, K. (1986). Subaerial volcanism in the western North Atlantic. In *The western North Atlantic region*, eds. P. R. Vogt and B. E. Tucholke, 69–86. Boulder, CO: Geological Society of America.

Samson, S. D. and Patchett, P. J. (1991). The Canadian Cordillera as a modern analogue of Proterozoic crustal growth. *Austral. J. Earth Sci.* 38:595–611.

Sarda, P., Moreira, M., Staudacher, T., Schilling, J.-G., and Allegre, C. J. (2000). Rare gas systematics on the southernmost Mid-Atlantic ridge: Constraints on the lower mantle and the Dupal source. *J. Geophys. Res.* 105:5973–96.

Saunders, A. D., Fitton, J. G., Kerr, A. C., Norry, M. J., and Kent, R. W. (1997). The North Atlantic igneous province. American Geophysical Union, Mon. 100:45–93.

Saunders, A. D., Tarney, J., Kerr, A. C., and Kent, R. W. (1996). The formation and fate of large igneous provinces. *Lithos* 37:81–95.

Schaefer, B. F., Parkinson, I. J., and Hawkesworth, C. J. (2000). Deep mantle plume osmium isotope signature from West Greenland Tertiary picrites. *Earth Planet. Sci. Lett.* 175:105–18.

Schiano, P., Birck, J. L., and Allegre, C. J. (1997). Os–Sr–Nd–Pb isotopic covariations in mid-ocean ridge basalt glasses and the heterogeneity of the upper mantle. *Earth Planet. Sci. Lett.* 150:363–79.

Schilling, J. G. (1973). Iceland mantle plume: Geochemical study of Reykjanes ridge. *Nature* 242: 565–71.

———. (1991). Fluxes and excess temperatures of mantle plumes inferred from their interaction with migrating mid-ocean ridges. *Nature* 352:397–403.

Schlanger, S. O. and Jenkyns, H. C. (1976). Cretaceous oceanic anoxic events: Causes and consequences. *Geol. Mijnbouw* 55:179–84.

Schoenberg, R., Kruger, F. J., Nagler, T. F., Meisel, T., and Kramers, J. D. (1999). PGE enrichment in chromitite layers and the Merensky reef of the western Bushveld Complex: A Re–Os and Rb–Sr isotope study. *Earth Planet. Sci. Lett.* 172:49–64.

Schubert, G. and Sandwell, D. T. (1989). Crustal volumes of the continents and of oceanic and continental submarine plateaus. *Earth Planet. Sci. Lett.* 92:234–46.

Schubert, G. et al. (1992). Geology and distribution of impact centers on Venus: What are they telling us? *J. Geophys. Res.* 97(13):257–303.

Schweitzer, J. and Kroner, A. (1985). Geochemistry and petrogenesis of Early Proterozoic intracratonic volcanic rocks of the Ventersdorp Supergroup, South Africa. *Chem. Geol.* 51:265–88.

Self, S., Thordarson, T., and Keszthelyi, L. (1997). Emplacement of continental flood basalt lava flows. American Geophysical Union, Mon. 100:381–410.

Semikhatov, M. A. and Raaben, M. E. (1996). Dynamics of the global diversity of Proterozoic stromatolites. Article II. *Stratigraphy Geol. Correlation* 4:24–50.

Sengor, A. M. C. (2001). Elevation as indicator of mantle plume activity. Geological Society of America, Spec. Pap. 352 (in press).

Sengor, A. M. C., Natal'in, B. A., and Burtman, V. S. (1993). Evolution of the Altaid tectonic collage and Paleozoic crustal growth in Eurasia. *Nature* 364:299–307.

Sepkowski, J. J., Jr. (1986). Pahanerozoic overview of mass extinction. In *Pattern and processes in the history of life*, eds. D. Raup and D. Japlonski, 277–95. Berlin: Springer-Verlag.

———. (1990). The taxonomic structure of periodic extinction. Geological Society of America, Spec. Pap. 247:33–44.

Sharma, M. (1997). Siberian traps. American Geophysical Union, Mon. 100:273–95.

Sharma, M., Basu, A. R., and Nesterenko, G. V. (1991). Nd–Sr isotopes, petrochemistry, and origin of the Siberian flood basalts, USSR. *Geochim. Cosmochim. Acta* 55:1183–92.

Shaw, H. R. (1973). Mantle convection and volcanic periodicity in the Pacific: Evidence from Hawaii. *Geol. Soc. Am. Bull.* 84:1505–26.

Shen, Y. and Forsyth, D. W. (1995). Geochemical constraints on initial and final depths of melting beneath mid-ocean ridges. *J. Geophys. Res.* 100:2211–37.

Sheth, H. C. (1999). A historical approach to continental flood basalt volcanism: Insights into pre-volcanic rifting, sedimentation, and early alkaline magmatism. *Earth Planet. Sci. Lett.* 168:19–26.

Sidorin, I., Gurnis, M., and Helmberger, D. V. (1999). Evidence for a ubiquitous seismic discontinuity at the base of the mantle. *Science* 286:1326–31.

Simonson, B. M. and Hassler, S. W. (1996). Was the deposition of large Precambrian iron formations linked to major marine transgression? *J. Geol.* 104:665–76.

Sleep, N. H. (1990). Hotspots and mantle plumes: Some phenomenology. *J. Geophys. Res.* 95:6715–36.

———. (1992). Hotspot volcanism and mantle plumes. *Annu. Rev. Earth Planet. Sci.* 20:19–43.

———. (1994). Lithospheric thinning by midplate mantle plumes and the thermal history of hot plume material ponded at sublithospheric depths. *J. Geophys. Res.* 99:9327–43.

———. (1997). Lateral flow and ponding of starting plume material. *J. Geophys. Res.* 102(10):1–12.

Sleep, N. H. and Windley, B. F. (1982). Archean plate tectonics: Constraints and inferences. *J. Geol.* 90:363–79.

Sloss, L. L. (1972). Synchrony of Phanerozoic sedimentary-tectonic events of the 4 North American craton and Russian platform. *Inter. Geol. Congress, 24th Montreal, Proc.*, Sec. 6, pp. 24–32.

Small, C. (1995). Observations of ridge–hotspot interactions in the Southern Ocean. *J. Geophys. Res.* 100(17):931–46.

Smith, R. B. and Braile, L. W. (1993). Topographic signature, space-time evolution, and physical properties of the Yellowstone–Snake River Plain volcanic system: The Yellowstone hotspot. Geological Survey of Wyoming, Mem. 5:694–754.

———. (1994). The Yellowstone hotspot. *J. Volcan. Geotherm. Res.* 61:121–87.

Smith, R. B. and Siegel, L. J. (2000). *Windows into the earth: The geologic story of Yellowstone and Grand Teton national parks*. Oxford: Oxford University Press.

Smrekar, S. E. and Phillips, R. J. (1991). Venusian highlands: Geoid to topography ratios and their implications. *Earth Planet. Sci. Lett.* 107:582–97.

Snavely, P. D. (1987). Tertiary geologic framework, neotectonics, and petroleum potential of the Oregon–Washington continental margin. In *Geology and resource potential of the continental margin of western North America and adjacent ocean basins*, eds. D. W. Scholl, et al., Houston, TX: Circum-Pacific Council for Energy and Mineral Resources.

Solomon, S. C. (1977). The relationship between crustal tectonics and internal evolution on the Moon and Mercury. *Phys. Earth Planet. Inter.* 15:135–45.

Sporli, K. B. and Balance, P. F. (1988). Mesozoic-Cenozoic ocean floor/continent interaction and terrane configuration, SW. Pacific area. *Oxford Monograph Geol. Geophys*. 8:176–90.

Stacey, F. D. and Loper, D. E. (1983). The thermal boundary layer interpretation of D″ and its role as a plume source. *Phys. Earth Planet. Inter.* 33:45–55.

Stefan, E. R., Smrekar, S. E., Bindschadler, D. L., and Senske, D. A. (1995). Large topographic rises on Venus: Implications for mantle upwelling. *J. Geophys. Res.* 100(23):317–27.

Stefanick, M. and Jurdy, D. M. (1984). The distribution of hotspots. *J. Geophys. Res.* 89:9919–25.

Stein, M. and Goldstein, S. L. (1996). From plume head to continental lithosphere in the Arabian–Nubian shield. *Nature* 382:773–8.

Stein, M. and Hofmann, A. W. (1992). Fossil plume head beneath the Arabian lithosphere? *Earth Planet. Sci. Lett.* 114:193–209.

———. (1994). Mantle plumes and episodic crustal growth. *Nature* 372:63–8.

Steinberger, B. (2000). Plumes in a convecting mantle: Models and observations for individual hotspots. *J. Geophys. Res.* 105(11):127–52.

Steinberger, B. and O'Connell, R. J. (1998). Advection of plumes in mantle flow: Implications for hotspot motion, mantle viscosity and plume distribution. *Geophys. J. Int.* 132:412–34.

Stewart, K., Turner, S., Kelley, S., Hawkesworth, C. J., Kirstein, L., and Mantovani, M. S. (1996). 3-D ^{40}Ar^{39}Ar geochronology in the Paraná flood basalt province. *Earth Planet. Sci. Lett.* 143:95–110.

Stofan, E. R., Smrekar, S. E., Bindschadler, D. L., and Senske, D. A. (1995). Large topographic rises on Venus: Implications for mantle upwelling. *J. Geophys. Res.* 100(23):317–27.

Storey, B. C. (1995). The role of mantle plumes in continental breakup: Case histories from Gondwanaland. *Nature* 377:301–8.

———. (1997). An active mantle mechanism for Gondwana breakup. *S. Afr. J. Geol.* 100:283–90.

Storey, B. C. and Kyle, P. R. (1997). An active mantle mechanism for Gondwana breakup. *S. African J. Geology* 100:283–90.

Storey, B. C., Mahoney, J. J., Kroenke, L. W., and Saunder, A. D. (1991). Are oceanic plateaus sites of komatiite formation? *Geol.* 19:376–9.

Storey, B. C., Leat, P. T., and Ferris, J. K. (2001). The location of mantle plume centers during the initial stage of Gondwana breakup. Geological Society of America, Spec. Pap. (in press).

Strom, R. G., Schaber, G. G., and Dawson, D. D. (1994). The global resurfacing of Venus. *J. Geophys. Res.* 99(12):899–926.

Strong, C. P. (1994). Late Cretaceous foraminifera from Hikurangi plateau, New Zealand. *Mar. Geol.* 119:1–5.

Sun, S. S. and McDonough, W. F. (1989). Chemical and isotopic systematics of oceanic basalts: Implications for mantle composition and processes. Geological Society of London, Spec. Publ. 42:313–45.

Sun, S. S., Nesbitt, R. W., and McCulloch, M. T. (1989). Geochemistry and petrogenesis of Archean and Early Proterozoic siliceous high magnesian baslats. In *Boninites and related rocks*, ed. A. J. Crawford, 149–73. London: Unwin Hyman.

Suzuki, A., Ohtani, E., and Kato, T. (1995). Flotation of diamond in mantle melt at high pressure. *Science* 269:216–18.

Swanson, D. A. and Wright, T. L. (1980). The regional approach to studying the Columbia River Basalt Group. Geological Society of India, Mem. 3:58–80.

Swanson, D. A., Wright, T. L., Hooper, P. R., and Bentley, R. D. (1975). Linear vent systems and estimated rates of magma production and eruption for the Yakima basalt on the Columbia Plateau. *Am. J. Sci.* 275:877–905.

Sweeney, R. J., Duncan, A. R., and Erlank, A. J. (1994). Geochemistry and petrogenesis of central Lebombo basalts of the Karoo igneous province. *J. Petrol.* 35:95–125.

Tackley, P. J. (1997). Effects of phase transitions on three-demensional mantle covection. In Earth's Deep Interior, ed. D. J. Crossley, 273–336. Amsterdam: Gordon and Breach.

Tackley, P. J. (2000). Mantle convection and plate tectonics: Toward an integrated physical and chemical theory. *Science* 288:2002–7.

Tackley, P. J., Stevenson, D. J., Glatzmaier, G. A., and Schubert, G. (1994). Effects of an endothermic phase transition at 670 km depth in a spherical model of convection in the earth's mantle. *J. Geophys. Res.* 99(15):877–901.

Taira, A., Tokuyama, H., and Soh, W. (1989). Accretion tectonics and evolution of Japan. In *The evolution of the Pacific Ocean margins*, ed. Zvi Ben Avraham, 100–23. Oxford: Oxford University Press.

Tanaka, K. L., Golombek, M. P., and Banerdt, W. B. (1991). Reconciliation of stress and structural histories of the Tharsis region of Mars. *J. Geophys. Res.* 96(15):617–33.

Tanaka, K. L., Scott, D. H., and Greeley, R. (1992). Global stratigraphy. In *Mars*, eds. H. H. Kieffer, B. M. Jakosky, and C. W. Snyder, 354–82. Tucson: University of Arizona Press.

Tarduno, J. A. and Gee, J. (1995). Large-scale motion between Pacific and Atlantic hotspots. *Nature* 378:477–9.

Tarduno, J. A. and Sager, W. W. (1995). Polar standstill of the mid-Cretaceous Pacific plate and its geodynamic implications. *Science* 269:956–9.

Taylor, S. R. and McLennan, S. M. (1985). *The Continental Crust: Its Composition and Evolution*. Oxford: Blackwell Scientific, 312 p.

Tejada, M., Mahoney, J. J., Duncan, R. A., and Hawkins, M. P. (1996). Age and geochemistry of basement and alkalic rocks of Malaita and Santa Isabel, Solomon Islands, southern margin of the Ontong Java plateau. *J. Petrol.* 37:361–94.

Thompson, R. N., Morrison, M. A., Dickin, A. P., and Hendry, G. L. (1983). Continental flood basalts – arachnids rule ok? In *Continental basalts and mantle xenoliths*, eds. C. J. Hawkesworth and M. J. Norry, 158–85. Nantwich, UK: Shiva Press.

Thurston, P. C. (1994). Archean volcanic patterns. In *Archean crustal evolution*, ed. K. C. Condie, 45–84. Amsterdam: Elsevier Scientific Publishers.

Thurston, P.C. and Chivers, K.M. (1990). Secular variation in greenstone sequence development emphasizing Superior Province, Canada. *Precamb. Res.* 46:21–58.

Tolan, T. L., Reidel, S. P., Beeson, M. H., Anderson, J. L., Fecht, K. R., and Swanson, D. A. (1989). Revisions to the estimates of the areal extent and volume of the Columbia River basalt group. Geological Society of America, Spec. Pap. 239, pp. 1–20.

Tomlinson, K. Y. and Condie, K. C. (2001). Archean mantle plumes: Evidence from greenstone belt geochemistry. Geological Society of America, Mem. (in press).

Tomlinson, K. Y., Hughes, D. J., Thurston, P.C., and Hall, R. P. (1999). Plume magmatism and crustal growth at 2.9 to 3.0 Ga in the Steep Rock and Lumby Lake area, western Superior Province. *Lithos* 46:103–36.

Tomlinson, K. Y., Stevenson, R. K., Hughes, D. J., Hall, R. P., Thurston, P. C., and Henry, P. (1998). The Red Lake Greenstone Belt, Superior Province: Evidence of plume-related magmatism at 3 Ga and evidence of an older enriched source. *Precamb. Res.* 89:59–76.

Trehu, A. M., Lin, G., Maxwell, E., and Goldfinger, C. (1995). A seismic reflection profile across the Cascadia subduction zone offshore central Oregon: New constraints on methane distribution and crustal structure. *J. Geophys. Res.* 100(15):101–16.

Turcotte, D. L. (1993). An episodic hypothesis for Venusian tectonics. *J. Geophys. Res.* 98(17):61–8.

Turcotte, D. L. and Oxburgh, E. R. (1978). Intra-plate volcanism. *Philos. Trans. Roy. Soc. Lond., Ser. A* 288:561–79.

Turncliffe, V. and Fowler, M. R. (1996). Influence of seafloor spreading on the global hydrothermal vent fauna. *Nature* 379:531–3.

Turner, S. P. and Hawkesworth, C. J. (1995). The nature of the subcontinental mantle: Constraints from the major element composition of continental flood basalts. *Chem. Geol.* 120:295–314.

Turner, S. P., Regelous, M., Kelley, S., Hawkesworth, C. J., and Mantovani, M. S. M. (1994). Magmatism and continental break-up in the South Atlantic: High precision $^{40}Ar/^{39}Ar$ geochronology. *Earth Planet. Sci. Lett.* 121:333–48.

Uken, R. and Watkeys, M. K. (1997). An interpretation of mafic dyke swarms and their relationship with major mafic magmatic events on the Kaapvaal craton and Limpopo belt. *S. Afr. J. Geol.* 100:341–8.

Ulmer, P. (1989). The dependence of the Fe^{+2}-Mg cation partitioning between olivine and basaltic liquid on pressure, temperature, and composition. *Contrib. Mineral. Petrol.* 101:261–73.

Unrug, R. (1997). Rodinia to Gondwana: The geodynamic map of Gondwana supercontinent assembly. *GSA Today* 7(1):1–5.

Vail, P. R. and Mitchum, Jr., R. M. (1979). Global cycles of relative changes of sea level from seismic stratigraphy. American Association of Petrological Geology, Mem. 29:469–72.

van der Hilst, R. D. and Karason, H. (1999). Compositional heterogeneity in the bottom 1000 km of Earth's mantle: Toward a hybrid convection model. *Science* 283:1885–8.

van der Hilst, R. D., Widiyantoro, R. S., and Engdahl, R. (1997). Evidence for deep mantle circulation from global tomography. *Nature* 386:578–84.

van Fossen, M. C. and Kent, D. V. (1992). Paleomagnetism and 122 Ma plutons in New England and the Mid-Cretaceous paleomagnetic field in North America; True polar wander of large-scale differential mantle motion? *J. Geophys. Res.* 97:19,651–61.

van Keken, P. (1997). Evolution of starting mantle plumes: A comparison between numerical and laboratory models. *Earth Planet. Sci. Lett.* 148:1–11.

Varsek, J. L., et al. (1993). Lithoprobe crustal reflection structure of the southern Canadian Cordillera 2: Coast Mountains transect. *Tectonics* 12:334–60.

Vaughan, A. P. M. (1995). Circum-Pacific mid-Cretaceous deformation and uplift: A superplume-related event? *Geol.* 23:491–4.

Veevers, J. J. (1990). Tectonic–climatic supercycle in the billion-year plate-tectonic eon: Permian Pangean icehouse alternates with Cretaceous dispersed-continents greenhouse. *Sed. Geol.* 68:1–16.

Veevers, J. J., Walter, M. R., and Scheibner, E. (1997). Neoproterozoic tectonics of Australia–Antarctica and Laurentia and the 560-Ma birth of the Pacific Ocean reflect the 400-My Pangean supercycle. *J. Geol.* 105:225–42.

Veizer, J. (1988). The evolving exogenic cycle. In *Chemical cycles in the evolution of the Earth*, eds., C. B. Gregor, R. M. Garrels, F. T. Mackenzie, and J. B. Maynard, 175–219. New York: John Wiley.

Veizer, J. (1989). Strontium isotopes in seawater through time. *Annu. Rev. Earth Planet. Sci.* 17:141–67.

Veizer, J. and Compston, W. (1976). $^{87}Sr/^{86}Sr$ in Precambrian carbonates as an index of crustal evolution. *Geochim. Cosmochim. Acta* 40:905–14.

Veizer, J. and Jansen, S. L. (1985). Basement and sedimentary recycling, 2. Time dimension to global tectonic. *J. Geol.* 93:625–43.

Veizer, J. et al. (1999). $^{87}Sr/^{86}Sr$, $d^{13}C$ and $d^{18}O$ evolution of Phanerozoic seawater. *Chem. Geol.* 161:59–88.

Vidale, J. E., Ding, X. Y., and Grand, S. P. (1995). The 410-km-depth discontinuity: A sharpness estimate from near-critical reflections. *Geophys. Res. Lett.* 22:2557–60.

Vink, G. E. (1984). A hotspot model for Iceland and the Voring plateau. *J. Geophys. Res.* 89:9949–59.

Vlastelic, L., Aslanian, D., Dosso, L., Bougault, H., Olivet, J. L., and Geli, L. (1999). Large-scale chemical and thermal division of the Pacific mantle. *Nature* 399:345–50.

Vogt, P. R. (1972). Evidence for global synchronism in mantle plume convection and possible significance for geology. *Nature* 240:338–342.

Vogt, P. R., et al. (1976). Subduction of aseismic oceanic ridges: Effects on shape, seismicity, and other characteristics of consuming plate boundaries. Geological Society of America, Spec. Pap. 172, 59 p.

Von Damm, K. L. (1990). Seafloor hydrothermal activity: Black smoker chemistry and chimneys. *Annu. Rev. Earth Planet. Sci.* 18:173–204.

Von Herzen, R. P. et al. (1989). Heat flow and the thermal origin of hotspot swells: The Hawaiian swell revisited. *J. Geophys. Res.* 94:13,783–800.

Walker, R. J., Morgan, J. W., and Horan, M. F. (1995). Os-187 enrichment in some plumes: Evidence for core–mantle interaction? *Science* 269:819–21.

Wang, Y., Weidner, D. J., Liebermann, R. C., and Zhao, Y. (1994). P–V–T equation of state of perovskite: Constraints on composition of the lower mantle. *Phys. Earth Planet. Inter.* 83:13–40.

Warner, M., et al. (1996). Seismic reflections from the mantle represent relict subduction zones within the continental lithosphere. *Geol.* 24:39–42.

Watts, A. B., Wessel, P., Duncan, R. A., and Larson, R. L. (1988). Origin of the Louisville ridge and its relationship to the Eltanin fracture zone system. *J. Geophys. Res.* 93:3051–77.

Weaver, B. L., Wood, D. A., Tarney, J., and Joron, J. L. (1987). Geochemistry of ocean island basalts from the South Atlantic: Ascension, Bouvet, St. Helena, Gough, and Trista da Cunha. Geological Society of London, Spec. Publ. 30:253–67.

Weis, D. et al. (1992). The influence of mantle plumes in generation of Indian oceanic crust. In *Synthesis of results from scientific drilling in the Indian Ocean*, eds. R. A. Duncan, et al., 57–89. Washington, DC: American Geophysical Union.

Wendlandt, E., DePaolo, D. J., and Baldridge, W. S. (1993). Nd and Sr isotope chronostratigraphy of Colorado Plateau lithosphere: Implications for magmatic and tectonic underplating of the continental crust. *Earth Planet. Sci. Lett.* 116:23–43.

Wessel, P. and Lyons, S. (1997). Distribution of large Pacific seamounts from Geosat/ERS-1: Implications for the history of intraplate volcanism. *J. Geophys. Res.* 102(22):459–75.

White, R. S. (1997). Mantle plume origin for the Karoo and Ventersdorp flood basalts, South Africa. *S. Afr. J. Geol.* 100:271–82.

White, R. S. and McKenzie, D. (1989). Magmatism at rift zones: The generation of volcanic continental margins and flood basalts. *J. Geophys. Res.* 94:7685–729.

White, R. S. and McKenzie, D. (1995). Mantle plumes and flood basalts. *J. Geophys. Res.* 100(17):543–85.

White, R. S., et al. (1987). Magmatism at rifted continental margins. *Nature* 330:439–44.

White, R. S., McKenzie, D., and O'Nions, R. K. (1992). Oceanic crustal thickness from seismic measurements and REE inversions. *J. Geophys. Res.* 97(19):683–715.

White, R. V., Tarney, J., Kerr, A. C., Saunders, A. D., Kempton, P. D., Pringle, M. S., and Klaver, G. T. (1999). Modification of an oceanic plateau, Aruba, Dutch Caribbean: Implications for the generation of continental crust. *Lithos* 46:43–68.

Whitehead, J. A. (1982). Instabilities of fluid conduits in a flowing earth – Are plates lubricated by the asthenosphere? *Geophys. J. Roy. Astron. Soc.* 70:415–33.

Whitehead, J. A. and Luther, D. S. (1975). Dynamics of laboratory diapir and plume models. *J. Geophys. Res.* 80:705–17.

Widom, E. and Shirey, S. B. (1996). Os isotope systematics in the Azores: Implications for mantle plume sources. *Earth Planet. Sci. Lett.* 142:451–65.

Wilde, P., Quinby-Hunt, M. S., and Berry, B. N. (1990). Vertical advection from oxic or anoxic water from the pycnocline as a cause of rapid extinction or rapid radiations. In *Extinction events in Earth history, lecture notes on earth science*, eds. E. G. Kauffman and O. H. Walliser, Vol. 30, 85–97. Heidelberg: Springer-Verlag.

Wilson, J. T. (1963). A possible origin of the Hawaiian Islands. *Can. J. Physics* 41:863–8.

Wilson, M. (1992). Magmatism and continental rifting during the opening of the South Atlantic Ocean: A consequence of Lower Cretaceous superplume activity? Geological Society of London, Spec. Publ. 68:241–55.

Windley, B. F. (1992). Proterozoic collisional and accretionary orogens. In *Proterozoic crustal evolution*, ed. K. C. Condie, 419–46. Amsterdam: Elsevier.

Winterer, E. L. (1976). Bathymetry and regional tectonic setting of the Line Islands chain. *Initial Repts. Deep Sea Drill. Proj.* 33:731–48.

Wood, B. J. (1995). The effect of water on the 410-km seismic discontinuity. *Science* 268:74–6.

Wood, R. A. and Davy, B. (1994). The Hikurangi plateau. *Mar. Geol.* 118:153–73.

Wooden, J. L. et al. (1995). Isotopic and trace-element constraints on mantle and crustal contributions to characterization of the Siberian continental flood basalts, Noril'sk area, Siberia. *Geochim. Cosmochim.* Acta 57:3677–704.

Worsley, T. R. and Nance, R. D. (1989). Carbon redox and climate control through earth history: A speculative reconstruction. *Palaeogeog., Palaeoclimat., and Palaeoecol.* 75:259–82.

Worsley, T. R., Nance, R. D. and Moody, J. B. (1984). Global tectonics and eustasy for the past 2 Ga. *Marine Geol.* 58:373–400.

———. (1986). Tectonic cycles and the history of the Earth's biogeochemical and paleoceanographic record. *Paleoceanog.* 1:233–63.

Wright, T. L. (1971). Chemistry of Kilauea and Mauna Loa lava in space and time. U.S. Geological Survey, Prof. Pap. 735, 40 p.

Wyession, M. E. et al. (1998). The D″ discontinuity and its implications. American Geophysical Union, Geodynamics Series, 28:273–98.

Wyllie, P. J. (1971). The role of water in magma generation and initiation of diapiric uprise in the mantle. *J. Geophys. Res.* 76:1328–38.

———. (1988). Magma genesis, plate tectonics and chemical differentiation of the Earth. *Rev. Geophys.* 26:370–404.

Xie, Q. (1996). Trace element systematics of mafic–ultramafic volcanic rocks from the Archean Abitibi greenstone belt, Canada: Implications for chemical evolution of the mantle and Archean greenstone belt development. Unpublished Ph.D. thesis, University of Saskatchewan, p. 222.

Xie, Q. and Kerrich, R. (1994). Silicate–perovskite and majorite signature komatiites from the Archean Abitibi greenstone belt: Implications for early mantle differentiation and stratification. *J. Geophys. Res.* 99:15,799–812.

Xie, Q., Kerrich, R., and Fan, J. (1993). HFSE/REE fractionations recorded in three komatiite-basalt sequences, Archean Abitibi greenstone belt: Implications for multiple plume sources and depths. *Geochim. Cosmochim. Acta* 57:4111–18.

Yanagi, T. and Yamashita, K. (1994). Genesis of continental crust under island arc conditions. *Lithos* 33:209–23.

Yang, H., Frey, F. A., Weis, D., Giret, A., Pyle, D., and Michon, G. (1998). Petrogenesis of the flood basalts forming the northern Kerguelen archipelago: Implications for the Kerguelen plume. *J. Petrol.* 39:711–48.

Young, G. M. (1991). The geologic record of glaciation: Relevance to the climatic history of Earth. *Geosci. Can.* 18:100–8.

Yuen, D. A., Cserepes, L., and Schroeder, B. A. (1998). Mesoscale structures in the transition zone: Dynamical consequences of boundary layer activities. *Earth Planets Space* 50:1035–45.

Yuen, D. A. and Fleitout, L. (1985). Thinning of the lithosphere by small-scale convective destabilization. *Nature* 313:125–8.

Yuen, D. A., Hansen, U., Zhao, W., Vincent, A. P., and Malevsky, A. V. (1993). Hard turbulent thermal convection and thermal evolution of the mantle. *J. Geophys. Res.* 98:5355–73.

Zhai, Y. and Halls, H. C. (1994). Multiple episodes of dike emplacement along the NW. Margin of the Superior province, Manitoba. *J. Geophys. Res.* 99(21):717–32.

Zhao, W. L., Yuen, D. A., and Honda, S. (1992). Multiple phase transitions and the style of mantle convection. *Phys. Earth Planet. Inter.* 72:185–210.

Ziegler, P. A. (1993). Plate-moving mechanisms: Their relative importance. *J. Geol. Soc. London* 150:927–40.

Zindler, A. and Hart, S. R. (1986). Chemical geodynamics. *Annu. Rev. Earth Planet. Sci.* 14:493–571.

Zolotukhin, V. V. and Al'mukhamedov, A. I. (1988). Traps of the Siberian platform. In *Continental flood basalts*, ed. J. D. MacDougall, 273–310. New York: Kluwer Academic Publishers.

Index